GIFT of MARY GLIDE GOETHE MEMORIAL FUND

APES OF THE WORLD

WOLVES OF THE WORLD: Perspectives of Behavior, Ecology, and Conservation.
Edited by *Fred H. Harrington* and *Paul C. Paquet*

IGUANAS OF THE WORLD: Their Behavior, Ecology, and Conservation.
Edited by *Gordon M. Burghardt* and *A. Stanley Rand*

HORSE BEHAVIOR: The Behavioral Traits and Adaptations of Domestic and Wild Horses, Including Ponies.
By *George H. Waring*

GAZELLES AND THEIR RELATIVES: A Study in Territorial Behavior.
By *Fritz R. Walther, Elizabeth Cary Mungall,* and *Gerald A. Grau*

THE MANAGEMENT AND BIOLOGY OF AN EXTINCT SPECIES: *PERE DAVID'S DEER.*
Edited by *Benjamin B. Beck* and *Christen Wemmer*

APES OF THE WORLD: Their Social Behavior, Communication, Mentality and Ecology.
By *Russell H. Tuttle*

APES
OF THE WORLD

Their Social Behavior, Communication, Mentality, and Ecology

by

Russell H. Tuttle

Department of Anthropology,
Committee on Evolutionary Biology,
The Morris Fishbein Center for the Study
of the History of Science and Medicine,
and The College,
The University of Chicago,
and
Yerkes Regional Primate Research Center,
Emory University

NOYES PUBLICATIONS
Park Ridge, New Jersey, U.S.A.

Library of Congress Catalog Card Number: 86-17960
ISBN: 0-8155-1104-3
Printed in the United States

Published in the United States of America by
Noyes Publications
Mill Road, Park Ridge, New Jersey 07656

10 9 8 7 6 5 4 3 2 1

Library of Congress Cataloging-in-Publication Data

Tuttle, Russell.
    Apes of the world.

    Bibliography: p.
    Includes index.
    1. Apes--Behavior. 2. Apes--Ecology. 3. Mammals--
Behavior. 4. Mammals--Ecology. I. Title.
QL737.P9T87 1986   599.88'0451   86-17960
ISBN 0-8155-1104-3

to
Mother
for much more than life
and
Khmātī Almas
for Marlene, Nicole and Matthew

Foreword

For several generations of primatologists, questions about virtually any general aspect of the biology of apes were referred to two books by Robert Mearns Yerkes. The first of these appeared in 1929 and was entitled *The Great Apes: A Study of Anthropoid Life.* It was co-authored with his wife Ada. The second book, *Chimpanzees: A Laboratory Colony* followed in 1943. The breadth of these two publications was remarkable though it now seems limited by comparison with the present work by Professor Russell Tuttle of the University of Chicago.

Yerkes himself listed a few of the research areas covered in the 1943 book which he felt to be significant ". . . because of the nature of our discoveries and their relations to human life." This list by Yerkes is worth considering, both for the common ground with *Apes of the World*, and for the vast amount of new information now available. It includes the following areas: (a) physical characteristics, growth, and maturation; (b) the sexual and reproductive cycles, and especially estrus, ovulation, gestation, and parturition; (c) aspects of sexual and reproductive behavior; (d) auditory and visual sensory and perceptual processes; (e) neural correlates of behavior; (f) behavioral adaptations, as in discrimination learning and other modes of habit formation; (g) factors, both internal and environmental, which affect behavioral adaptations; (h) memory and imagination; (i) capacity for the modification of environment as in the shaping and use of tools; (j) symbolism, ideation, insight, or their counterpart; (k) linguistic expression and capacity; (l) suggestibility; (m) emotional traits and their modes of expression; (n) social relations and organization; (o) drug addiction and drug susceptibility; (p) parasite control and related problems of health and hygiene.

With the exception of husbandry and drug addiction, these subjects all find a place in Tuttle's book, but with a data base so different that a reader might be forgiven for momentarily forgetting that the same subjects are under discussion. In some ways the transformation in the corpus of facts known to Yerkes that is made available to us in Tuttle's book brings to mind the revolution brought about in biology during the same period by molecular genetics.

The contents of *The Great Apes* was derived largely from research on comparative anatomy and physiology, and from studies of the behavior of animals in captivity, for comparative psychology was then at its most vigorous and creative, and contemporaries with whom Yerkes interacted regularly and productively included E.B. Holt, Karl Lashley, Edward Thorndike, and John B. Watson. There was an emphasis then on sensory abilities, perception, intelligence testing, and sexual behavior, on all of which Yerkes was an authority. He pioneered as chairman of the National Research Council Committee for Research in Problems of Sex from 1921 to 1947, a story in itself, for there was oppressive censorship of research and publications on sexual behavior in those days: Frank Beach and Alfred Kinsey were among those whose ground-breaking research was made possible by Yerkes's support. But you would search in vain for much of interest on the natural history of sexual behavior in *The Great Apes*. Nothing was known of the extraordinary diversity of their mating systems and patterns of social organization. Despite the efforts of H. Bingham, C. Carpenter, H. Nissen and others, such topics were still largely terra incognita.

As will be immediately evident to readers of *Apes of the World*, the situation today is vastly different. With the possible exception of the bonobo or pygmy chimpanzee, we now have a very complete picture of the ecology and behavior of all of the anthropoid apes, creating a new frame of reference for almost all other aspects of their behavior. Hundreds of thousands of observation hours in the field over the past 20 years have yielded data so richly that non-human primates are now among the best known of all vertebrates from the viewpoint of their social behavior and its variation, both within and between species. Much contemporary theorizing in evolutionary biology employs primate field data, whether the issues are the nature and evolutionary significance of infanticide, or comparative studies of the relationships between diet and patterns of social organization.

By the mid-1980's, several long-term field projects had reached the crucial point of development at which radically new syntheses began to emerge. With the common chimpanzee, the names of J. Goodall, A. Kortlandt, J. Itani and T. Nishida come to mind;

with the mountain gorilla, D. Fossey and G. Schaller; with the orangutan, B. Galdikas, J. MacKinnon, and P. Rodman; with gibbons, D. Chivers and others.

The time is clearly ripe for a major reassessment of where things now stand with regard to the biology of the anthropoid apes, taking special account of the revelations that have only begun to make sense in recent years about their social behavior, tool use and cognitive abilities. Radically new lines of evidence have emerged, both from field research, and from the remarkable explorations into the abilities of captive apes to work with natural and language-based systems of communication, initiated by B. and A. Gardner, and sustained with such vigor and imagination by E. Menzel, D. Premack, D. Rumbaugh, E.S. Savage-Rumbaugh, and H. Terrace.

As a physical anthropologist, Russell Tuttle is ideally suited to the task. Avid and meticulous as a scholar, he has accomplished a truly monumental achievement with this new synthesis on the social behavior, communication, mentality and ecology of the apes. The emphasis is radically different from *The Great Apes*, but had Yerkes survived to see this synthesis of the new flowering of research in primatology, I have no doubt that it would have received his full blessing.

Some of the new developments he had already anticipated. In a famous and uncannily prescient insight, he concluded " . . . from the various evidences that the great apes have plenty to talk about, but no gift for the use of sounds to represent individual, as contrasted with racial, feelings or ideas. Perhaps they can be taught to use their fingers, somewhat as does the deaf and dumb person, and thus helped to acquire a simple, nonvocal, sign language."

Tuttle acknowledges this and many other debts to Yerkes fully and graciously, and yet this is a book that Yerkes himself could never have written. It is no dry compendium of facts and quotations, but a personal statement about the current state of our understanding of the biology and behavior of the apes. Without sacrificing comprehensiveness or reliability it is leavened with wit and word play. When issues are unresolved, and controversies rage, all viewpoints are fairly represented. Where major lacunae still exist, they are firmly indicated, and no punches are pulled.

Delicate questions of sexual comportment are handled with a forthrightness that would have warmed the hearts of Yerkes's committee for research on sex. The bizarre homosexual "hunching" behavior of female pygmy chimpanzees, for example, first observed in captivity, and now well documented in the wild, earns appropriate comment. Noting the suggestions by Rumbaugh and Savage-Rumbaugh that distinctive hand positions, some apparently iconic, may facilitate this physically intricate interaction, Tuttle remarks that "though falling short of the *Kama Sutra*, the remarkable variety of copulatory positions that were enjoyed by the bonobos seemed to require a flexible system of communication." Yet they achieved the necessary coordination successfully, usually without signs of psychological friction. As Tuttle puts it, "they rarely rub each other the wrong way."

The text is decorated with puns and humorous asides that go a long way to make this a book that, compendium though it is, can also be read for interest and pleasure. It is no mean achievement to have made judicious use of a light touch in a treatise that will surely serve as the major source book on the anthropoid apes for the next generation of primatologists.

The Rockefeller University   Peter Marler
Field Research Center
Millbrook, New York
October, 1986

Preface

"Perhaps no greater tribute could be rendered to the memory of Dr. Yerkes than to fill the great gaps which he and Mrs. Yerkes astutely noted in *The Great Apes* and to augment the field studies of apes which he foresaw as essential to all aspects of modern primatology." (Tuttle, 1977, p. 293)

Nearly six decades ago, Robert M. and Ada W. Yerkes published *The Great Apes, a Study of Anthropoid Life* (1929). It is a thick compendium of information on gibbons, orangutans, chimpanzees and gorillas. The authors hoped that it would enhance the value of apes as subjects for psychobiological research. They originally intended the work to be primarily historical and objective and only secondarily original and critical (Yerkes and Yerkes, 1929, *x*). But they candidly confessed that while their treatment of gibbons remained largely an historical compilation, they could not resist critical commentaries on literature about the orangutan and sharing their own psychobiological research on common chimpanzees, a bonobo, an orangutan, and a mountain gorilla.

The Yerkeses believed that they could present hypotheses and support theses only incidentally (Yerkes and Yerkes, 1929, *ix*). In 1929, systematic psychobiological research was in its early infancy. It was gestated during the first third of the twentieth century chiefly due to the efforts of Wolfgang Köhler (1887-1967) in Tenerife and Berlin, Mrs. Nadezhda Nikolaevna (Ladygina) Kots (1890-1963) of Moscow, and Robert M. Yerkes (1876-1956) and his numerous students and associates in the United States. Field studies of apes were virtually nonexistent then.

xi

The early experiments of psychobiologists and the pioneer field studies of Clarence Ray Carpenter (1934, 1935, 1938, 1940) and other naturalists quickly rallied the imaginative intellectual cannibals of the sciences and humanities—anthropologists. In 1929, Earnest Albert Hooton, the doyen of American physical anthropology, was nearly midway in his flourishing career at Harvard University. Although his research was focused primarily on human variation (then termed "race") and the interfaces between cultural, archeological and physical anthropology, Hooton astutely recognized the value of primate studies for illuminating aspects of the hominid career. The titles and substance of his books, *Up from the Ape* (1931, 1946), *Apes, Men and Morons* (1937), *Why Men Behave like Apes and Vice Versa* (1940), and *Man's Poor Relations* (1942), reflect his passion for primatology. Hooton aimed to inform and *sui generis* to amuse lay readers as well as students and specialists. He was overwhelmingly successful. He interested some of the best graduate students at Harvard University in careers as physical anthropologists and secured a place for primate studies in American anthropology. Indeed, because of Hooton, Louis S.B. Leakey, Adolph H. Schultz, Sherwood L. Washburn, and other anthropologists, for many years primate studies assumed a greater importance in anthropology than in psychology or biology.

The Great Apes is divided into six parts, treating I. historical discoveries of the apes and "terms, types, and relations" among them; II-V. details on what was known about the geographic distributions, lifeways, sociality, mentality, affective behavior, communication, neural systems and captive maintenance of gibbons, orangutans, chimpanzees and gorillas; and VI. comparisons and conclusions.

In order to survey and to synthesize newer information on apes, I have departed markedly from the organizational plan of *The Great Apes*. Some topics have been passed over, others have been expanded upon, and new ones have been introduced. Still I have tried to touch base with the Yerkeses' volume whenever it was practicable so that the novelty of current questions and approaches (or lack thereof) may be compared with the state of the science 57 years ago.

Like *The Great Apes*, *Apes of the World* is meant to serve as a source book for a wide spectrum of biological and social scientists, most particularly those who would draw upon knowledge of apes to model human behavioral evolution. It is also designed to serve as a textbook in classes on human evolution which contain substantive units on primatology.

Unlike the Yerkeses, I have written the text largely in my own words, without including extensive quotations from primary sources.

It is detailed, but I trust not overly dry. Indeed a warning must be issued here. Unless the reader agrees that wordplay is almost as much fun as foreplay, she or he had best not proceed with this book. Regardless of the irresistible puns, I have endeavored to write a precise, scholarly work that will not only engage readers but also stimulate them to explore the primary literature, attend the nearest zoological park, and perhaps make their own synthetic or theoretical contributions, ranging from term papers to novel models of hominoid evolution.

Where I have erred factually, please correct me altruistically in correspondence or more publicly in reviews.

The inspiration for this project came from the apes. But the strength to persevere over the past few years came from quite a different source, viz. through services at the Rockefeller Memorial Chapel. This was especially true during periods of grief following the deaths of May Elizabeth Tuttle, Charles King, and Richard N. Tuttle.

Aunt May was an epitome of Christian grace. Early on, Grandfather King instilled in me a fascination for primates and, with my father, Richard Tuttle, conveyed an earthy wit, which is occasionally manifest in this work. The apes regularly provide ripe material for that art. I could not resist exploiting bits of it.

Three choral directors, Thomas Peck, William Peterman, and Rodney Wynkoop unknowingly contributed greatly to this enterprise by providing refreshing respites from the stacks of reprints and books that crowded my workbenches and desk.

Ben Beck and George Narita provided impetus for the project from midway to the elusive end. Richard Klein, Cornelia Wolf and Samuel Wilson helped me to computerize the index.

The new John Crerar Library at the University of Chicago and its crack reference librarians, Yvonne Glanze, Christa Modschiedler, and Kathleen Zar, were godsends that enormously expedited the pursuit of major and minor bits of information. The head typist, James Bone, interacted with the text and its author in a highly professional manner. This allowed speedy culmination of what had at times seemed to be an endless adventure. Mrs. Katherine Barnes, Lois Bisek and Queen Ward typed earlier drafts.

The apes at Yerkes Regional Primate Research Center, Brookfield Zoo and Lincoln Park Zoo have given me firsthand acquaintance with their capacities. It has been a privilege to work with them, their caretakers and fellow students over the years. I regret never having seen apes in the wild and am most grateful to the many persons who have sent me accounts of their field research. This holds doubly for colleagues in Asia and Africa, whom I hope to have repaid by fully presenting their contributions herein.

Sherry Washburn arranged my initial study at the Yerkes Regional Primate Research Center, which was then at Orange Park, Florida. There I met Charles Rogers, Emil Menzel, Irwin Bernstein, Richard K. Davenport, William Mason, W.C. Osman-Hill, their families and numerous visitors, who generously shared their homes and knowledge of primate behavior and ecology. Dr. Geoffrey Bourne and his successor Frederick King have always welcomed us to the modern facilities in Atlanta and Lawrenceville, Georgia. Duane Rumbaugh and Sue Savage-Rumbaugh were most gracious hosts and stimulating colleagues when we visited their language research laboratories in Atlanta.

The table of contents, which follows, tells what is in the book so I will not belabor the matter here. Readers might profit also from browsing in the index.

I have not delved deeply into the literature on captive maintenance because of space limitations and the fact that Terry Maple (1980; Maple and Hoff, 1982) has accepted the task of summarizing available information on the topic. Neither could I deal at length with problems of ape conservation in the wild. Undoubtedly, this is the most vital issue confronting all scientists who would draw upon knowledge of the apes in order to understand the human career and condition and search for general biological principles among a spectrum of animals.

Without immediate moves to preserve the apes *and their habitats*, we will lose forever a wealth of information about nature and ourselves. This would be especially tragic now that we have many well formulated hypotheses to test, have developed incisive research tools and approaches, and possess the technology to process great quantities of data. We are on the threshold to move from considering apes as amusements to creatures which truly reveal things about ourselves. Hopefully, this book will stimulate reflections that will preserve the apes and profit humankind.

The following persons and publishers graciously provided photographs and/or permitted us to reprint figures from their publications or collections: C. Boesch, G.H. Bourne, R.L. Ciochon, D.J. Chivers, N. Creel, Edinburgh University Press, R.A. and B.T. Gardner, R. Garrison, C.P. Groves, T. Kano, S. Karger AG, K. Kawanaka, W.C. McGrew, Oxford University Press, R.E. Passingham, D. Premack, H. Preuschoft, H.J. Rijksen, D.M. Rumbaugh, E.S. Savage-Rumbaugh, R. Tenaza, R.L. Tilson, C.E.G. Tutin, D.P. Watts, and the Zoological Society of San Diego.

I am most grateful also to the following persons who helped me to identify the first names of persons in the references and the addresses of colleagues from whom figures were requested: P.J. Andrews, B.B. Beck, J. Erwin, I.M. Friedman, J. Kruijt, W.C. McGrew, K. Nagatoshi, D.M. Rumbaugh, R. Singer, P. Ulinski, and F. deWaal.

My laboratory and museum studies on apes were supported in part by NIH grant RR-00165 from the Division of Research Resources of the Yerkes Primate Research Center; the Marian and Adolph Lichtstern Fund of the University of Chicago; the Wenner-Gren Foundation for Anthropological Research, Inc.; NSF grants GS-834, GS-1888, GS-3209, and SOC75-02478; and PHS Research Career Development Award 1-K04-GM16347-01.

The book was completed while I was a Fellow of the John Simon Guggenheim Memorial Foundation and with partial support from NSF grant BNS-8504290.

October 1986   Russell H. Tuttle

Contents

1

Terminology, Taxonomy, Distribution, and Phylogeny

"The student of behavior need not be also systematic and morphologist, but to ignore these disciplines with their highly developed techniques and their invaluable assemblages of information would be inexcusable." (Yerkes and Yerkes, 1929, p. 36)

"Because behavior may vary with type (genus, species, variety) as with individuality, sex, stage of development, and physiological condition of the organism, it is essential that an animal which serves as object of study be so carefully and accurately described that it may be classified and safely compared with other individuals in the several respects mentioned." (Yerkes and Yerkes, 1929, p. 42)

If these insightful caveats had been heeded by the majority of psychobiologists and students of primate behavior, the primatological literature would be more useful for current comparative purposes and there would have been greater potential for communication between behavioral scientists and evolutionary biologists. Robert Yerkes worked assiduously to identify his subjects systematically and, as inaugural director of the Yerkes Laboratories, to ensure that the biological components of psychobiology received equal consideration with the proximate research goals of individual behavioral scientists.

The scientific and vernacular terms which I will use are listed in Table 1 and are elaborated upon in this chapter. Basically, I have followed the conventions suggested by Simpson (1961) for rendering taxonomic nomina into vernacular forms.

Living apes and humans and certain fossil forms belong to the superfamily Hominoidea of the suborder Anthropoidea of the order Primates. In 1929, the terms *anthropoid*(s) and *anthropoid ape*(s) were applied commonly to any and all of the living apes. Anthropoid means man-like (Yerkes and

Yerkes, 1929, p. 40). Before the twentieth century, the term ape, (like its German cognate *Affen* and French equivalent *singe*) embraced the monkeys as well as the great and lesser apes. Now that the term ape has achieved a narrower connotation, excluding the monkeys, the adjective anthropoid is no longer needed to discriminate between gibbons, siamang, orangutans, common chimpanzees, bonobos and gorillas, on the one hand, and monkeys, on the other. In contemporary scientific writing, anthropoid specifically refers to the suborder Anthropoidea, which encompasses the living and fossil New and Old World monkeys, the apes, and the hominid primates. Thus, in modern scientific contexts, the phrase anthropoid ape(s) is redundant and verbose and the term anthropoid(s) cannot be used nominally to indicate only the apes.

Experts differ about how many families of hominoid primates there are and which apes, humans, man-apes and ape-men belong in each family. Traditionally, the extant apes and humans are placed in either two or three families. I will use the three family scheme (Table 1). The Hylobatidae includes all gibbons and the siamang. The Pongidae is comprised of the organgutans, common chimpanzees, bonobos and gorillas. Humans are in their own family, the Hominidae. Curiously, *The Great Apes* was mistitled anent the inclusion of the gibbons and siamang. The "great apes" only include the orangutans, bonobos, common chimpanzees, and gorillas. The phrase great apes is synonymous with pongid apes as the term pongid is used herein. Lesser apes and hylobatid apes are the proper designations for the siamang and gibbons.

Disagreement exists over the superfamilial, familial, and subfamilial placement of the siamang and gibbons; familial rearrangement of the African apes to reflect certain close biomolecular similarities with humans; and the classification of many fossils, particularly fragmentary and relatively early ones (Ciochon and Corruccini, 1983).

Chiarelli (1968, 1972, 1975) advocated that the siamang and gibbons should be removed from the Hominoidea and placed in the Cercopithecoidea because many species of *Hylobates* have 44 chromosomes and other karyological resemblances with the leaf-eating monkeys (Colobinae) of Asia and Africa. This suggestion proved to be indigestible to primatologists because it is inconsistent with dental, dermatoglyphic, and other morphological data and with the bulk of the evidence from comparative studies of DNA and proteins (Groves, 1972; Simons, 1972; Schultz, 1973; Frisch, 1973; Darga et al., 1973, 1984; Andrews and Groves, 1976; Tuinen and Ledbetter, 1983; Cronin et al., 1984; Creel and Preuschoft, 1984). Further, a gene-mapping study indicates that the Hylobatidae furcated from the common stem leading to the Pongidae after the Cercopithecoidea had diverged (Turleau et al., 1983).

Thenius (1981) recommended that the Hylobatidae constitute their own superfamily, the Hylobatoidea. At the opposite extreme, some experts (e.g.

Simpson, 1963; Clark, 1971; Groves, 1972)* who focused on morphological complexes, particularly those related to suspensory behavior, concluded that the siamang and gibbons constitute a subfamily (Hylobatinae) of the Pongidae. If this scheme were followed, vernacularly the lesser apes would be termed the hylobatine apes and the great apes would be called the pongine apes.

On the basis of comparative studies of serum proteins, Goodman (1962, 1963) concluded that the gorillas and chimpanzees should join the humans in the Hominidae. Among living great apes, only the orangutans would remain in the Pongidae. The lesser apes would still be in the Hylobatidae.

Goodman's data and taxonomic scheme are particularly attractive to cladists. Some current cladists place all apes and hominids in the Hominidae (Delson and Andrews, 1975; Delson, 1977) or exclude only the Hylobatidae among the living Hominoidea (Szalay and Delson, 1979; Goodman and Cronin, 1982; Groves, 1984). In cladistic taxonomy, the recency of common ancestry, as evinced by shared derived (synapomorphic) character complexes, is weighted over other factors in the establishment of sister groups that will be given common taxonomic nomina. Because the protein and DNA structures are assumed to reflect the genetic relatedness between organisms more clearly than gross morphological and behavioral features do, molecular primatologists argue that they are the best indicators of phylogenetic relationships in the Hominoidea.

The more traditional, evolutionary taxonomists argue that we must take account of the degrees of divergence, due to adaptive changes, since speciation from common stocks. Simpson (1963) argued cogently that since the hominid adaptive zone is so very different from that of the apes, humans should be placed in a discrete family. The hominid adaptive zone is characterized by habitual terrestrial bipedal positional behavior, techno-logical sophistication and versatility, and elaborate symbolically mediated vocal communication and ideology.

Most biomolecular data indicate that humans and the African apes are much closer to one another than any of them are to the orangutans (Goodman and Cronin, 1982; Goodman et al., 1983; Mai, 1983; Cronin, 1983). However, one study on the genetic distances between humans, common chimpanzees, bonobos, gorillas, and orangutans revealed that the orangutans are no further distant from humans than the African apes are (Bruce and Ayala, 1978). It will take a good deal more research to clarify which branches in hominoid taxonomic schemes and evolutionary trees will collapse from eccentric loadings of cladistic versus traditional factors and what will sprout in their steads.

---

*Apparently, Groves later changed his mind and agreed that the lesser apes belong in their own family, the Hylobatidae (Andrews and Groves, 1976, p. 174).

Table 1: Nomina and Common Names of Major Primate Groups and Anthropoid Taxa

		Common Names
Order	Primates	primates
Suborder	Prosimii	prosimians (lemurs, lorises and bush babies)
	Anthropoidea	anthropoids (humans, apes and monkeys)
Superfamily	Ceboidea	New World (ceboid) monkeys
	Cercopithecoidea	Old World (cercopithecoid) monkeys)
	Hominoidea	apes and humans
Family	Hylobatidae	lesser apes
	Pongidae	great apes
	Hominidae	humans, *Australopithecus* and intermediate species

Genus	Subgenus	Species	Subspecies	Common Names
Hylobates	*(Symphalangus)*	*syndactylus*	*syndactylus*	Sumatran siamang
			continentis	Malayan siamang
	(Nomascus)	*concolor*	*concolor*	Tonkin black gibbon
			lu	Laotian black gibbon
			hainanus	Hainan black gibbon
			leucogenys	Northern white-cheeked gibbon
			siki	Southern white-cheeked gibbon
			gabriellae	Red-cheeked gibbon
	(Bunopithecus)	*hoolock*	*hoolock*	Western hoolock gibbon
			leuconedys	Eastern hoolock gibbon
	(Hylobates)	*pileatus*		Pileated or capped gibbon
		klossii		Kloss gibbon, beeloh
		moloch		Javan, silvery, or moloch gibbon

(continued)

Table 1: (continued)

Genus	Subgenus	Species	Subspecies	Common Names
Hylobates	*(Hylobates)*	*agilis*	*muelleri*	Müller's gibbon
			albibarbis	Bornean agile gibbon
			agilis	agile gibbon
		lar	*lar*	Malayan lar or white-handed gibbon
			carpenteri	Chiengmai lar gibbon
			entelloides	Tenasserim lar gibbon
			vestitus	Sumatran lar gibbon
Pongo		*pygmaeus*	*pygmaeus*	Bornean orangutan
			abelii	Sumatran orangutan
Pan	*(Pan)*	*troglodytes*	*verus*	Upper Guinea, pale-faced or masked chimpanzees
			troglodytes	Lower Guinea, black-faced or western chimpanzee
			schweinfurthi	Eastern or long-haired chimpanzee
			?koolokamba	Koolokamba or gorilla-like chimpanzee
		paniscus		Bonobo, pygmy chimpanzee
	(Gorilla)	*gorilla*	*gorilla*	Western or lowland gorilla
			beringei	Mountain gorilla
			graueri	Grauer's gorilla
Homo		*sapiens*	*sapiens*	Modern humans, people

THE HYLOBATID APES

After humans, the hylobatid apes are the next most numerous hominoid primates. They are widely distributed in the forests of southeastern Asia extending from Bangladesh eastward through Assam and Burma to southern China and Hainan Island in the South China Sea. They range from Yunnan Province, China, southward through Vietnam, Laos, Thailand, Kampuchea (Cambodia) and Peninsular Malaysia onto the large islands of Sumatra, Java and Borneo [now politically divided into East Malaysia (Sarawak and Sabah) and Kalimantan, Indonesia] and several of the Mentawai Islands off the west coast of Sumatra (Figure 1). Much of their current distribution is on insular and peninsular sections of the Sunda Shelf, which joined to form more continuous and extensive land masses during periods of lowered sea level in the Pleistocene period.

Hylobatids are the smallest of the living apes. They are unique among the Hominoidea in their minimal expression of sexual dimorphism, particularly in body size, cranial features, and canine tooth dimensions (Frisch, 1963, 1973; Schultz, 1956, 1962a, 1973; Wolpoff, 1975; Gaulin and Sailer, 1984). However, Oxnard (1983, p. 19) found notable sexual dimorphism in some of their body proportions.

Unlike the pongid apes and humans (and like the Old World monkeys), all lesser apes have special sitting pads of cornified, hairless skin attached firmly to their ischial tuberosities. These pads are called ischial callosities. Hylobatids are the only apes that do not build nests.

Taxonomic splitters have recognized as many as twenty different species of hylobatid apes while the lumpers have reduced their number to between six (Groves, 1972) and nine (Chivers, 1977a). There is also disagreement over the number of genera in the Hylobatidae. Schultz (1933, 1973) steadfastly maintained that the siamang belongs in its own genus, *Symphalangus*. After considering dental, dermatoglyphic and biomolecular evidence, Frisch (1973), Biegert (1973), and Darga et al. (1973), respectively, concurred with Schultz. Per contra, on the basis of thoroughgoing multivariate statistical studies of the cranial morphology of lesser apes, Creel and Preuschoft (1976) and Corruccini (1981) concluded that *Symphalangus* is probably only a subgenus of *Hylobates*. Accordingly, the siamang should be formally designated *Hylobates* (*Symphalangus*) *syndactylus*. Groves (1972) and Andrews and Groves (1976) reviewed available evidence pertaining to the problem and concluded that *Hylobates* (*Symphalangus*) *syndactylus* is the preferred nomen for the siamang.

With one exception we will follow the latest classification of Groves (1984), who periodically conducted a comprehensive taxonomic revision of the lesser apes. He diagnoses 8 species in 3 subgenera (*Symphalangus*, *Nomascus* and *Hylobates*) of the genus *Hylobates*. Because hoolocks have

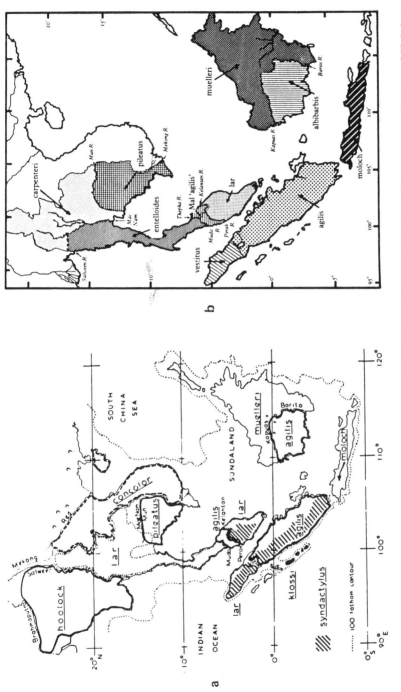

Figure 1: (a) Distribution of the Hylobatidae (Courtesy of D.J. Chivers). (b) Distribution of lar-group populations of *Hylobates* (*Hylobates*) sspp. (Courtesy of N. Creel and H. Preuschoft, 1984).

38 chromosomes a fourth subgenus (*Bunopithecus*) is now recognized within *Hylobates* (Prouty et al., 1983a,b; Marks, 1983; Groves, 1984).

The siamang, *Hylobates* (*Symphalangus*) *syndactylus*, is the largest of the lesser apes. Siamang are approximately twice as heavy as most other hylobatid species. Adult males weigh between 20 and 28 lb (9.1-12.7 kg) and adult females weigh between 20 and 27.5 lb (9.0-12.5 kg) (Schultz, 1973). Adults of both sexes are black, with long shaggy hair. They possess glabrous laryngeal or gular air sacs, which are quite conspicuous when inflated (Figure 22). The second and third toes of the siamang are commonly webbed together, sometimes as far distally as the terminal phalanges (Schultz, 1973). It is on the basis of this feature that the scientific names *Symphalangus* (one with joined digital bones) and *syndactylus* (one with joined digits) were devised. Siamang are distinct among the Hylobatidae in having a diploid chromosome number of 50. Like other hylobatid apes, siamang live in monogamous family groups (Chivers, 1977a).

Siamang are found on Sumatra and Peninsular Malaysia. Groves (1972) once considered the Sumatran and Malaysian populations to be separate subspecies, and then abandoned the distinction (Groves, 1984). They are extensively sympatric with other lesser apes. During the Pleistocene, siamang probably lived on Java as witnessed by fossils in the Djetus and Trinil faunas and from two fissure deposits at Punung (Hooijer, 1960). Fossil siamang have not been recovered from Borneo, though Hooijer (1962) speculated that future excavations at the Niah Caves in Sarawak might produce them. Because of the somewhat larger size of Sumatran siamang, including subfossil ones (Hooijer, 1960), and their more primitive dentition (Kitahara-Frisch, 1971), Groves (1972) suggested that the Peninsular Malaysian populations may be descendents of comparatively recent invaders from a southern homeland.

Hylobates (*Nomascus*) *concolor* is found in Vietnam (Tien, 1983), Laos, Kampuchea, southern China (Zhang et al., 1981; Zhixiang and Zhengyu, 1983) and on Hainan Island. They are colloquially referred to as crested, black, Indochinese and concolor gibbons. Groves (1972) stated that they are the most distinctive of living gibbons because of their slender builds, hooked noses, special ability to stand in a fully erect and humanoid posture, and certain other features. He recognized six subspecies of *Hylobates concolor* (Table 1). Concolor gibbons are distinct among the Hylobatidae in possessing a diploid chromosome number of 52. According to data compiled by Schultz (1933), they exhibit little or no sexual dimorphism in body weight. Ten adult males ranged between approximately 11 and 15 lb (4.9-6.8 kg) and eleven adult females ranged between approximately 10 and 15 lb (4.6-6.8 kg). But in some other features the sexes are strikingly dimorphic. Only the adult males possess relatively small laryngeal air sacs. And the adults are sexually dichromatic. Both sexes are fawn colored at birth. They become black at about 6 to 9 months of age. Male concolor gibbons remain black. As the females mature (at between 5 and 7 years old), they revert back to a fawn coat color and retain a black crown patch (Groves, 1972).

After the siamang, the next largest lesser ape is the hoolock or white-browed gibbon, *Hylobates (Bunopithecus) hoolock*, of eastern Bangladesh (Green, 1978; Gittins, 1984a; Gittins and Akonda, 1982; Gittins and Tilson, 1984; Sarker and Sarker, 1984), Assam, Burma (Southwick and Southwick, 1985), and west Yunnan, China. Five adult males weighed between 13 and 17.4 lb (6.0-7.9 kg) and 2 adult females weighed around 14 lb (6.4-6.6 kg) (Schultz, 1933). Both sexes probably have small laryngeal air sacs. The hoolock gibbons exhibit sexual dichromatism, which develops like that of the concolor gibbons. Groves (1984) and Haimoff et al. (1984) recognize 2 subspecies of *Hylobates hoolock*, the western hoolock (*H. h. hoolock*) of Assam and Bangladesh and the eastern hoolock (*H. h. leuconedys*) of Burma.

Until fairly recent times, hoolock (Groves, 1972) or concolor (Chivers, 1977a) gibbons were fairly widely distributed in China (Gulick, 1967). The fossil form (*Bunopithecus sericus*), attributable to *Hylobates hoolock*, was recovered from Pleistocene deposits in Sichuan, China (Matthews and Granger, 1923; Colbert and Hooijer, 1953; Groves, 1972).

The subgenus *Hylobates* contains 5 species. All of them have a diploid chromosome number of 44 (Groves, 1984).

The pileated or capped gibbon, *Hylobates (Hylobates) pileatus*, is a rather small-bodied species which lives in southeastern Thailand and Kampuchea, west of the Mekong River (Brockelman, 1975). In Thailand, there is some overlap between pileated gibbons and the closely related species *Hylobates lar* (Marshall et al., 1972). Hybrids have been observed in the Khao Yai Park (Chivers, 1977a; Brockelman, 1978, 1981; Brockelman and Gittins, 1984). Pileated gibbons lack laryngeal air sacs and infrequently exhibit interdigital webbing. Like the concolor and hoolock gibbons, the species is sexually dichromatic. But the pelage patterns develop differently in the pileated gibbons. Young male and female capped gibbons are buffy, with black spots on their crowns and chests. Ontogenetically, the spots spread until the males are totally black, except for a white brow band or facial ring, a white ring around the crown cap, white hands and feet, and a white pubic tuft. The head, hands and feet of the female are like those of the male. But the female only acquires a dark triangular ventral patch which extends from her cheeks to her groin. Both the Dagwoods and the Blondies can have long tufts of hair drooping over their ears (Groves, 1972).

Hylobates (Hylobates) klossii inhabit Selatan, Utara, Sipora, and Siberut of the Mentawai Islands, Indonesia. Colloquially, they have been called the beeloh, Kloss gibbon and dwarf siamang. Like the siamang, the species is monochromatically black. And it may possess a higher frequency of extensive webbing between the second and third toes than other species of

Hylobates do (Schultz, 1932, 1973). But here, really close resemblance to the siamang ceases. For instance, like other species of the subgenus *Hylobates, H. klossii* possess a diploid chromosome number of 44 (Hösli and Lang, 1970). And beelohs lack laryngeal air sacs (Chasen and Kloss, 1927; Schultz, 1933; Tenaza and Hamilton, 1971). They are about the same size as the concolor and many subspecies of the lar gibbons. Two adult males weighed 11.4 and 13.4 lb (5.2 kg, 6.1 kg) and 4 adult females weighed between 11.4 and 14.3 lb (5.2-6.5 kg) (Schultz, 1933).

The silvery, grey, or Javan gibbon, *Hylobates (Hylobates) moloch*, is a monochromatic species confined to Java (Groves, 1984; Kappeler, 1984a). The young have creamy coats which quickly change to grey with maturity. Adults usually have at least a trace of black capping and commonly sport black ventral thoracic markings. The facial ring is poorly defined (Groves, 1972, 1984). They lack laryngeal sacs. Twelve males weighed between 11 and 14.5 lb (5.0-6.6 kg) and 6 females were between 10 and 14.1 lb (4.5-6.4 kg) (Schultz, 1973).

The agile gibbons, *Hylobates (Hylobates) agilis* consist of 3 subspecies which are widespread in Indonesia and Malaysia (Groves, 1984). None of them have laryngeal sacs. Müller's Bornean gibbon, *H. (H.) agilis muelleri*, is widely distributed in Borneo, except for southwestern Kalimantan (Chivers, 1977a; MacKinnon, 1984) and the west coast of the Kapus River. Haimoff et al. (1984) consider them to constitute a distinct species, *H. (H.) muelleri*.

Müller's gibbons are polychromatically brown to grey, with black capping and breast patches which sometimes extend cranially onto the cheeks and distally along the medial surfaces of the forelimbs and hind limbs. Their hands and feet can be lighter than the rest of their bodily pelage and they sport a light brow band and sometimes also light cheek-whiskers (Groves, 1972, 1984).

The Bornean agile gibbon, *Hylobates (Hylobates) agilis albibarbis*, lives in southwestern Kalimantan, north of the Kapus River. It is distinguished from the other Bornean gibbons more by its calls than its pelage, though some individuals sport well developed facial rings and have the lightest caps and breast patches among Bornean gibbons (Groves, 1984).

The proper agile gibbon, *Hylobates (Hylobates) agilis agilis* lives on Sumatra, south of the Aceh region; in Peninsular Malaysia, north of the Perak River; and in southernmost isthmian Thailand (Chivers, 1977a). Three females weighed between 12.5 and 12.8 lb (5.7-5.8 kg) and 10 males weighed between 11 and 16.3 lb (5.0-7.4 kg) (Schultz, 1973). Agile gibbons exhibit dramatic asexual dichromatism. In most populations, and often within families, there are light phase and dark phase individuals. Adults of both sexes may be dark or light. Data compiled by Groves (1970a) support the hypothesis that the light phase is due to a simple autosomal recessive factor, influenced by modifiers. The lighter individuals can be buff to golden brown or fawn to greyish and sport dark ventrums and caps. Dark individuals are black with some regions tending toward maroon. All proper agile gibbons have white brow bands and some also sport white cheek-whiskers (Groves, 1972, 1984).

Groves (1984) lumped all other populations of gibbons into *Hylobates* (*Hylobates*) *lar*, within which he recognized 4 subspecies. Collectively, they can be referred to as the lar gibbons. Schultz (1973) compiled body weights for 47 adult male and 38 adult female *Hylobates lar*. The males weighed between 9 and 16 lb (4.1-7.3 kg) and the females weighed between 8.6 and 13.4 lb (3.9-6.1 kg). Lar gibbons lack laryngeal sacs and generally do not have marked interdigital webbing (Schultz, 1973). They have white hands and feet and a white facial ring. All 4 subspecies are asexually dichromatic (Groves, 1984).

The Malayan white-handed gibbon, *Hylobates* (*Hylobates*) *lar lar*, lives in Peninsular Malaysia, except the region between the Perak and Muda Rivers, and in southern Thailand as far north as the Isthmus of Kra. Pale individuals have a creamy pelage while dark phase individuals are medium to dark brown (Groves, 1972, 1984). Gittins (1978a) observed mixed groups of *Hylobates lar lar* and *Hylobates agilis agilis* in Ulu Muda, Kedah, Peninsular Malaysia (Gittins and Raemaekers, 1980, p. 99; Brockelman and Gittins, 1984).

The Chiengmai white-handed gibbon, *Hylobates* (*Hylobates*) *lar carpenteri*, occurs in northwestern Thailand, west of the Mae Nam Ping River, and neighboring parts of Burma. Light phase individuals are pale creamy white and dark phase individuals are chocolate brown (Groves, 1968a,b, 1969, 1970a, 1972, 1984).

The Tenasserim white-handed gibbon, *Hylobates* (*Hylobates*) *lar entelloides*, is found in southern and isthmiam Burma and adjacent areas of Thailand northward from the Isthmus of Kra and west of the Mae Nam Ping River. Pale individuals are fawn or honey colored; dark phase individuals are blackish brown (Groves, 1984).

Sumatran white-handed gibbons, *Hylobates* (*Hylobates*) *lar vestitus*, live in the Istimewa and Aceh regions of northern Sumatra. Pale phase individuals are light golden brown or fawn and dark individuals are greyish brown (Groves, 1971a, 1972, 1984).

The striking vocalizations and eye-catching arm-swinging displays of the lesser apes are also proving to be invaluable characteristics for distinguishing the different species. Initial studies on captives (Lamprecht, 1970; Dang et al., 1969; Goustard and Demars, 1971, 1973; Demars and Goustard, 1972; Dupette and Goustard, 1978; Tembrock, 1974; Goustard, 1976, 1979a-c, 1984a,b; Schilling, 1984) and free-ranging populations (McCann, 1933; Ellefson, 1968, 1974; Chivers, 1972, 1974, 1976, 1977; Haimoff, 1984, 1985; Haimoff and Gittins, 1985; Haimoff et al., 1984; Marshall and Marshall, 1976; Marshall et al., 1984; Tenaza, 1976; Tenaza and Hamilton, 1971; Marler and Tenaza, 1977; Tenaza and Tilson, 1977;

Chivers and MacKinnon, 1977; Gittins, 1978, 1984b,c; Kappeler, 1984b; Whitten, 1984a; Mitani, 1984, 1985a,b; Caldecot and Hamilton, 1983) indicate that, like birds and certain forest monkeys (Struhsaker, 1970, 1975), the lesser apes differ markedly from place to place in the kinds of sounds that they make, which sex makes them, how the sounds are combined, and which individuals display athletically and chase after calling (Chapter 7).

THE PONGID APES

The only extant Asian great ape is the orangutan, *Pongo pygmaeus*. Among the apes, orangutans are as unique for their solitude as the hylobatid apes are for their monogamy. Orangutans share with the African great apes a common diploid chromosome number of 48, the habit of nest building, and many morphological features (Tuttle, 1969a, 1975; Kluge, 1983).

Orangutans are extensively sympatric with the siamang and white-handed gibbons (*Hylobates lar vestitus*) in northern Sumatra and with Müller's and Bornean agile gibbons in Borneo. Attesting to a once wider distribution, Pleistocene fossil and subfossil teeth of *Pongo* have been recovered from Trinil deposits in central Java and from caves in the Padang Highlands of Sumatra, Barat, in the Yunnan and Guangxi (Kwangsi) provinces of southern China, and in northern North Vietnam. Dental and cranial remains from the Niah caves of Sarawak, are outside the present geographic range of Bornean orangutans (Hooijer, 1948, 1961; Simonetta, 1957; Medway, 1966; Groves, 1971b; de Vos, 1983).

Recent students of the taxonomy of orangutans unanimously concur with the intuition of Yerkes and Yerkes (1929, p. 108) that there is only one living species, perhaps composed of two insular subspecies: *Pongo pygmaeus pygmaeus* of Borneo and *Pongo pygmaeus abelii* of Sumatra (Jones, 1969; Bemmel, 1969; Groves, 1971b; MacKinnon, 1974, 1975; Rijksen, 1978, Röher-Ertl, 1982, 1983, 1984).

Although it is commonly stated (e.g. Groves, 1971b) that Sumatran orangutans are larger than Bornean orangutans, according to MacKinnon (1975) and Eckhardt (1975a) there is probably no major difference between the body weights of the two subspecies. Reliable weights for large samples of wild orangutans are not available. Seven Sumatran males weighed between 75 and 190 lb [34-86 kg; $\bar{x}$ = 145 lb (66 kg)] while 9 Bornean males weighed between 75 and 200 lb [34-91 kg; $\bar{x}$ = 160 lb (73 kg)]. Five Sumatran female orangutans weighed between 74 and 98 lb [34-44 kg; $\bar{x}$ = 82 lb (37 kg)] and 10 Bornean females weighed between 72 and 100 lb [33-45 kg; $\bar{x}$ = 82 lb (37 kg)] (Eckhardt, 19795a). Schultz (1956, p.

895) also calculated that male orangutans are on average twice as heavy as females. Thus, *Pongo pygmaeus* is quite sexually dimorphic in body size.

MacKinnon (1975) characterized adult Bornean orangutans as generally exhibiting somewhat darker body hair (chocolate or maroon); more glabrous gular air sacs; broader and less hairy faces, with prominent round muzzles; less plantigrade feet; and a more rotund body build than Sumatran orangutans. Some male Sumatran orangutans sport long, pointed, yellow or white beards, whereas those of Bornean males are generally darker hues of orange and red (MacKinnon, 1975). Some old female Sumatran orangutans also grow respectable beards (Rijksen, 1978).

Most adult males of both subspecies develop striking fibrofatty cheek pads or flanges that project laterally beneath the facial skin and musculature (Fooden and Izor, 1983). The cheek pads provide male Sumatran orangutans with an ovate facial configuration (Figure 23). The somewhat floppier flanges of Bornean males are heavier over the upper face and brows, giving them a more quadrangular countenance (MacKinnon, 1975).

MacKinnon (1974, p. 77) commented that "the Bornean and Sumatran races are so similar that subspecific status is doubtful." Further, Rijksen (1978, pp. 27-30) discovered that in Sumatra two quite different types of orangutan coexist and even occur in the same pedigree. The dark-haired, long-fingered type invariably has dark hair and their skin is dark brown to blackish. They are the smallest of the two Sumatran types. They appear to be rather delicately built, with slender forelimbs and hind limbs and long fingers and toes. Their well developed thumbs and first toes are always nailed.

The light-haired, short-fingered type of Sumatran orangutan has hair of reddish hues. And their skin is either light brown or dark greyish brown. Their forelimbs and hind limbs are stout and their fingers and toes are relatively short and thick. Their pollices and halluces are small. Their first toes always lack nails and their thumbs can also be nailless.

Rijksen (1978) was acquainted with more than 80 wild and rehabilitant Sumatran orangutans. Among them, he observed many individuals that were intermediate between the two extremes sketched hereabove. But he noted that while the lighter colored orangutans variably exhibited some features of the dark form, such as long fingers, the characteristics of the dark form were quite consistent. All dark individuals had long slender extremities and nailed thumbs and first toes (Rijksen, 1978, p. 27).

The African great apes include several subspecies of common chimpanzees, the bonobo (Sarich, 1984; Socha, 1984), and two or more subspecies of gorillas. Because they all share a unique mode of quadrupedal positional behavior—termed *knuckle-walking*—for which the forelimbs are specially adapted (Figure 7), all of the extant African apes should be placed in the genus *Pan*, with two subgenera and three species (Tuttle, 1967, 1969a, 1970, 1974, 1975; Groves, 1970b, p. 15). Accordingly, the common chimpanzees are formally designated *Pan (Pan) troglodytes*; the bonobos and *Pan (Pan) paniscus*; and the gorillas are *Pan (Gorilla) gorilla.*

The common chimpanzees are widely distributed in the forest belt of tropical Africa, from The Gambia and Guinea-Bissau eastward through West and northern Central Africa to the Great Lakes region of western Uganda and northwestern Tanzania (Figure 2a). Ecologically, they are the most versatile of the extant apes, inhabiting primary and secondary forests, deciduous woodlands, riverine forests and forest-savanna mosaic regions.

Common chimpanzees are sexually dimorphic, but much less so than orangutans and gorillas (Schultz, 1956, 1962a, 1969; Gavan, 1971; Johanson, 1974a, Wolpoff, 1975). Schultz (1940) compiled data on the weights of 9 adult male and 19 adult female wild and captive common chimpanzees. The males averaged around 100 lb (31.8-60.0 kg). The female average weight is 88 lb (31.3-49.8 kg).

On the basis of hair patterning, skin pigmentation and cranial morphology, Hill (1967, 1969) recognized four subspecies of *Pan troglodytes*. The westernmost subspecies is the masked, pale-faced or Upper Guinea chimpanzee, *P. t. verus*. They have been found in Senegal, The Gambia, Guinea-Bissau, Guinea, Sierra Leone, Liberia, Ivory Coast and Ghana as far as Lake Volta (Hill, 1969; Bournonville, 1967; Dupuy, 1970; Dupuy and Verschuren, 1977; Robinson, 1971; Myers et al., 1973; Monfort and Monfort, 1973; Geerling and Bokdam, 1973; Bourlière et al., 1974). Their distribution is interrupted by the Dahomey Gap. But they recur in Nigeria west of the Niger River. At birth, *P. t. verus* have flesh-colored faces, with or without bluish-brown pigmentation around the eyes and on the bridge of the nose. As they mature, the face darkens, especially if exposed to sunlight, but the mask usually remains discernible. Their large, pale ears protrude laterally. They possess white beards and pallid palms and soles.

The black-faced, Lower Guinea or western chimpanzee, *P. t. troglodytes*, lives east of the Niger River in Nigeria, Cameroon, Equatorial Guinea, Gabon, westernmost Central African Republic west of the Sangha River, Congo, perhaps Cabinda (Angola), and Zaire north of the mouth of the Zaire River (Jones and Sabater Pí, 1971; Sabater Pí and Groves, 1972; Gibbons, 1970; Gartlan and Struhsaker, 1972; Emmons et al., 1983; Spinage, 1981; Tutin and Fernandez, 1984a,b). They are probably recently extinct south of the Zaire River in northwestern Angola. Another common name for *P. t. troglodytes* is the tschego.

Neonatal *P. t. troglodytes* are pale-faced. Later, tan blotches appear, spread and darken until, by adolescence, the face and ears are black. In adults, the rest of the skin, including the palms and soles, is also black. White hairs are absent from or sparse on their chins.

The long-haired or eastern chimpanzee, *P. t. schweinfurthi*, lives in Zaire, north of the Zaire River and east of the Ubangi River. Thence they extend eastward into the southwestern extremity of Sudan, western Uganda,

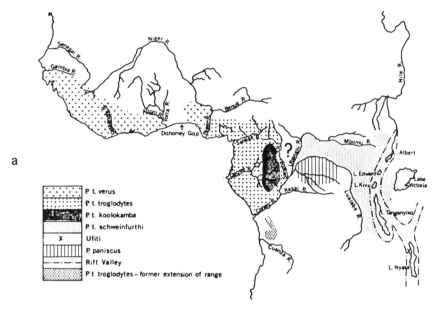

a

© S. Karger AG, Basel

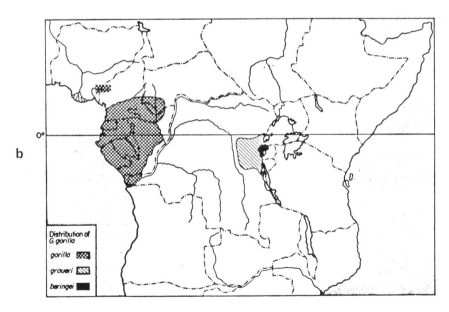

b

Figure 2: (a) Distribution of bonobos and common chimpanzees in 1970 (Courtesy of G.H. Bourne and S. Karger AG, Basel). (b) Distribution of gorillas in 1970. Nigerian populations may now be extinct. (Courtesy of C.P. Groves.)

Rwanda, Burundi, and along the northeastern shore of Lake Tanganyika in the Kigoma and Tabora Regions of western Tanzania (Butler, 1966; Hill, 1969; Albrecht, 1976; Verschuren, 1978; Kano, 1971, 1972).

The faces of newborn *P. t. schweinfurthi* are pale. Then they darken uniformly to bronze or coppery hues. Their ears, palms and soles are also hues of bronze. If aged individuals are exposed to sunlight, their faces may become quite dark. *P. t. schweinfurthi* sport long side-whiskers and body hair. Their chins bear relatively few white hairs (Hill, 1969). The *Pan troglodytes schweinfurthi* of Gombe National Park are lightweights compared to the averages which Schultz (1940) compiled for the species. In a sample of 15 adults, the average Gombe male would weigh 87 lb and the average Gombe female would weigh 66 lb (Zihlman and Cramer, 1978, p. 90). Wrangham and Smuts (1980) found that the average weight of 6 Gombe females [66 lb (29.8 kg); r = 59-72 lb (26.9-32.9 kg)] was about 76% of the average weight of 9 Gombe males [87 lb (39.5 kg); r = 80-95 lb (36.2-43.0 kg)]. Rahm (1967, p. 205) calculated the average weights of 3 adult males to be 94 lb (42.8 kg) and of 9 adult females to be 75 lb (34.2 kg). The animals had been captured recently near the western shore of Lake Kivu in Zaire.

The koolokamba or gorilla-like chimpanzee, *P. t. koolokamba*, is the least well known subspecies of common chimpanzee, which Hill (1967, 1969) recognized. Captives have been confused with gorillas and some authors have suggested that they are in fact hybrids between gorillas and chimpanzees (Shea, 1984). To my knowledge, the question of hybridization between chimpanzees and gorillas has not been systematically tested experimentally. Rumors about the production of chimpanzee-gorilla hybrids (and even pongid-human ones) sometimes buzz through primatological conferences. Alas, these chimrillas, hupanzees, and gomans have the credibility of human clones. Yet the recent arrival of viable siabons should temper our skepticism regarding future possibilities (Myers and Shafer, 1979; Wolkin and Myers, 1980; Shafer et al., 1984; Kortlandt, 1981; Rumbaugh, 1981a).

Solitary and small groups of koolokambas are supposed to live sympatrically with Lower Guinea chimpanzees (*P. t. troglodytes*) in the inland montane rain forests of Cameroon, Gabon and perhaps the Congo. Following an initial brown blotchiness, the prognathous faces of koolokambas become ebony black. Their ears are smaller than those of other common chimpanzees and they lie close to their heads. As in gorillas, the nasal alae of koolokambas are padded, giving their noses some resemblance to "a squashed tomato" (Hill, 1967). They have prominent side-whiskers and hirsutulous chins. Their body hair is long and coarse. Their black hands and feet are massive (Hill, 1969).

After the elusive koolokamba, the bonobo, *Pan (Pan) paniscus*, is the least well known of the African apes. It has received notable attention recently because of suggestions that it might be a close counterpart of *Ramapithecus* in the flesh or of the precursor of *Australopithecus* (Zihlman et al., 1978; Zihlman, 1979). Both of these fossil forms have been proffered as the stem Hominidae.

Bonobos are also commonly called pygmy or dwarf chimpanzees because of their presumed smaller overall body size. But if they are indeed a unique species, terming them pygmy or dwarf chimpanzees is about as misleading as calling the common chimpanzees pygmy gorillas.

Bonobos are confined to the highly humid and swampy (Phillips, 1959) forests of central Zaire, south and west of the Zaire and Lualaba Rivers and north of the Kasai River (Nishida, 1972a; MacKinnon, 1976; Badrian and Badrian, 1977; Kano, 1979, 1984a; van den Audenaerde, 1984; Figure 2a). While most observers agree that bonobos are generally less heavy than the common chimpanzees, Hill's statement (1969, p. 44) that the body weight of *Pan paniscus* is less than half that of *Pan troglodytes* surely exaggerates the diminuitive nature of the beast (Coolidge and Shea, 1982; Shea, 1983a, 1984b). The first thoroughly described wild-shot specimen, a pregnant female, was "unusually small for an adult chimpanzee" (Coolidge, 1933, p. 7). No body weight was given for it. But its bones were around half the average weight of counterparts from *Pan troglodytes* (Coolidge, 1933). Schultz (1969, p. 32) argued that *Pan paniscus* is not smaller than many individuals among the several "races" cf. subspecies) of *Pan troglodytes*. Schultz's view is strongly supported by the data which Zihlman and Cramer (1978) compiled on 18 adult *Pan paniscus*, consisting of 13 wild-shot specimens and 5 captives. The average body weight of the females is 69 lb and that of the males is 86 lb (range, 25-48 kg; overall average, 35.5 kg). These weights are remarkably similar to those of the Gombe chimpanzees.

Bonobos have been characterized as having black faces (except around the mouth), ears, palms and soles from an early age (Hill, 1969; Badrian and Badrian, 1977). However, some captives exhibit lighter pigmentation (Plates 8-16 in Savage-Rumbaugh and Wilkerson, 1978). Bonobos sport long thick side-whiskers and have long fine body hair (Figure 6). Some adults retain the infantile white perianal tuft of hair. Unlike common chimpanzees and some humans, they appear to escape the distinguishing devastation of alopecia. Their ears are smaller than those of common chimpanzees, falling more into the size range of humans and gorillas (Coolidge, 1933).

Adults of both sexes have gracile crania, with minor development of bony superstructures, like the supraorbital ridges. The muzzles of bonobos are less prognathic than those of common chimpanzees. Their foreheads are more conspicuous and their skulls are more rounded than those of common chimpanzees. All in all, the crania of adult *Pan paniscus* are strikingly similar to those of juvenile *Pan troglodytes* (Coolidge, 1933; Hill, 1969; Fenart and Deblock, 1973, 1974; Cramer, 1977; Zihlman and Cramer, 1978; Shea, 1983a,b, 1984b, 1985). Cranial capacities of the two species overlap considerably though the average cranial capacity of *Pan paniscus* is lower than that of *Pan troglodytes* (Fenart and Deblock, 1973; Cramer, 1977; Holloway, 1978; Chapter 6).

Although bonobos cannot be distinguished from all subspecies of common chimpanzees on the basis of gross body weight, they do seem to exhibit several characteristic postcranial features and proportions. Observers have commented particularly on their narrow shoulders and slender, long-limbed physiques (Coolidge, 1933; Hill, 1969; Badrian and Badrian, 1977; Zihlman and Cramer, 1978; Susman, 1979; Coolidge and Shea, 1982; Shea, 1981, 1983b, 1984b). Narrowness of the shoulders in bonobos is probably due to the unusual shortness of their collar bones and the narrowness of their shoulder blades (Coolidge, 1933; Zihlman and Cramer, 1978). When standing quadrupedally, the backs of bonobos often appear to be less inclined caudally than those of quadrupedal common chimpanzees (Susman, 1979, Figure 9). This may be related to the fact that the thigh bones are generally longer in bonobos than in common chimpanzees (Zihlman and Cramer, 1978).

Johanson's (1974) extensive odontometric study on bonobos and common chimpanzees revealed that *Pan paniscus* have smaller teeth than *Pan troglodytes*. The two species can also be distinguished by shortness of the mandible in *Pan paniscus* (Cramer, 1977). In the dentition of bonobos, only the permanent canine teeth are sexually dimorphic. The amount of canine dimorphism is less in bonobos than in common chimpanzees (Fenart and Deblock, 1973; Johanson, 1974; Almquist, 1974; Kinzey, 1984). Indeed in many cranial and postcranial features, *Pan paniscus* exhibits less sexual dimorphism than *Pan troglodytes* does (Fenart and Deblock, 1973; Cramer, 1977; Zihlman and Cramer, 1978; Susman, 1979; Laitman and Heimbuch, 1984).

Unlike *Pan troglodytes*, in sexually mature male *Pan paniscus* the glans penis is well developed and quite conspicuous during penile erections (Savage-Rumbaugh and Wilkerson, 1978). In this feature the bonobo more closely resembles the gorilla than the common chimpanzee. The clitoris of the female bonobo is quite discernible even during the maximal monthly sexual swelling of her labia minora when it become reoriented to point ventrally between the thighs (Dahl, 1985). Like the penis, it enlarges when the female is libidinous or generally excited. In common chimpanzees, the clitoris is much less visible, especially during the monthly estrous tumescence (Savage-Rumbaugh and Wilkerson, 1978; Savage and Bakeman, 1978; Dahl, 1985).

Gorillas, *Pan (Gorilla) gorilla*, are the largest of the extant apes. They now inhabit two widely separated areas of tropical Africa, one in West Africa and the other in northeastern Central Africa and southwestern Uganda (Figure 2b). In some areas, they are sympatric with common chimpanzees. But, unlike chimpanzees, the gorillas show a marked proclivity for particular types of montane and secondary forests (Schaller, 1963).

In many features, gorillas are among the most sexually dimorphic animals (Schultz, 1962, 1973; Tobias, 1971, 1975; Wolpoff, 1975). Because large male gorillas were much prized as hunting trophies (by museum curators as well as by "sportsmen") more information is available on their body weights than on those of wild adult females (Cousins, 1972). Groves (1970c) computed the average adult male weights of 32 western gorillas, 6 mountain gorillas, and 4 eastern lowland gorillas to be 307, 343, and 360 lb, respectively. Schaller (1963, p. 76) cited weights of 214.5 and 159.5 lb (Gyldenstolpe, 1928; Grzimek, 1957) for individual wild-shot adult female mountain and western gorillas, respectively. Like *Pongo pygmaeus*, adult females are around half the weight of adult male *Pan gorilla*.

Until approximately 10 years of age, male gorillas are difficult to distinguish from female gorillas (Figure 28). Then the male secondary sexual characteristics develop rather rapidly (Schaller, 1963). They include massive cranial superstructures (supraorbital ridges and sagittal and nuchal crests), elongate palates and mandibles, large canine teeth, and a distinctive subcutaneous fibrofatty pad over the posterior part of the sagittal crest. Adult male gorillas are called "silverbacks" (Fossey, 1979; Harcourt, 1979) and "silverbacked males" (Schaller, 1963) because of stunning patches of greyish-white hair which develop over their lumbar regions.

The skin of newborn gorillas is pinkish-grey. Their hair is brown or black or mixtures of both colors. By one year of age, the skin has darkened to charcoal grey or black and, like the chimpanzees, they have developed a perianal tuft of white hair. It disappears sometime after their fourth year (Schaller, 1963; Groves, 1970b; Fossey, 1979). The ears of gorillas lie close to their heads and, like those of orangutans, appear to be relatively small. One of the most distinguishing features of the gorilla face is the "squashed tomato" nose. The noses of gorillas are so variable that Schaller (1963) could use them to identify the individuals in his study population.

In a classic taxonomic study, Coolidge (1929) recognized only 2 subspecies of gorillas. More recent and thorough studies (Vogel, 1961; Groves, 1967, 1970b,c) demonstrated that there are probably 3 subspecies of *Pan gorilla*.

The western or lowland gorilla, *Pan gorilla gorilla*, lives in the forests of Cameroon, Equatorial Guinea, Gabon, Congo, Cabinda (Angola), the western extremity of Central African Republic (on both sides of the Sangha River), the bit of Zaire which lies north of the mouth of the Zaire River, and perhaps a small isolated area in the Cross River district of southeastern Nigeria (Bützler, 1980; Coolidge, 1936; Groves, 1971c; Jones and Sabater Pi, 1971; Emmons et al., 1983; Tutin and Fernandez, 1984a,b; Fossey, 1982). The term lowland gorilla is misleading since *Pan gorilla gorilla* also inhabit plateau regions in Cameroon and a quite hilly region of Nigeria (Groves, 1967).

The western gorilla is the smallest of the 3 subspecies. Among gorillas, they have a relatively short palate and tooth rows. Brown pelage is more common in western gorillas than in the eastern gorillas. The "saddle" of males may extend onto their thighs. The "tip" of the nose overhangs the median nasal septum, forming a distinct lipped structure, which is apparently unique to *Pan gorilla gorilla* (Napier and Napier, 1967; Groves, 1967, 1970b,c).

The mountain gorilla, *Pan gorilla beringei*, is restricted to two enclaves. The best known population lives on the 6 dormant volcanoes and intervening saddle areas of the Virunga Volcanoes, which lie astride the international borders of Zaire, Rwanda and Uganda. Gorillas are absent from the 2 westernmost Virunga Volcanoes, which are still active (Schaller, 1963, p. 32). The second population of mountain gorillas lives in the Bwindi Forest Reserve (formerly, the Kayonza Forest and Impenetrable Central Forest Reserve), Kigezi District, southwestern Uganda, 15 miles north of the easternmost Virunga Volcano (Groves and Stott, 1979).

Pan gorilla beringei is the second largest subspecies of *Pan gorilla*. They have very large jaws and teeth and exceptionally long palates and toothrows (Groves, 1970c). Their hair is black, quite long and fine. In adult males, the "saddle" is restricted to the back. The vertebral borders of their shoulder blades are sinuous. Their arms are short and their long great toes seem to be more aligned with the lateral 4 toes than is the case in the other gorillas (Schultz, 1934; Groves, 1970c).

The largest subspecies is probably the eastern lowland or Grauer's gorilla, *Pan gorilla graueri*. It inhabits lowland and certain highland forests of northeastern Zaire, east of the Lualaba River and west of the section of the Western (Albertine) Rift in which Lake Edward, Lake Kivu and the northern tip of Lake Tanganyika lie (Groves and Stott, 1979). Because *Pan gorilla graueri* has been seen at altitudes up to 2,400 m (8,052 ft) in the Tshiaberimu highlands west of Lake Edward, Stott (1981) suggested that the subspecies be termed Grauer's gorilla instead of the eastern lowland gorilla.

In many features, Grauer's gorillas are intermediate between the western and the mountain gorillas. The hair of Grauer's gorillas is black, like that of mountain gorillas, but short, like that of western gorillas. Like *Pan gorilla beringei*, the saddle is restricted to the back in adult male *Pan gorilla graueri* (Groves, 1970c). The subspecific classifications of some populations of eastern gorillas remain equivocal (Groves and Stott, 1979).

TEMPERMENT AND TAXONOMY

Both as a topic of basic scientific interest and in order to implement informed selections of subjects for particular investigations on mentality and sociality, Yerkes and Yerkes (1929) explored the possibility that hereditary interspecific and interracial (cf. subspecfic) variations exist in the temperaments of the apes. In special chapters of *The Great Apes*, they carefully assembled available published information and their own personal observations and impressions on affectivity and its expressions in each form.

Yerkes and Yerkes (1929) could cite no affective traits by which the various species and subspecies of gibbons, including the siamang, could be distinguished from one another. They elected to emphasize their good-tempered, affectionate, timid and gentle nature while warning that particular captives will bite savagely and otherwise reject the effusive attentions of humans. They considered the data on affectivity in the Hylobatidae to be "crude, incomplete, unreliable, merely glimpses thorugh a clouded glass, and well-nigh valueless . . . " (Yerkes and Yerkes, 1929, p. 80).

On the basis of literary anecdotes and personal observations of a 5 year-old male, a 10 year-old male, and 2 females between 6 and 8 years old, Yerkes and Yerkes (1929) concluded that orangutans are reserved, cautious, timid, apprehensive, curious and exploratory. They also stressed that at sexual maturity, the males become dangerously aggressive toward human keepers. The oldest male orangutan with which they were personally acquainted was slain during a donnybrook with his attendant at the Abreu primate colony in Havana, Cuba.

In 1929, the paucity of reliable information on the temperaments of gorillas was rivalled only by the dearth on gibbons. Therefore, Yerkes and Yerkes were reduced to rag picking in the popular works of great white hunters, like Du Chaillu (1861, 1867) and Akeley (1922, 1923a,b), and sharing their impressions of an adolescent female mountain gorilla, named Congo, with which they had worked intensively. They sketched the temperament of the "generalized gorilla" as shy and retiring instead of annoying, aggressive or inquisitive; sly and cunning versus direct and frank in its actions; and stolid, stoic, independent, introverted, and above all, self-dependent. They disagreed with claims that the gorilla is a particularly savage and ferocious beast, though they granted that unusually fearful, provoked, old, injured and ill individuals can be hazardous to human health (Yerkes and Yerkes, 1929, p. 455).

Yerkes and Yerkes (1929) concluded that in contrast with orangutans and gorillas, chimpanzees are extraordinarily emotional, extroverted creatures that generally exhibit sanguine, good-natured, sociable and amicable temperaments unless they are ill or have been mistreated by humans. Because of their firsthand acquaintance with a good number of chimpanzees in various age-sex categories and the excellent descriptions by Köhler (1921, 1925) and Kots (1921), Yerkes and Yerkes (1929, p. 284) presented a more detailed and reliable account of temperamental changes from infancy to senility in the chimpanzee than was possible for the other apes.

They recognized close parallels between chimpanzees and humans in the ontogenetic transformations that occur in their temperaments and affective expressions. But they repeatedly warned readers of the terrible inadequacy of systematically collected information about temperament in any ape. Further, they pointed out that the divergence in affective behavior between the chimpanzee and the other pongid apes is perhaps no greater than that which exists between chimpanzee youngsters and adults (Yerkes and Yerkes, 1929, p. 276).

The mountain gorilla, Congo, was not the only pongid rarity that Robert Yerkes was privileged to study in the early years of primate psychobiology. By chance, one of his first 2 pongid subjects was a black-faced juvenile male bonobo, which he named Chim. Yerkes and Yerkes suspected that he was not an ordinary chimpanzee, as witnessed by the legend under a picture of him, which states "He resembled the gorilla more closely than do most examples of his genus" (1929, Figure 126, p. 394). Nevertheless, because his engaging personality had made him a family favorite (Yerkes, 1977; Blanshard, 1977) Chim seems to have heavily influenced their profile of the chimpanzee temperament. Panzee, the white-faced juvenile female common chimpanzee, purchased with Chim, was described as distrustful, retiring, lethargic and stupid (Yerkes and Learned, 1925, p. 30; Yerkes and Yerkes, 1929, p. 281). She was also sick and soon dead.

Yerkes and Yerkes did not include temperamental traits in their discussions on the taxonomy of the apes. They left for future scientists the systematic studies that might reveal naturalistic differences between the temperaments of various hominoid taxa. The task stayed undone. It would appear now that such studies are largely impracticable. Indeed, mindful of the morasses into which scientists who worked on human race and temperament have bumbled, some would claim that questions about hereditary temperamental differences among our complex collaterals are unanswerable. For those who would step in where modern anthropologists fear to tread, I recommend that they first ponder Hooton's horrific epitome of a twelve-year study on incarcerated persons, *Crime and the Man* (1939).

Since 1929, we have merely accumulated more anecdotes on the temperaments of individual apes. Nevertheless, some taxonomists have enumerated temperamental and other poorly understood affective traits in their classifications of pongid apes. For instance, Hill (1969, pp. 45-46) included certain "behavioural peculiarities" to justify the specific status of *Pan paniscus*. These were based on Coolidge (1933), who had relied upon the description of Chim by Yerkes and Learned (1925). Wisely, Hill did not repeat the statement that "Whereas Panzee in both facial appearance and manner suggested the Irish type, Chim similarly suggested the Negro" (Yerkes and Learned, 1925, p. 24; Coolidge, 1933, p. 49). It smartly illustrates the state of the art and social prejudice then. Paradoxically, Yerkes and Yerkes (1929) had chosen a bonobo to exemplify the temperament of the common chimpanzee and, only 4 years later, Coolidge (1933) cited descriptions of the same animal to characterize a new species. Coolidge (1933, p. 51) was careful to point out that Chim probably exhibited "greater intelligence" than Panzee because he was older than she was. Nevertheless, the exemplary reputation that Chim's benefactors bestowed on the bonobo has held up well (Palmans, 1956; Rempe, 1961; Hill, 1969; Gijzen, 1974; Savage-Rumbaugh, 1984a). Hill (1969, p. 45) stated that "Bonobos are of gentle disposition; they rarely bite, scratch or strike, but protect themselves by kicking . . . "

Schaller (1963) bolstered the reputation of gorillas as stolid creatures. Marler (1976) wrestled with this stereotype in a comparison between the vocal behaviors of Virunga gorillas and Gombe chimpanzees. He noted that distressed chimpanzees of all ages and both sexes whimper while only infant gorillas emit counterpart cries. However, if sheer vocal output were used as an index of temperament, the silverbacked male gorilla could not be viewed as less emotional and expressive than common chimpanzees are (Marler, 1976, p. 257).

In classifying orangutans, Groves (1971b) and MacKinnon (1974, 1975) avoided discussions of temperament, except that MacKinnon (1974, p. 70) remarked on the greater sociability of the Sumatran subspecies. Per contra, in a section (tellingly titled "typology") of his monograph on Sumatran orangutans, Rijksen (1978, p. 29) portrayed the short-fingered type as more amicable, extroverted, playful, fearful of new stimuli and inclined to exhibit bluff-behavior than the long-fingered type of orangutan. He described the latter as introverted, with a tendency to threaten and attack other orangutans and humans instead of displaying bluff-behavior. He further added that "It was common knowledge among the former local orangutan hunters that the dark (long-fingered type) orang utans made less desirable pets than the lighter ones, because they were assumed to be unpredictable, unmanageable and even dangerous animals in captivity." (ibid., p. 29). It appears that Rijksen reached these conclusions on the basis of his interactions with relatively few rehabilitants near his camp. Such as been the stuff of human racism. These passages must count among the least substantive in his informative monograph.

Maple (1980, pp. 86-94) attempted to assess the temperaments of 37 pongid apes objectively by systematically recording a variety of facial expressions, vocalizations and body movements as they reacted to the sound of food carts that convey their daily sustenance at the Yerkes Primate Center. All of the gorillas (n = 8) and chimpanzees (13) and most (12 of 16) of the orangutans were adults. The chimpanzees and orangutans were well represented by both sexes. There were 6 female and only 2 male gorillas. Maple (1980, p. 87) determined that chimpanzees are much more expressive than the other apes that he measured and that the gross amounts of expressiveness by orangutans and gorillas are approximately equal though they are qualitatively different. The gorillas were notable for the infrequency of facial expressions; low grunts and growls; chest-beating; and digit-sucking. The orangutans moved little as the food cart approached, showed considerable interest in Maple's recording device, and were still, except for long calls by adult males. Hungry chimpanzees are noisy and active. Like Yerkes and Yerkes (1929), Maple (1980, p. 83) concluded that in order of increasing emotionality are orangutans, gorillas and chimpanzees.

Clearly, before further attempts are made to generalize the temperaments of the apes and to utilize such information in taxonomic classifications and the selection of subjects for scientific research, much more thoroughgoing studies will have to be conducted. The form that these studies should take is hard to envision. Projects on monozygotic twins, raised under different controlled conditions, would be difficult to implement because of the rarity of multiple births and their uncertain mortality, especially in wild populations of apes (Goodall, 1979; Rosenthal, 1981).

FOSSILS

The comments of Yerkes and Yerkes (1929, p. 44) on the phylogeny of apes and humans occupy only one page of their opus. They adopted Smith's (1924) dendrogram of hypothetical evolutionary relationships among the primates and reiterated his conclusion that morphologically the gorilla is more like the human than the other living apes are. *Homo* is arrogantly central and culminant in Smith's dendrogram (Figure 3a). In view of historical human behavior, perhaps it is fitting that Smith drew the Pleistocene and Recent Hominidae as the blackest branch in a tapering tree.

In 1929, hominoid fossils were quite scanty and poorly dated. Hominoid specimens that have survived 5 decades of scientific scrutiny (though not always in the taxa of the twenties) include the fragmentary jaws and teeth of several species from Oligocene deposits in the Fayum region of Egypt (U.A.R.) and from Miocene deposits in Europe, Pakistan, and India; the juvenile skull and natural endocast from Pliocene or Pleistocene deposits at Taung, South Africa; mid Pleistocene cranial remains and perhaps the thigh bones of *Homo erectus* from Java; the robust cranium and other remains of 3 or 4 individuals from Broken Hill (Kabwe), Zambia, and fairly good examples of Neanderthal and later human skeletons from Europe. The "human" branch of Smith's phylogenetic tree, labeled *Hesperopithecus*, was pruned when the tooth upon which it was premised proved to be that of a peccary. The Piltdown forgery, which warped British theories of hominid evolution between 1915 and 1955, also should be footnoted either as a prank gone awry or a grave put-down aimed at prominent members of the British academic establishment.

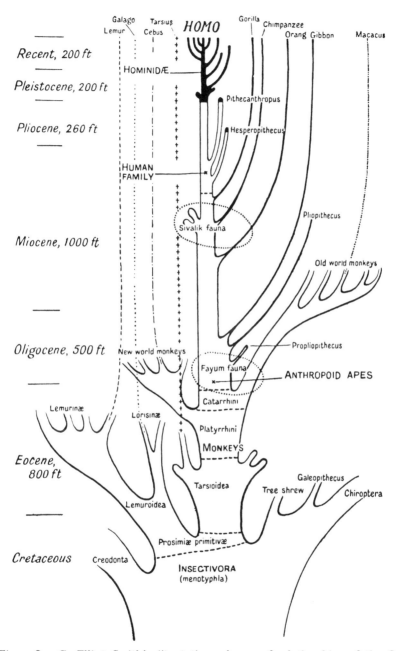

Figure 3a: G. Elliot Smith's "tentative scheme of relationships of the Order Primates." See text (p. 24) for discussion. (From G. Elliot Smith, *The Evolution of Man*, 1927, Oxford University Press.)

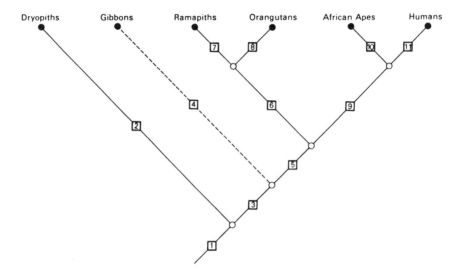

Figure 3b: Cladogram of hominoid relationships. The numbered boxes represent the following extinct morphotypes: 1. proto-hominoid; 2. dryopith; 3. stem extant hominoid; 4. proto-gibbon; 5. pre-great ape/human; 6. stem ramapith; 7. ramapith; 8. proto-orangutan; 9. pro-African ape/human; 10. proto-African ape; 11. proto-human. (© Russell L. Ciochon, 1983.)

During the period between 1929 and 1985, numerous fossil apes were recovered from Early and Middle Miocene deposits in the Turkana and Kavirondo Districts of western Kenya and the Karamoja and Sebei Districts of eastern Uganda. Potassium/argon dates bracket most of these hominoid fossils between 14 and 18 million years. Europe and Asia have also yielded numerous fossil apes. They are not associated with minerals amenable to radiometric dating. But faunal correlations indicate that the Eurasian hominoid specimens span a period of the Middle and Late Miocene between 6-9 and 16 million years.

Although for many years Europe ranked well behind Africa and South Asia in the number of Miocene hominoids it produced, it should no longer be regarded as a backwater of Miocene hominoid evolution. At the current rate of discovery, Europe and Western Asia could rival South Asia and even Africa in considerations of hominoid evolution. Thus, we will have come full circle back to the continent which provided the first specimens of *Dryopithecus* and *Pliopithecus* in the nineteenth century.

Paleoanthropological research over the past 50 years in China, Ethiopia, Java, Kenya, South Africa, and Tanzania has produced a prodigious quantity of Pliocene and Early-Middle Pleistocene hominid remains that are bracketed radiometrically and by faunal inferences between 3.5 million and 300,000 years. Europe and northwestern Africa have produced fewer, but nonetheless quite interesting, Middle Pleistocene hominids which severally evince morphological affinities with Asian and African *Homo erectus* and archaic *Homo sapiens* (Clark and Campbell, 1978; Howell, 1978; Sigmon and Cybulski, 1981; Smith and Spencer, 1984; Delson, 1985).

Before 1965, there was a superfluity of taxonomic nomina, giving the impression that there had been an extensive Miocene radiation of Eurasion and African apes. Then, in a major taxonomic revision, Simons and Pilbeam (1965) lumped many of the Miocene hominoid primates into two genera of dryopithecine pongid apes (*Dryopithecus* and *Gigantopithecus*) and one hominid genus (*Ramapithecus*). They recognized 3 subgenera and 7 species of *Dryopithecus*: *D.* (*Dryopithecus*) *fontani*; *D.* (*D.*) *laietanus*; *D.* (*Sivapithecus*) *indicus*; *D.* (*S.*) *silvalensis*; *D.* (*Proconsul*) *major*; *D.* (*P.*) *nyanzae*; and *D.* (*P.*) *africanus*. *Gigantopithecus blacki* and *Ramapithecus punjabicus* were monospecific in Simons and Pilbeam's (1965) scheme.

Dryopithecus (*Proconsul*) *africanus*, *D.* (*P.*) *nyanzae* and *D.* (*P.*) *major* are respectively small, medium and large species known only from eastern African Early and Middle Miocene deposits. *Dryopithecus fontani*, a medium-sized species, had a fairly wide distribution in Europe during the Middle and Late Miocene. Contemporaneously, the small *Dryopithecus laietanus* was supposed to have lived in northern Spain and South Asia (Simons and Pilbeam, 1965). *Dryopithecus indicus* is a large, later Miocene species from South Asia, Turkey, and perhaps China. Contemporaneously, *Dryopithecus sivalensis*, a medium-sized species, lived in South Asia and western Kenya (Simons and Pilbeam, 1965).

Gigantopithecus blacki is known only from southern China. Though generally thought to date from the Early or Middle Pleistocene, it is of uncertain provenance. Simons and Pilbeam (1965) concluded that *Ramapithecus punjabicus* was widely distributed in Europe, Kenya, South Asia and China during the later Miocene.

Simons (1963) also suggested that certain European and eastern African Miocene species should be considered as congeneric in the hylobatid genus *Pliopithecus*. The species thus lumped were *Pliopithecus* (*Epipliopithecus*) *vindobonensis* from Czechoslovakia and *Limnopithecus macinnesi* and a smaller form, *Limnopithecus legetet*, from Kenya.

These lumpophilic exercises were much applauded by the scientific community because they were in keeping with modern principles of evolutionary biology. They laid the groundwork for better focused studies and discussion on the hypothetical functional morphology and adaptations of the Miocene Pongidae, Hylobatidae and (perhaps) Hominidae. Refined dating techniques, paleoecological studies, and neonontological behavioral and anatomical studies of primates and other mammals also gave new impetus to paleoprimatology.

Additional discoveries of fossil apes since 1965, have inspired another era of taxonomic splitting and concomitant phylogenetic ramifications, though nowise comparable to that of earlier decades. The field is somewhere between flux and fulmination now. Andrews (1978a,b) and Kay and Simons (1983) have tackled the formidable task of reassessing the classic Neogene Hominoidea in light of many new specimens from Africa and Eurasia. Simons and Fleagle (1973), Andrews and Simons (1977), and Fleagle and Simons (1978) have revised the taxonomy of the Neogene hylobatid apes.

Based on a study of 700 specimens, Andrews (1978a) recognized 3 genera and 7 species of apes from eastern African Miocene deposits. He placed 6 of the species in the subfamily Dryopithecinae of the Pongidae and one species in the Hylobatidae.

Proconsul (Proconsul) major, P. (P.) nyanze, and *P. (P.) africanus* are well represented in the collection by gnathic specimens. Andrews (1978a) diagnosed two new species of *Proconsul* largely on the basis of dental features. *Proconsul (Rangwapithecus) gordoni* is similar in size to *Proconsul (Proconsul) africanus. Proconsul (Rangwapithecus) vancouveringi* is the smallest species of *Proconsul.* It is poorly represented by only 10 fragmentary specimens.

Andrews (1978a) removed *Limnopithecus legetet* from the Hylobatidae and placed it in the Dryopithecinae. Although Andrews and Simons (1977; Andrews, 1978a) left the hypodigm of *Limnopithecus macinessi* in the Hylobatidae, they renamed it *Dendropithecus macinessi.* Previously, Simons and Fleagle (1973) had resurrected the generic status of *Limnopithecus* because of postcranial skeletal differences between Kenyan *Limnopithecus macinnesi* and Czechoslovakian *Pliopithecus vindobonensis.*

Fleagle and Simons (1978; Fleagle, 1984) erected another new species of eastern African hylobatid ape on the basis of about 20 fragmentary specimens from Napak, Uganda. As this newcomer's nomen implies, *Micropithecus clarki* is the smallest hominoid primate recognized by Fleagle and Simons (1978). Andrews (1978a) provisionally considered these specimens to belong in *Limnopithecus legetet.*

Assuming that these hominoid species are valid, during the Early and Middle Miocene period, a considerable adaptive radiation of apes may have occurred in eastern Africa. This radiation included 2 species of hylobatid apes and 6 species of pongid apes. In craniodental size, *Micropithecus clarki* is similar to the capuchin monkey (*Cebus capucinus*) and *Dendropithecus macinnesi* resembles the siamang. Craniodentally, the smallest African dryopithecine ape (*Limnopithecus legetet*) was somewhat smaller than the common gibbon and the largest African dryopithecine ape (*Proconsul major*) was about the size of the female gorilla. Between 3 and 5 of these species were at least broadly sympatric at the most productive fossil localities in eastern Africa. Today only 2 or 3 species of apes are commonly sympatric in African, Southeast Asian and Indonesian localities.

Unlike some of his predecessors (Simons and Pilbeam, 1965, 1972; Pilbeam, 1969a, 1972), Andrews (1978a,b) concluded that the ancestors of extant apes could not be discerned among the eastern African Miocene specimens. He suggested that *Dendropithecus macinnesi* may have given rise to the mid-Miocene European hylobatid species of *Pliopithecus*; *Proconsul nyanzae* may be ancestral to mid-Miocene *Dryopithecus fontani* and *Dryopithecus laietanus*; and *Proconsul major* or *Proconsul africanus* may have direct phylogenetic links with *Sivapithecus* spp. and *Ramapithecus* spp. Andrews (1978a,b) proposed no descendents for *Limnopithecus legetet* and the 2 species of *Proconsul* (*Rangwapithecus*).

Both hylobatid and dryopithecine apes are known from the European Miocene. Except for a few controversial specimens, only dryopithecine apes are known from the Asian Miocene. In 1973, Simons and Fleagle commented that no discoveries of *Pliopithecus* had ever been made at the same European sites which yielded specimens of *Dryopithecus*. They inferred from this that species of the two genera were possibly adapted to different ecologies. But later the Hungarian site of Rudabanya provided specimens of *Pliopithecus* in addition to *Dryopithecus* and a form like *Ramapithecus* (Kretzoi, 1975).

Andrews's (1978b) study of the classic Eurasian dryopithecine specimens and recently discovered fossils from Greece, Hungary, Pakistan, and Turkey (Bonis et al., 1974, 1975; Bonis and Melentis, 1976, 1977a,b, 1978; Kretzoi, 1975; Simons and Chopra, 1969; Pilbeam, 1976; Pilbeam, Barry, et al., 1977; Pilbeam, Meyer, et al., 1977; Pilbeam et al., 1980; Pilbeam and Smith, 1981; Pilbeam, 1982; Andrews and Tobien, 1977; Andrews, 1982; Tekkaya, 1974; Andrews and Tekkaya, 1976) led to a major revision in the taxonomy of Eurasian Miocene apes. Although he lumped some specimens with exotic nomina like *Ankarapithecus meteai*, *Graecopithecus freybergi* and *Bodvapithecus altipalatus*, on balance, his scheme is that of a splitter. In addition to recognizing the probable validity of the 2 European species of *Dryopithecus*, he resurrected *Sivapithecus* to full generic status and proposed that it contains 4 species, at least 3 of which are found in Europe as well as in Asia. The chief criterion for assigning specimens to one of the 4 species of *Sivapithecus* is relative size. Simons (1976) and Pilbeam, Meyer, et al. (1977) concurred that *Sivapithecus* is generically distinct from *Dryopithecus*.

Sivapithecus meteai is the largest species, gnathodentally approximating the size of female gorillas and sometimes exceeding dimensions of *Proconsul major*. The smallest species is *Sivapithecus darwini*, the jaws and teeth of which approximate those of the siamang. *Sivapithecus indicus* and *Sivapithecus sivalensis* are the second and third largest species, respectively. Andrews (1978b) was especially enthusiastic about certain Turkish specimens of *Sivapithecus darwini*. He suggested that because they are morphologically and geochronologically intermediate between Early Miocene

Proconsul and Late Miocene *Sivapithecus* they probably represent an emigrant population from Africa which links the African dryopithecine apes with the Eurasion ones.

Initially, Andrews (1978b) did not connect any of the Middle-Late Miocene Eurasion apes to extant species. He suggested that *Sivapithecus meteai* might have given rise to *Gigantopithecus blacki* via *Gigantopithecus bilaspurensis* of the Indian Late Miocene (Andrews, 1978b). Since the Himalayan *yeti* has yet to be validated, this lineage too must be considered dead without issue.

Recently, Andrews and Tekkaya (1980) detected orangutan-like features in the maxilla, dentition, and lower face of a specimen of *Sivapithecus meteai* from the Sinap deposits near Yassorien, Turkey. Andrews (1982) and Preuss (1982) argued that *Sivapithecus*, especially as evidenced by a partial skull of *S. indicus* from the Potwar Plateau, Pakistan (Pilbeam and Smith, 1981; Pilbeam, 1982), is related to *Pongo pygmaeus*. Pilbeam (1982; Lipson and Pilbeam, 1982) doubts that *S. indicus* was ancestral to *Pongo* but he considers it likely that a similar late Miocene Asian species of *Sivapithecus* founded the lineage leading to *Pongo*.

The alpha taxonomy of European Miocene hylobatid apes is in disarray. According to Zapfe (1960) there are 4 or 5 species and 3 subgenera of *Pliopithecus* represented in collections from Austria, Czechoslovakia, and France. Localities in southern Germany and Switzerland have also produced specimens of *Pliopithecus*. Kretzoi (1975) reported an additional species from Rudabanya, Hungary. Simons and Fleagle (1973), Andrews and Simons (1977) and Fleagle and Simons (1978) proposed that *Pliopithecus* probably represents primitive descendents of the Oligocene hylobatid precursors and that among the African fossil species, *Dendropithecus macinnesi* is the most likely ancestor of modern hylobatid apes.

Unearthing the roots of the Hominidae has been as arduous as discerning the origins of the Pongidae and Hylobatidae. Among Miocene primates, there have been 3 main contenders for the coveted superlative, first hominid. They are, in order of increasing popularity, *Oreopithecus*, *Gigantopithecus*, and *Ramapithecus*. Hürzeler's (1958, 1968) suggestion that *Oreopithecus bambolii*, a chimpanzee-sized form (Straus, 1963) from late Miocene deposits in Tuscany, Italy, represents a stage in hominid evolution, has not been accepted. Indeed, some primatologists (Simons, 1972, p. 264; Delson, 1979) have been impressed by cercopithecoid affinities in the teeth and skeleton of *Oreopithecus*. Like certain monkeys of the New World, *Oreopithecus* might have developed postcranial skeletal features like those of extant apes via parallel evolution. Or the features in common could merely represent chance similarity among these forms (Tuttle, 1975).

As a potential ancestral hominid, *Gigantopithecus* has fared somewhat better than *Oreopithecus*. In 1945, Weidenreich, best known for his invaluable monographs on Chinese *Homo erectus* (alias Peking Man), proclaimed *Gigantopithecus blacki* to be a hominid on the basis of 3 molar teeth from a "dragon bone" shop in Hong Kong. He further suggested that *Gigantopithecus* was ancestral to *Homo erectus* of Java and perhaps China. In 1952, von Koenigswald, an expert on Javanese *Homo erectus*, concluded that the teeth belonged to a highly specialized hominid that was not ancestral to *Homo.*

Between 1956 and 1960, Chinese paleontologists recovered 3 mandibles and over 1000 isolated teeth of *Gigantopithecus* from caves in Kwangsi Province (Woo, 1962). In 1965, Simons and Pilbeam removed *Gigantopithecus blacki* from the Hominidae and placed it in the Pongidae. What he took from its status, Simons (1972, p. 254) returned in size, declaring that *Gigantopithecus blacki* might have been 8 feet tall and weighed 600 pounds.

Early in the 1970's several paleoanthropologists, who were unimpressed by the hominid credentials of *Ramapithecus*, renominated *Gigantopithecus* as ancestral to Pleistocene Hominidae (Eckhardt, 1972, 1973, 1975b; Frayer, 1973; Robinson, 1972). Robinson (1972) erected a new hominid subfamily, the Paranthropinae, to accommodate *Gigantopithecus* with *Paranthropus*, a form which other authorities on early Hominidae consider to be a robust species of *Australopithecus.*

Simons (1961, 1964, 1968, 1972) was the first full-time champion for *Ramapithecus punjabicus* as the earliest hominid and ancestor of *Australopithecus.* The hypodigm consisted of less than a dozen jaws and teeth, previously classified under different nomina, from the Siwalik Hills of India, the Salt Range of Pakistan, and Ft. Ternan, Kenya (Simons, 1964; Simons and Pilbeam, 1965). Additional gnathic and dental specimens were assigned to *Ramapithecus* as scientists restudied existing museum collections (Pilbeam, 1969b; Simons, 1976, 1978) and resumed field work in Kenya, India, Turkey, Hungary, and Pakistan (Leakey, 1962; Prasad, 1969; Tekkaya, 1974; Kretzoi, 1975; Andrews and Tobien, 1977; Pilbeam, Meyer, et al., 1977).

Andrews (1978b) concluded that there are at least 2 species of *Ramapithecus* represented in the greatly increased sample: *Ramapithecus punjabicus* from Late Miocene localities in Pakistan and India and *Ramapithecus wickeri.* *Ramapithecus wickeri* from Ft. Ternan, Kenya, and specimens like it from Candir, Turkey are probably Middle Miocene in age (Andrews, 1978b). The fauna from Candir indicates that the Turkish locality is older than Ft. Ternan. This suggests that *Ramapithecus* could have originated in Asia and then immigrated to Africa. This form is also a likely ancestor for the South Asian *Ramapithecus punjabicus* and its Hungarian counterpart (Andrews, 1978b). Andrews (1978b) refrained from linking *Ramapithecus* to Pliocene and Pleistocene species.

Support for the hominid status of *Ramapithecus* has collapsed. Greenfield (1979) and Andrews (1982) would lump it into *Sivapithecus.* Further, Andrews (1982) would have *Sivapithecus punjabicus* in the lineage with

Pongo pygmaeus (Figure 3b). This would leave us with no recognized ancestral Hominidae prior to *Australopithecus afarensis* from early Pliocene deposits at Hadar, Ethiopia (Johanson, et al., 1978; Johanson and White, 1979) and the bipedal creatures that left their footprints at site G, Laetoli, northern Tanzania (Leakey, 1979, 1981; Leakey and Hay, 1979; Hay and Leakey, 1982; Tuttle, 1981, 1986).

We must conclude that because the fossil record is so badly broken, puzzles about where, when, and how the extant apes and humans evolved cannot be resolved now. Although the Miocene fossils are numerous, they were obtained from a very small segment of our planet. Western and central Africa, the Arabian peninsula (Andrews et al., 1978; Hamilton et al., 1978), southeastern Asia and other major regions that might have been hotbeds in the hylobatid, pongid, and hominid careers remain very poorly sampled. Unless these huge gaps are filled, we may never be able to document thoroughly the origins and deployment of gibbons, orangutans, chimpanzees, gorillas, and humans.

2

Positional Behavior

It flies through the air with the greatest of ease, that daring young gibbon on the wisp of a breeze. (Anon.)

Long before the mid-nineteenth century evolution revolution, philosophers and natural historians had noted similarities and differences in the locomotor systems of apes and humans (Eiseley, 1958; Greene, 1959). Lamarck (1809) had even toyed with hypothetical transformations from one form to another. Huxley (1863), Haeckel (1866, 1868, 1874) and Darwin (1871) were quite vague about the nature of changes from apes to humans because little was known about the naturalistic locomotor habits of the apes (Tuttle, 1974). Most captives were sickly youngsters that died soon after importation into Europe.

By 1929, a lively controversy had developed over the role that a specific pattern of arboreal locomotion—termed brachiation—had played in the evolution of hominid bipedalism (Tuttle, 1974, 1977). The chief architect of the theory, Keith (1891, 1903, 1912, 1923, 1927, 1934, 1940), and some later contributors to hominoid evolutionary theory, including Schultz (Carpenter, 1940), Clark (1968), Gregory (Gregory and Raven, 1937), and Washburn (Carpenter, 1940) had opportunity to observe one or more species of apes in the wild, often as they hunted them for anatomical studies. Between 1929 and the mid 1960s, *The Great Apes* served as the best general reference on the positional behavior of apes. During this period several alternative models on the evolution of bipedalism were devised and aired (Tuttle, 1969a, 1974, 1977).

Positional behavior is a broad term, encompassing both locomotion (movements from place to place) and more or less static postures (i.e., situations in which the relations between an animal and its external

33

environment are relatively stable) (Prost, 1965). Yerkes and Yerkes (1929) focussed primarily on the locomotor behavior of apes. Following Pocock (1905), they devoted special attention to the relative facility with which the different species of apes walk quadrupedally and bipedally on the ground and climb in trees. They summarized information about captives and free-ranging subjects. Generally they were careful to distinguish between them. Yerkes and Yerkes (1929) discussed preferred sitting, resting, and sleeping postures but presented virtually no information about foraging behavior and feeding postures in natural settings (Tuttle, 1977).

On the basis of anecdotes from natural historians and travelers, Yerkes and Yerkes (1929, pp. 53-56, 555-556) inferred the following about the positional behavior of the lesser apes: *Hylobates* spp. move rapidly by arm-swinging from branch to branch instead of jumping or springing. They are remarkably adept, even bird-like, in the ability to change course during rapid arboreal flight. Siamang are less agile than the smaller gibbons. On the ground, the bipedal ability and predisposition of hylobatid apes are surpassed only by those of humans among catarrhine primates. Pongid-like quadrupedal knuckle-walking was undescribed for the gibbon or siamang. They sleep upright in crotches of trees assisted by hand and footholds (Tuttle, 1977).

Schlegel and Müller (1839-1844), Wallace (1869), Hornaday (1879, 1885) and other nineteenth century hunter-naturalists provided the main corpus of anecdotes used by Yerkes and Yerkes (1929, pp. 112-122, 537, 555-556) to portray the positional behavior of *Pongo pygmaeus*. They concluded that the "strictly arboreal" orangutan "differs conspicuously from the gibbon in its slow, deliberate, cautious, almost slothful movements" (ibid., p. 112). They rarely brachiate (i.e., travel by arm-swinging unassisted by pedal grasps), spring, or jump. Instead of brachiating, orangutans climb with free use of the hind limbs and forelimbs. They commonly use bridging to transfer from tree to tree. Yerkes and Yerkes (1929) stated that grounded captive orangutans are rather slow quadrupeds, whose weight rests on the outer edges of the feet and "on knuckles" (ibid., p. 555). They did not distinguish the terrestrial hand postures of orangutans from true knuckle-walking hand postures of the African apes (Tuttle, 1967).

Yerkes and Yerkes (1929, p. 537) ranked orangutans as the least bipedal of all apes. They described the occasional bipedal stances and steps of captives as "extremely difficult and fatiguing" (ibid., p. 116) and cited no instances of arboreal bipedal locomotion by free-ranging orangutans (Tuttle, 1977).

Yerkes and Yerkes (1929) concluded that crouching is a common daytime resting posture of orangutans and that, unlike gibbons, weight rarely rests on their ischial regions. They reported that orangutans sleep in nests, lying on their sides or supine. They regarded the elongate forelimbs of orangutans as greatly advantageous for reaching food on branches that could not support their full weight (Tuttle, 1977).

Yerkes and Yerkes (1929, pp. 213-216, 555-556) inferred that common chimpanzees are eminently arboreal and especially cited their climbing skills. While acknowledging that chimpanzees "can swing themselves rapidly from branch to branch and may on occasion as it were throw themselves or fall through the air from one handhold to another," they also remarked that their mode of arboreal flight falls far short of that of gibbons (ibid., p. 214). Terrestrially, chimpanzees normally progress quadrupedally on knuckled hands and plantigrade feet. Although they are "capable of standing and walking erect for short times and distances," bipedalism is relatively difficult for them (ibid., p. 215).

Yerkes and Yerkes' accounting of jumping by chimpanzees is somewhat ambiguous (Tuttle, 1977). In the text they stated that "whereas jumping by leg propulsion when on a tree is an exceptional occurrence . . . similar jumping on the ground is commonly observed" (Yerkes and Yerkes, 1929, p. 215). In their synoptic table, they indicated that the chimpanzee "Jumps freely and skillfully in trees or on the ground" (ibid., p. 556).

Yerkes and Yerkes (1929, p. 556) noted that chimpanzees sleep supine or lying on their sides in arboreal nests. They made no definitive statements about other postural behavior of chimpanzees (Tuttle, 1977).

Yerkes and Yerkes (1929, p. 537) concluded that the gorilla is ill-adapted structurally for arboreal existence and of all the anthropoid types most closely approaches man in terrestrialness and in locomotor habits. They speculated that the decreased agility, consequent upon terrestrial existence, might be related to decreased responsiveness and slower neuromuscular processes in gorillas versus the more arboreal apes.

They presented the following additional points on the positional behavior of gorillas (ibid., pp. 406-409, 536-537, 555-556). They walk quadrupedally on the ground with their hands knuckled and their feet plantigrade. Young gorillas can be outrun by humans. And gorillas are much slower runners than chimpanzees are. Arboreally, they are slow, cautious, and clumsy climbers that jump only when forced to do so (ibid., p. 556). Their sleeping postures include lying on one side, supine, and sitting upright with the back against a tree.

Yerkes and Yerkes made contradictory statements about the bipedal abilities and predispositions of gorillas. In the text (p. 409) they noted that chimpanzees walk bipedally erect and plantigrade on the ground more willingly and readily than gorillas and orangutans do. Elsewhere (p. 537) they ranked the gorilla highest among the three great apes in "ease, skill, and efficiency in bipedal walking." In the synoptic comparison of anthropoid apes, they stated that gorillas "Stand erect and walk bipedally somewhat more naturally and readily than the chimpanzee" (p. 555).

HYLOBATID APES

The greatest advance in documenting the naturalistic positional behavior of apes has come through field studies of the Hylobatidae. Equivalent studies on the great apes have limped sorely behind those on the lesser apes.

Carpenter (1940) and Ellefson (1968, 1974) provided detailed descriptions on the locomotor, foraging, feeding, sleeping, and display behavior of Thai and Peninsular Malaysian lar gibbons. Fleagle (1976, 1977) augmented their accounts with quantitative data on Peninsular Malaysian lar gibbons and provided a thorough report on sympatric siamang.

Gittins (1983) performed a similar quantitative study on the positional behavior of Peninsular Malaysian agile gibbons. Tidbits on positional behavior in hylobatid apes occur in papers and monographs by Chivers (1972, 1974, 1977), McClure (1964), MacKinnon (1974b), Miller (1942), Papaiouannou (1973), Rijksen (1978), Srikosamatara (1984), and Tenaza and Hamilton (1971). Studies on captives (Avis, 1962; Baldwin and Teleki, 1976; Carpenter, 1976; Carpenter and Durham, 1972; Cortright, 1976; Fleagle, 1974; Fleagle et al., 1981; Hollihn, 1984; Hollihn and Jungers, 1984; Ishida et al., 1975, 1978, 1984, 1985; Jungers and Stern, 1980, 1984; Kimura, 1985; Kimura et al., 1977, 1979, 1983; Okada, 1985; Okada and Kondo, 1982; Okada et al., 1983; Preuschoft and Demes, 1984; Prost, 1967; Tuttle, 1972; Tuttle et al., 1979; Yamazaki, 1985; Yamazaki and Ishida, 1984) have added to our knowledge of the versatility, adaptability and mechanics of hylobatid locomotion.

Gibbons are brachiators par excellence (Figure 4). Because of light weight and special musculoskeletal mechanisms in their forelimbs, they can propel themselves upwards and outwards over considerable distances between supports in the forest canopy (Tuttle, 1969b, 1972, 1975; Jungers and Stern, 1980). This dramatic mode of forelimb propelled locomotion is termed ricochetal arm-swinging (Tuttle, 1969b) and ricochetal brachiation (Fleagle, 1976). Ellefson (1974) argued that brachiation has been over-emphasized in descriptions of gibbon locomotion since most of their movements over more than a few yards include variable amounts of bipedal running, dropping to lower strata, vertical climbing, bridging, hoisting, and hauling, in addition to arm-swinging. Fleagle (1976) found that white-handed gibbons used brachiation during 56% of travel bouts and 45% of bouts in which they shifted positions within a feeding site. Their travel bouts included much less climbing (21%), leaping (15%) and bipedalism (8%) than brachiation. He concluded that "Brachiation is the predominent activity observed during travel and accounts for the majority of distance travelled" (Fleagle, 1976, p. 259).

Agile gibbons also employ brachiation predominantly to move in the forest canopy (Gittins, 1983). They climb (14% of locomotor activity), walk

Figure 4: Male hoolock gibbon brachiating. (Photo courtesy of R.L. Tilson.)
(continued)

Figure 4: (continued)

bipedally (7%) and leap (6%) much less frequently than they brachiate (74%).

Thai pileated gibbons brachiated 88%, climbed 6%, leaped 5%, and ran bipedally 1 percent of locomotor bouts in the tropical semi-evergreen forest (Whitmore, 1975) of the Khao Soi Wildlife Sanctuary. They preferred branches (80%) to boughs (10%) and twigs (10%) as locomotor substrates. They travelled in the middle (45%), upper (34%) and lower canopies (22%; n = 754 bouts) (Srikosamatara, 1984).

Like the smaller gibbons, siamang employ brachiation (51%) much more than climbing (37%), bipedalism (6%) and leaping (6%) during travel bouts. But siamang climb (74%) much more than they brachiate (23%) during feeding translocations. In all 3 species the four modes of locomotion tend to be concentrated on different substrates with brachiation occurring beneath boughs and branches greater than two centimeters in diameter, bipedalism on sturdy horizontal branches, and climbing and leaping on a variety of structures, notable among which are twigs less than two centimeters in diameter (Carpenters, 1940; Fleagle, 1976; Gittins, 1983).

Lar gibbons and siamang execute approximately four-fifths of their feeding translocations on branches and twigs (Fleagle, 1976). Fruit, young leaves, flowers, shoots, and other preferred foods are commonly located at the peripheries of trees in the middle canopy. Often hylobatid apes climb directly from one feeding spot to another without having to return to stabler regions near the core of a tree. Their elongate forelimbs enable them to reach foods in precarious places. They can climb smooth trunks and vines with their hands and feet (Carpenter, 1940; Tuttle, 1969b, 1972; Van Horn, 1972).

The hylobatid apes employ a wide spectrum of sitting and suspensory feeding postures. They tend to hang from small supports and to sit on larger supports while feeding (Chivers, 1974; Fleagle, 1976; Gittins, 1983). The capacity to feed while hanging beneath branches doubles the hylobatid ape's potential feeding arena. Unlike a seated feeder, the suspended ape does not deform peripheral branches so that food thereupon is harder to reach (Grand, 1972).

Although agile gibbons foraged about equally in seated and hanging postures, they seemed to prefer sitting on firm supports to hanging while they fed, as attested by the former postures occurring twice as often as the latter during feeding bouts (Gittins, 1983).

Thai pileated gibbons sat (62%) more than they hung (38%) to feed. They used branches (73%) much more than twigs (22%) and boughs (5%) as feeding supports (Srikosamatara, 1984).

Carpenter (1940) confirmed that lar gibbons commonly sleep sitting up, though they also sleep supine or on a side on wide branches. Ellefson (1974) found that resting gibbons usually sat hunched on a branch and held neighboring branches with their hands. Sometimes they lay supine or on a side for long periods during the day.They slept seated on large horizontal branches. A variety of supine and one-sided postures preceded the sitting posture as they settled for the night)Ellefson, 1974). Agile gibbons also prefer seated to other postures during rest (95% of bouts) and sleep (88% of bouts) (Gittins, 1983). Siamang sleep sitting in the terminal branches of tall emergent trees (Chivers, 1974, pp. 23, 49; Gittins and Raemaekers, 1980).

ORANGUTANS

Although Bornean and Sumatran orangutans have been surveyed and studied by a respectable number of modern investigators (Carpenter, 1938; Davenport, 1967; Galdikas, 1979; Galdikas-Brindamoor, 1975; Harrisson, 1962; Horr, 1972, 1975, 1977; MacKinnon, 1971, 1973, 1974a, 1974b, 1979; Okano, 1965; Rijksen, 1975, 1978; Rijksen and Rijksen-Graatsma, 1975; Schaller, 1961; Yoshiba, 1964), there is meager quantitative description of their positional behavior. Even detailed qualitative descriptions of the sort that Carpenter (1940) provided on lar gibbons are rare. The chief mode of locomotion by orangutans is versatile climbing (Figure 5). This includes clever, cautious bridging transfers, vertical ascents and descents, hoisting and occasional pedally assisted arm-swinging and quadrupedal suspensory movements beneath branches (Davenport, 1967; Galdikas, 1979; Horr, 1977; MacKinnon, 1971, 1974a; Rijksen, 1978; Schaller, 1961).

When orangutans cannot reach a new support across a gap in the canopy directly, they may sway the base support until it closes the distance and then grab onto the new support. Definite arboreal "highways" are used in at least some orangutan habitats (Horr, 1977; MacKinnon, 1974b). As a mode of travel, brachiation, unassisted by pedal grasps, is secondary to climbing and occurs only over short distances. Rijksen (1978, p. 175) noted that brachiation is a common component of certain displays. MacKinnon (1974a, p. 42) stated that "orangutan brachiation lacks the speed and flow of the specialized ricocheting brachiation of gibbons and siamangs and also differs in that the arms are swung overhead rather than underarm." Sugardjito (1982) confirmed that orangutans are overarm brachiators. Jumping and vertical dropping over notable distances are also uncommon among orangutans. Fleeing subjects sometimes leap (Rijksen, 1978, p. 174). MacKinnon witnessed "tumble descents" in which the fleeing subject fell rapidly through the foliage while briefly grasping and releasing supports with hands and feet (MacKinnon, 1971, 1974a). Schaller (1961) observed one jump over 3 feet to a lower branch. Davenport (1967) and Sugardjito (1982) saw no leaping or jumping in free flight. However, 2 of Davenport's Bornean subjects executed dramatic "dives" in which they held onto a branch with their feet and lunged or fell forward so that ultimately they

Figure 5: Adult female Sumatran orangutan travelling by suspensory locomotion while her infant clings ventrolaterally to hair of her body. (Courtesy of H.D. Rijksen, 1978.)

hung by their feet alone. Rijksen (1978, p. 183) also witnessed display diving by orangutans in the Gunung Leuser Reserve, Aceh, Sumatra.

Orangutans walk quadrupedally atop large horizontal and slightly inclined boughs and branches. Sumatran orangutans hold onto the supports instead of fist-walking as they commonly do on the ground (Rijksen, 1978, p. 175). Arboreal bipedalism has been reported by Davenport (1967) and Rijksen (1978, pp. 194-195) among recent observers. Rijksen (1978) noted that bipedal stances were most commonly executed by females and sub-adult males.

Sugardjito (1982) quantified the locomotor behavior of adult Sumatran orangutans during 219 hours of observation at Ketambe in the Gunung Leuser National Park. He considered 5 categories of locomotor activity: quadrumanous scrambling, including dives (41% of locomotor bouts); brachiation (21%); tree swaying (15%); quadrupedal walking (13%) and vertical climbing (10%). Females employed quadrumanous scrambling (< 50% of their locomotor bouts) and walking (16%) more than males did (35% and 10%, respectively). Per contra, the hefty males were more disposed to sway base trees so they could transfer directly into another (> 20% of male bouts) and to brachiate (>20%) than females were (9% and 18%, respectively). Males and females engaged in similar frequencies of vertical climbing (Sugardjito, 1982).

In brief, the new observations of locomotion by *Pongo pygmaeus* basically confirm the conclusions of Yerkes and Yerkes (1929) about their aboreal modes of travel. But it has been discovered that the orangutan, especially the Bornean subspecies, is not strictly arboreal. Large adult males are particularly disposed to walk quadrupedally on the ground (deSilva, 1971; Galdikas, 1979; Galdikas-Brindamoor, 1975; Horr, 1975, 1977; MacKinnon, 1971, 1974a,b; Rijksen, 1978; Rodman, 1979). In some cases, terrestrial locomotion is used to cross man-made discontinuities in the forest (deSilva, 1971; Galdikas-Brindamoor, 1975). But in other contexts, terrestriality seems to be wholly naturalistic. For instance, Rodman (1979, p. 241) found that one adult male in the Kutai Nature Reserve of Kalimantan Timur, Indonesia, spent 20% of his travel time on the ground. Galdikas (1979, p. 223) discovered that as large males became fully habituated to the presence of human observers, "almost all of their long distance traveling was done on the ground." In the Ulu Segama Reserve of Sabah and the Ranun River region of eastern Sumatra, MacKinnon (1974a) was chased by unhabituated male orangutans, some of which climbed down from the trees in order to pursue him. He also saw female and juvenile orangutans traveling briefly on the ground, especially during wet weather (MacKinnon, 1974a). Galdikas (1979, p. 222) found that females in the Tanjung-Puting Reserve, Kalimantan Tengah, Indonesia, foraged on the ground much less frequently than adult males did. Rijksen (1978, p. 314) and Sugardjito (1982) concluded that wild orangutans at Ketambe, in the Genung Leuser National Park, Sumatra, rarely descend to the ground. They do so to cross gaps in the forest, to feed, and to flee from other orangutans.

The feeding and resting postures of *Pongo pygmaeus* have been described by Schaller (1961), Harrisson (1962), Davenport (1967), MacKinnon (1971, 1974a,b), and Rijksen (1978). The best general summary on feeding postures was provided by MacKinnon (1974a, p. 44):

> Feeding animals usually sit on a firm branch holding on above with one arm whilst gathering and handling food with the other. At other times they hang out from, or beneath, a branch suspended by one arm and one leg. The free arm and leg are used for gathering and holding food. Sometimes animals hang by both hooked feet and use both hands to collect and open fruit.

Rijksen (1978, p. 43) noted that 2 point suspension by an ipsilateral hand and foot was used frequently by orangutans to harvest fruit-laden twigs in strangling fig trees. Both young and adult orangutans occasionally hang unimanually as they feed on fruit (Harrisson, 1962, p. 78; MacKinnon,

1974b, p. 121). The highly mobile hip, shoulder, elbow and knee joints, and powerful, prehensile feet and hands, which are endowed with individual digital control (Tuttle, 1967, 1969b, 1970, 1975), enable orangutans to distribute their weight among several small branches and lianas, any one of which probably could not support them. Rijksen (1978) noted that 3 point suspension was especially common during feeding and locomotion by his subjects. Sometimes orangutans collect large fruits or clusters of smaller foods and carry them to large branches or nests where they can feast in less perilous pavilions (Harrisson, 1962; MacKinnon, 1971; Rijksen, 1978). Their long reach and great strength enable them to break off and to bend terminal twigs towards themselves, thereby obtaining foods that are inaccessible to many other non-volant animals (Rijksen, 1978).

Orangutans sleep in arboreal nests at night and commonly build more and less elaborate daytime structures for protection against heavy rainfall and siestas, respectively (Chapter 4). Davenport's (1967, p. 256) subjects characteristically slept on their sides or supine. MacKinnon (1974a, p. 43) reported the following:

> Animals resting in trees, nests, or on the ground often sat upright, sometimes holding an overhead branch with one arm. Some animals lie on their backs or fronts with their limbs dangling over branches. Orangutans can remain suspended between two upright supports with their legs taking much of their weight.

COMMON CHIMPANZEES

Over the past half century the common chimpanzees of West, Central, and especially, eastern Africa have been extensively surveyed and sometimes intensively studied in the wild. None of the observers focused systematically on their positional behavior. However, several scientists have provided solid qualitative descriptions that complement and occasionally clarify ambiguities in the brief compendium of Yerkes and Yerkes (1929).

Chimpanzees are much more consistently terrestrial than the Asian apes are. Although they spend a considerable portion of their waking hours foraging, feeding, and resting in trees, and although they sleep in trees at night, they generally move from one site to another by knuckle-walking on the ground (Albrecht and Dunnett, 1971; Bournonville, 1967; Ghiglieri, 1984; Goodall, 1962, 1963a,b, 1965; Itani, 1979; Itani and Suzuki, 1967; Izawa, 1970; Izawa and Itani, 1966; Jones and Sabater Pi, 1971; Kano, 1972; Kortlandt, 1962, 1968, 1972; Lawick-Goodall, 1968; Nissen, 1931; Rahm, 1967; Reynolds, 1964, 1965; Reynolds and Reynolds, 1965; Sugiyama, 1968, 1969, 1973; Suzuki, 1969; Teleki, 1973a, p. 110; Wrangham, 1979).

All observers attest to the adeptness with which chimpanzees climb vertical tree trunks and boughs. Kortlandt (1968) emphasized that the elongate forelimbs and large hands of chimpanzees enable them to obtain food from isolated and emergent trees that do not have lianas extending to lower strata. The trunks of these trees cannot be climbed by sympatric monkeys.

Goodall (1965, p. 437) commented that as subadults, chimpanzees "move easily and quickly in trees but as they attain maturity arboreal locomotion becomes slow and careful unless the animals are frightened or excited." Chimpanzees grip some branches with hands and feet. If the branches are wide, knuckle-walking may be employed (Albrecht and Dunnett, 1971, p. 18; Kortlandt, 1972, p. 15; Lawick-Goodall, 1968, pp. 177-178; Reynolds, 1964, 1965, p. 62). Bridging is a common mode for transferring from tree to tree (Goodall, 1965, p. 439; Izawa and Itani, 1966, p. 127). Travel through the canopy is limited to short distances (Tuttle, 1977).

Chimpanzees sometimes brachiate, jump or leap, and engage in bipedalism in trees. It is impossible to discern actual frequencies of these positional behaviors from available reports (Tuttle, 1977). Goodall (1965, p. 440) stated that brachiation for short distances is common. She observed that it is sometimes a prelude to bridging (Lawick-Goodall, 1968, p. 178). Reynolds and Reynolds (1965, p. 384) listed brachiation along branches as common. But elsewhere Reynolds (1965, p. 59) stated that he saw brachiation occasionally.

Most observers have seen chimpanzees execute short arboreal jumps. They do not always mention whether the jumps were propelled chiefly by forelimbs or hind limbs. Schaller and Emlen (1963, p. 371) saw chimpanzees jump from branch to branch, sometimes over distances exceeding 8 ft. Goodall (1965, p. 440) described some jumps that were powered by the hind limbs. Frisky youngsters, displaying adults, and fugitives of all ages execute longer leaps, at least some of which are probably vertical drops (Albrecht and Dunnett, 1971, p. 20; Izawa and Itani, 1966, p. 127; Jones and Sabater Pi, 1971, p. 61; Reynolds, 1965, pp. 58-61; Reynolds and Reynolds, 1965, p. 384; Schaller and Emlen, 1963, p. 371). Chimpanzees grasp landing sites quadrupedally or with their hands alone (Goodall, 1965, p. 440).

Terrestrial chimpanzees at Gombe National Park, Tanzania, usually leap quadrupedally or bipedally across streams and small gulleys (Lawick-Goodall, 1968, p. 177). Kortlandt (1962, p. 8) observed chimpanzees jumping bipedally over 6 feet from a standing start to cross a stream on a plantation in Zaire. Reynolds and Reynolds (1965, p. 439) saw chimpanzees jumping quadrupedally and occasionally bipedally across streams in the Budongo Forest, Uganda. Izawa and Itani (1966, p. 148) inferred that chimpanzees in the Kasakati Basin, Tanzania, ordinarily waded streams instead of jumping over them.

Reynolds and Reynolds (1965, p. 384) noted that Budongo chimpanzees occasionally walked short distances bipedally on branches while grasping supports overhead with their hands. Reynolds (1965, p. 62) also reported that they would often walk bipedally along stout branches. Rahm (1967, p. 203) stated that fugitives from net hunters in Zaire very often ran upright like gibbons over branches and jumped distances "they would not even try under normal conditions" (Tuttle, 1977).

Gombe chimpanzees "frequently stand upright in order to look over long grass and other vegetation"; "frequently walk bipedally for short distances" in tall grass, on wet ground, or when carrying food; and run bipedally during terrestrial branch waving displays (Lawick-Goodall, 1968, p. 177). The frequency of bipedalism may have been increased by the artificial feeding situation at Gombe National Park (Tuttle, 1977).

Kortlandt (1962, 1967, 1968, 1972; Kortlandt and Kooij, 1963) was impressed by the variety and frequency of terrestrial bipedal behavior in wild chimpanzees. But his observations were made in the vicinity of plantations, provisions, and experimental objects (Tuttle, 1977). Reynolds (1965, p. 62) regarded terrestrial bipedal walking as "rather common" in Budongo chimpanzees. Bipedal standing to get a better view was more common than bipedal movement on the ground. Tutin et al. (1983, p. 169) noted that Senegalese chimpanzees "frequently stood bipedally to scan their surroundings" when they travelled in open areas.

Yerkes and Tomilin (1935, p. 336) reported that bipedal locomotion develops readily during the second half of the infant chimpanzee's first year. They concluded that " . . . in frequency and amount, probably also in ease and facility, it is inversely related to age." Riesen and Kinder (1952) found that infant chimpanzees can stand with manual assistance during the second half of their first year and master unassisted bipedalism early in the second year.

Feeding chimpanzees sit, squat, recline, or stand on branches and hold onto neighboring (commonly overhead) branches with one hand as they draw in foods with the other hand (Figure 14b). Chimpanzees inhabitating the Okorobiko Mountains of Equatorial Guinea ate approximately 80% of their food while seated, 15% while standing up, and 5% while lying down, generally in a supine position (Sabater Pí, 1979). Sometimes they use their feet to bend, break, or hold branches as they feed from them (Goodall, 1963, p. 41; Nissen, 1931, p. 67). Chimpanzees also hang unimanually or from a hand and a foot in order to reach food at the ends of branches (Lawick-Goodall, 1968, p. 185; Reynolds, 1965, p. 59).

During daytime rest periods, chimpanzees sit or recline on the ground, in arboreal or terrestrial nests, on branches, or in the crotches of trees. When they recline supine or on a side in trees, they commonly hold overhead branches with hands or feet (Lawick-Goodall, 1968, p. 201; Nissen, 1931, p. 31). In arboreal night nests, they sleep supine, on one side with flexure of the hind limbs, and occasionally prone (Goodall, 1965, pp. 448-449; Izawa and Itani, 1966, p. 147; Lawick-Goodall, 1968, p. 201; Nissen, 1931, p. 45) During diurnal rests, they sat upright 59% of time on the ground and 41% of time in trees and lay supine 86% of time on the ground and 14% of time in trees (Galdikas and Teleki, 1981).

BONOBOS

Bonobos have been studied at several localities in Zaire (Badrian and Badrian, 1977; Badrian et al., 1983; Horn, 1975; Kano, 1979; Kuroda, 1979; MacKinnon, 1976; Nishida, 1972). Only one team focussed on their locomotor behavior (Susman et al., 1980; Susman, 1984). In general, the studies are characterized by brief duration and limited contact with the skittish subjects.

Most observers concluded that bonobos are somewhat more adventurous and adept at arboreal locomotion than common chimpanzees are (Badrian and Badrian, 1977; Horn, 1975; Kano, 1979; MacKinnon, 1976; Susman, 1980, 1984; Susman et al., 1980; Figure 6). Badrian and Badrian (1977) found that they sometimes travelled arboreally over more than a kilometer between food sites. Still, like common chimpanzees, bonobos usually travel from place to place by knuckle-walking on the ground. And they sometimes nest (Chapter 4) and obtain a considerable portion of their food on the ground (Horn, 1975; Kano, 1979; Kuroda, 1979; MacKinnon, 1977; Susman et al., 1980) (Chapter 3).

Arm-swinging is an important component of the arboreal locomotor repertoire of bonobos, especially when they are excited. Initially Susman et al. (1980) found that adult and younger bonobos in the Lomako Forest of the Equateur Region, Zaire, employed arm-swinging along (and occasionally between) boughs and branches during 15% of their arboreal locomotor bouts. They engaged in "quadrumanous climbing," scrambling and trans-ferring behavior during 20% of arboreal locomotor bouts. Vertical drops (5%) and hind limb propelled leaping and diving (18%) also occurred. They are part of the flight behavior of bonobos, which are quite shy of humans in the forest. Once away from terrestrial intruders, bonobos descend quickly and flee on the ground.

In calmer moments, bonobos walked quadrupedally atop branches and boughs (34% of arboreal locomotor bouts). Usually, they gripped the substrate with hands and feet. But on large horizontal boughs, bonobos knuckle-walked (Susman et al., 1980).

Susman et al. (1980) found that, like the hylobatid apes, bonobos move bipedally during a notable amount (9%) of arboreal locomotor bouts. Most arboreal bipedal walking was confined to horizontal boughs. It is particularly practiced during displays of excited subjects (Susman, 1984).

Figure 6: Bonobo in a tree. Note long slender limbs and tufts of hair on its head. (Photo by Ron Garrison. © Zoological Society of San Diego.)

Foraging bonobos often stand bipedally on horizontal boughs and then may walk bipedally while carrying recently picked food. Susman et al. (1980) once encountered an adult male bonobo which was walking bipedally on the ground with both hands full of stalks.

On the basis of a longer study (n = 1,722 locomotor bouts observed over a span of 20 months), Susman (1984) calculated that, arboreally, the Lomako bonobos engaged in quadupedal locomotion (mostly palmigrade on branches and boughs) during 31% of bouts and quadrumanous (including vertical) climbing (mainly on trunks) and scrambling in foliage for another 31% of bouts. Arm-swinging (21%), leaping and diving (10%), and bipedalism (6%) were less common than arboreal quadrupedalism. Susman (1984) sensed that leaping, diving, and bipedalism were less frequent in the later study because the subjects were somewhat less inclined to spook and threaten the observers.

Badrian et al. (1981) mentioned that bonobos sometimes hung unimanually and drew branches toward themselves with a free hand or foot, stood bipedally to pull down food laden foliage, or broke off branches and moved to a spot where they could sit and eat. Kano and Mulavwa (1984) also noted that the bonobos at Wamba carried harvested fruits in their hands, feet, mouths and groins to sitting places. Indeed 90% of 132 bouts of arboreal fruit-eating was conducted while the bonobos were seated. Hanging (5%), quadrupedal standing (2%), and reclining (3%) postures were much less common during feeding bouts.

GORILLAS

Gorillas are predominantely terrestrial (a fact for which many trees might be grateful). Mountain gorillas and eastern gorillas spend 80 to 97% of their time on the ground. They generally feed in the herb and shrub stratum of the forest. Most of their diurnal resting spots and night nests are located on or very near the ground (Bingham, 1932; Donisthorpe, 1958; Jones and Sabater Pi, 1979; Kawai and Mizuhara, 1959-60; Schaller, 1963; Tuttle and Watts, 1985). Like chimpanzees, gorillas move from locality to locality by knuckle-walking (Figure 7). On average, in Virunga gorillas it accounted for 940 m/km travelled (Tuttle and Watts, 1985).

Virunga gorillas spent only about 3% of the average daily activity period in trees. Most of their arboreal tenure was devoted to feeding (Chapter 3): in adults, 91%, and in juveniles, 73% of their non-locomotor arboreal activity. Youngsters played in trees (20% of their non-locomotor arboreal time). The remainder of diurnal arboreal time was spent in unnested rest (Tuttle and Watts, 1985, p. 265).

Watts categorized the arboreal climbing of Virunga gorillas into 3 modes: vertical, horizontal, and on vines. Vertical climbing accounted for 71% of

Figure 7: Knuckle-walking juvenile gorilla. (Inset shows knuckle pads of friction skin over the mid phalanges.)

the total distance climbed, while horizontal climbing and climbing on masses of vines constituted only 22% and 6% of the distance, respectively. Adults were particularly inclined to remain near the main trunk and relatively vertical boughs and to use them as their primary arboreal substrate. When they did venture from the core of a tree, they preferred stout horizontal boughs. Hefty males kept off vines (Tuttle and Watts, 1985).

Although cautious at climbing trees, gorillas can manage quite capably in them (Figure 8). Youngsters are particularly inclined to explore, play, feed, and rest in trees (Tuttle and Watts, 1985). They enter them by climbing step-wise up the trunk and utilizing existent irregularities in the bark, branches, and other structures as handholds and footholds (Schaller, 1963, p. 83). In trees, they are basically quadrupedal climbers. Brachiation between and along branches and boughs is very rare and never ricochetal (Tuttle and Watts, 1985). Further, gorillas rarely jump in trees. The distances that they have traversed by jumping in trees are less than 10 feet (Schaller, 1963, p. 84). Occasionally they jump from atop low branches to the ground (Donisthorpe, 1958). They land either quadrupedally or feet first in a manner that permits quick exits by knuckle-walking (Donisthorpe, 1958; Schaller, 1963; Tuttle and Watts, 1985).

Unlike chimpanzees, gorillas refuse to jump from tall trees. They descend tree trunks feet first. If the tree has protrusions suitable for handholds and footholds, the movements of descent are basically the reverse of those used for ascension. If the tree lacks suitable handles, gorillas slide hand over hand down the trunk while using the soles of the feet as brakes (Donisthorpe, 1958; Schaller, 1963; Tuttle, 1970; Tuttle and Watts, 1985). To descend directly from low branches to the ground, gorillas usually hang "first by both arms, then only by one arm as they scan vegetation below for several seconds before finally releasing their hold." (Schaller, 1963, p. 85).

Virunga gorillas sometimes grasped herbaceous plant stems overhead and hauled themselves forwards while ascending steep slopes. The basic pattern of limb movements is reminiscent of versatile arboreal climbing. During descents of steep ravines, the subjects occasionally moved rump first while holding onto the stems of large herbs in order to modulate their speed. Youngsters, young adults, and even adult females also somersaulted for short distances downhill (Tuttle and Watts, 1985).

Gorillas rarely engage in bipedal positional behavior (Bingham, 1932; Schaller, 1963; Tuttle and Watts, 1985). Schaller and Emlen (1963, p. 370) observed bipedal running only thrice over distances between 15 and 60 ft. Bipedal standing and running are brief but important components in their chest-beating displays (Chapter 7; Figure 25). At Visoke, silverbacks exhibited the highest frequencies of bipedal running because they most often executed chest-beating displays (Tuttle and Watts, 1985).

Schaller (1963) also noted instances of bipedalism during object transport

Figure 8: Postures of mountain gorillas. (a) young adult female sitting on the ground; (b) blackback squatting while eating *Galium*; (c) 3-year-old female standing bipedally on a vine; (d) 3-year-old female standing bipedally in front of a prone silverback on the ground. (Photos by David P. Watts.)

and a precopulatory display. Jones and Sabater Pí (1971, p. 60) figured the bipedal postures of a young western gorilla feeding in a tree and Goodall (1977; p. 457) noted that Kahuzi gorillas sometimes stood bipedally to pull down vines from shrubs.

Terrestrially feeding gorillas generally sit (61% of time) or squat (32% of time) and reach for food in all directions (Tuttle and Watts, 1985). Jones and Sabater Pí (1971, p. 59) noted that "whether feeding, resting, or just sitting and looking around, gorillas in arboreal situations usually supported themselves by grasping branches with both feet and at least one hand." Unlike chimpanzees, mountain gorillas do not use their feet commonly to secure branches or directly to collect foods. Their feeding is almost entirely manual and oral (Tuttle, 1970, pp. 190-191).

At Visoke, one silverback never fed in trees, but other individuals spent up to 9% of their feeding time arboreally. Like climbing, the frequency of arboreal feeding varied inversely with body size. Mostly they squatted (86% of arboreal feeding time). Arboreal bipedal feeding and feeding while reclining or suspended were rare in Virunga gorillas. The suspensive feeders were usually youngsters and never were silverbacks (Tuttle and Watts, 1985).

Gorillas sleep either on a side with snug hind limb flexure and one or both forelimbs folded across the chest or prone with the limbs tucked under the body (Schaller, 1963, p. 198).

DISCUSSION

The modern field studies of apes surveyed hereabove generally confirm the continuum from gorillas to gibbons in the relative frequencies of terrestriality and arboreality. Gorillas are the most terrestrial apes, followed by common chimpanzees, bonobos, and orangutans. The Hylobatidae are the only exclusively arboreal apes. The observations that orangutans are more terrestrial than they had been thought to be and the curious mixture of terrestrial and arboreal feeding, nesting, and locomotor habits of the bonobos are having considerable theoretical impact among evolutionary modelers and scenarists.

We still lack sufficient data from systematic field studies of apes to resolve questions about the relative importance of suspensory postures and locomotion, including brachiation, in their positional behavioral repertoires. Fleagle's (1976) study of the siamang is singular. We must have similar long-term quantitative studies on the positional behavior of many other species and populations of lesser apes, orangutans, common chimpanzees, bonobos, and gorillas in a variety of habitats in order to conduct controlled comparisons and to interpret the postcranial structures of living and fossil

Hominoidea functionally. The most informative approach would be to study 2 or more species of apes concurrently in the same locality.

A survey of the largely impressionistic literature on pongid and hylobatid arboreal positional behavior suggested that extensive, rapid, bimanual suspensory progression is only characteristic of the lesser apes. It is a dramatic component of their displays during intraspecific altercations (Tuttle, 1975, p. 454). The rapid brachiation that has been attributed to chimpanzees and other anthropoid primates is probably vertical dropping and leaping between supports instead of true ricochetal arm-swinging like that which the hylobatid apes exquisitely execute.

Arm-swinging along horizontal branches or between juxtaposed limbs and vines seems to be a very rare mode of locomotion in the gorillas. It is employed more frequently by orangutans and common chimpanzees, and perhaps even more often by bonobos (Tuttle, 1975; Susman and Badrian, 1980). Sumatran orangutans evince brief bouts of "brachiation" during "show off" displays (Rijksen, 1978, p. 175).

Transfers between adjacent areas of small flexible branches, wherein the feet grasp base supports and the hands reach ahead carefully to test and to secure new vantage points, is most common as a mode of progression in the orangutan. Common chimpanzees, bonobos, and lesser apes occasionally employ transferring behavior but it is not a major component in their locomotor repertoires. Gibbons and orangutans have been observed to bridge gaps in the canopy with their bodies while youngsters climbed over them (Tuttle, 1975, p. 453).

When apes climb vertically on tree trunks and vines and maneuver versatilely in the peripheries of trees during feeding translocations, hoisting and hauling actions by the forelimbs are conspicuous. Like arm-swinging, these activities subject the forelimbs to tensile stresses and are facilitated by elongation of the forearms and hands. Similarly, suspensory feeding, in which all apes, except some gorillas, engage to an extent, subjects the forelimbs to tensile stresses. And the elongate forelimbs of heavy apes, like the siamang and orangutans, enable them to reach most of the arboreal foods that the lightweight gibbons can reach from more precarious vantage points. Thus, it is overly simplistic to invoke brachiation, with narrow connotations of rapid arm-swinging locomotion, to explain many of the morphological similarities beween extent apes and other living and fossil anthropoid primates that have been designated brachiators (Tuttle, 1975, p. 464).

Whereas our knowledge of the suspensory behavior of apes is penumbral, that on their jumping and leaping, particularly in trees, is obscure indeed. Students of hominoid locomotion have shown various levels of remissness in documenting whether the forelimbs or the hind limbs are the principal propellent organs during "jumps" and "leaps," how the animals land, and whether these actions truly are different from what other observers have labeled vertical drops and dives. Similar confusion arises when one tries to discern precisely how hylobatid and pongid subjects were oriented in space and how they propelled themselves on the basis of naturalistic accounts wherein observers scored a wide spectrum of movements under the catchall, climbing. For theorists interested in the evolution of orthograde positional behaviors, including terrestrial bipedalism, it is especially important to have vertical climbing scored discretely.

Quantitative and contextual information on the bipedal behavior of apes is still spotty and ambiguous. Bipedal hylobatid apes, bonobos, and common chimpanzees progress rapidly along horizontal boughs and branches, especially when they are excited by human intruders. Bipedal locomotion also occurs during arboreal chasing displays between hylobatid groups and certain terrestrial displays of chimpanzees (Bygott, 1979; Lawick-Goodall, 1968; Nishida, 1979) and gorillas. Some of the more stationary displays of orangutans also have bipedal components. There is no data in recent naturalistic literature that would allow us to rank-order the pongid apes in regard to their bipedal potentialities. Observations on captives indicate no clear-cut differences between chimpanzees, gorillas, and orangutans in their fatiguability during terrestrial bipedalism. No experiments have been conducted to test this factor.

Like numerous other mammals (some of which appear quite ungainly), the apes readily stand bipedally to see over objects in their way. Passing comments have been made about bipedal feeding postures in apes, but their frequencies of occurrence have not been recorded.

3

Feeding Behavior

"The gorilla is a strict vegetarian like the elephant and buffalo—three of the four most dangerous animals in Africa. It behooves one to walk softly with vegetarians!" (Bradley, 1922, p. 131).

"When man is his prey, he devours him as he does animals that he can catch, though his sluggishness prevents his taking many animals as prey." (Ford, 1852-53, p. 32).

Because they considered knowledge of the nutritional requirements of apes to be vital for their maintenance and breeding in captivity, Yerkes and Yerkes (1929) searched the literature and corresponded with persons who had observed apes in the wild. They collected but a handful of morsels. A few of their hunter-informants had directly observed animals eating their last meals. Others had examined the stomach contents of their bag or fecal deposits. Little information was available about the species and plant parts that were eaten and ignorance was perfect concerning the quantities and variety of foods consumed daily by any wild ape.

Yerkes and Yerkes (1929) concluded that the apes are primarily vegetarians and that, with the likely exception of gorillas, most species also sometimes consume small quantities of eggs, insects, and perhaps small vertebrates. They fairly presented opinions of protagonists and antagonists in the controversy over the extent to which chimpanzees and gorillas were naturalistically carnivorous. But they felt that arguments for regular meat-eating by African apes were biased by observations of the easy acceptance of raw and cooked meat by many captives. Indeed the accommodation of most apes, particularly young ones, to a wide selection of humanoid menus led Yerkes and Yerkes (1929) to challenge the idea that dietary preferences are instinctive and to eschew facile use of the word instinct throughout the book.

HYLOBATID APES

Carpenter's (1940) classic study of *Hylobates lar* in Thailand served the first entree of facts on naturalistic foods and feeding habits of apes. Over a period of 4 months, he often observed gibbons feeding. He also examined the stomach contents of 23 specimens, which had been shot by members of the Asiatic Primate Expedition of 1937, and collected botanical specimens for scientific identification and biochemical analysis. He did not publish the nutritional contents of the plants.

Carpenter (1940) characterized gibbons as highly selective hand (and foot) to mouth feeders that chiefly consumed figs, grapes, plums, mangoes and other fleshy fruits. He estimated that 80% of their diet consisted of fruits. Leaves, buds, flowers, avian eggs, nestlings and insects comprised the remaining 20% of their fare. Though fruits, raindrops and dewdrops constituted the principal sources of water for the gibbons, they would also dip a hairy hand into tree bowls and streams and then drink the imbibed liquid by holding the hand overhead or sucking on it.

Carpenter (1940) noted that during daily activity cycles, the gibbons generally had 2 major feeding periods; one falling between 0830 h and 1100 h and the other between 1430 h and 1630 h or 1700 h. Their food sources were often concentrated so that observers could predictably locate them.

Three decades later, Ellefson (1974) made the next substantive contribution on the feeding ecology of white-handed gibbons based primarily on a 16-month intensive study at Tanjong Triang in Peninsular Malaysia. He retrieved some items that were dropped by feeding subjects and picked others himself. Altogether Ellefson (1974) collected 140 specimens, of which 80% were identified by botanists. No nutritional analyses were conducted on them. Ellefson also examined fecal deposits. In keeping with modern conservationist aims, he did not transform subjects into specimens in order to analyze their stomach contents.

The majority of Peninsular Malaysian food plant species are located in the middle stratum of the forest. Like Chiengmai lar gibbon diets, the bulk of Peninsular Malaysian lar gibbon fares consist of fruits and berries (Ellefson, 1974). Nine of the 22 plant items that Carpenter (1940) listed were also eaten by gibbons at Tanjong Triang. But by 2 different measures—percent of ingested species collected and percent of the time spent feeding on various classes of food—Ellefson found that approximately two-thirds of his subjects' vegetal diet consisted of fruits and berries and one-third was new leaves, buds, flowers and shoots. This differs notably from Carpenter's estimate of 80% frugivory for Chiengmai lar gibbons.

Ellefson (1974) concluded that in terms of the number of plant species eaten, white-handed gibbons should be considered generalized feeders.

The gibbons at Tanjong Triang ate a wide variety of arthropods, including large cryptic stick insects, termites, ants, spiders and centipedes. Some adults ate arthropods nearly every day. There was no evidence that eggs, nestlings or other vertebrate prey were consumed (Ellefson, 1974). They drank water almost daily though one week passed during which Ellefson (1974) observed no drinking. Hand-dipping was much more common than direct application of the mouth to the sources of water.

Ellefson (1974) calculated that his subjects spent 80% of an average daily activity period feeding (25%) and foraging (55%). He presented 2 different sets of figures, however. In a table (p. 9) he showed that of the average 9.75 hour waking period away from the sleeping locality, 4.5 hours were spent foraging and 2.75 hours were spent feeding. In the text he indicated that 5.33 hours/day were spent foraging (p. 64) and 2.5 hours/day were spent feeding (p. 55). The percentages hereabove are based on the second set of figures. An average day contained 5 feeding bouts, each lasting about 30 minutes (Ellefson, 1974).

Ellefson also noticed that gibbons seemed to search for specific kinds of food during certain periods of the activity cycle. For instance, arthropod hunting, which required considerable diligence, often took precedence over other foraging behavior around midday. And it was fairly common for the gibbons to locate and to eat new leaves after long bouts of frugivory (Ellefson, 1974, p. 65).

In an intensive 12-month study of one group of lar gibbons in the Krau Game Reserve of Peninsular Malaysia, Raemaekers (1978a,b, 1984) soundly established that they select and seek figs again before settling in for the night. Overall, they concentrated on figs for 22% of their feeding time. The remainder of their diet consisted of other fruit (28%), young leaves and shoots (25%), flowers (7%), mature greenery (4%), and insects (13%) (Raemaekers, 1984).

The predilection of western Thai lar gibbons for fruits has been documented by Fooden's (1971) examination of the stomach contents of 21 wild-shot specimens. All of them consisted exclusively of fruit pulp and seeds. Like Carpenter (1940), Fooden (1971) noted that the fruit fragments were coarse, indicating that gibbons do not chew pulpy fruit thoroughly.

Based on a 2-year study of agile gibbons (*Hylobates agilis agilis*) at Sungai Dal, a lowland and hill dipterocarp forest in the Gunong Bubu Forest Reserve, Perak, Peninsular Malaysia, Gittins (1980, 1982; Gittins and Raemaekers, 1980) calculated that one group devoted 41% of its feeding time to fruits other than figs, 17% to figs, 39% to young leaves, 3% to flowers, and 1% to arthropods other than those in the figs. Mature leaves were quantitatively inconsequential in their diet. On average, agile gibbons are active for 9 hours per day (r = 6.5 - 11.0 hours) of which time 53% is spent foraging and feeding, 11% travelling, 29% resting and 8% calling (Gittins and Raemaekers, 1980; Gittins, 1982). They obtain fruit mainly in the smaller trees of the middle canopy (20-40 m [66-132 ft]) and generally forage higher for leaves (Gittins and Raemaekers, 1980). The home ranges of 2 groups were 29 and 25 ha. The daily ranges of Malaysian agile gibbons ($\bar{x}$ = 1.22 km; r = 0.65-2.20 km) were intermediate between those of Malaysian white-handed gibbons ($\bar{x}$ = 1.49 km; r = 0.45-2.90 km) and siamang ($\bar{x}$ = 0.74 km; r = 0.20-1.70 km), but were closer to the former species (Gittins and Raemaekers, 1980, p. 78).

Rijksen (1978, p. 108) commented that figs constitute the major dietary item of North Sumatran white-handed gibbons (*Hylobates lar vestitus*). MacKinnon (1974a, p. 45) noted that *Hylobates lar vestitus* ate sodium and potassium-rich soil from caves that had been dug by elephants in the Ranun area of northern Sumatra.

Rodman (1978) found that Müller's Bornean gibbons (*Hylobates agilis muelleri*) in the Kutai Nature Reserve of Kalimantan Timur, Indonesia, have approximately equal amounts of fruit and leaves in their diet.

In the Khao Soi Dao Wildlife Sanctuary, East Thailand, a group of 6 *Hylobates pileatus* that ranged over 36 ha (27 ha of which they defended) was predominantely frugivorous (Srikosamatara, 1984). As measured by time spent feeding on various classes of food, their diet consisted of 26% figs, 45% other soft skinned and hard rinded fruits, 11% young leaves, 2% young shoots, and 15% insects. They generally became active around dawn (0600 h) and, on average, settled for the night at 1427 h (n = 27; r = 1245-1550 h). The average 8.2-hour daily activity period was spent as follows: 26% feeding, 25% travelling, 37% resting, 4% calling, 5% grooming and 3% playing (Srikosamatara, 1984).

Pileated gibbons fed mostly in the middle (49%) and upper (39%) canopies and much less often (12%) in the lower story of the forest. They concentrated on fruits, including figs, at the outset and end of their activity period and continued to eat them throughout the day. Young leaves and shoots were eaten most intensively between 1000 h and 1400 h. Insects were taken more evenly throughout the day (Srikosamatara, 1984, p. 248).

Tenaza and Hamilton (1971) briefly observed *Hylobates klossii* feeding twice. On Pagai Utara (North Pagai Island) a group fed on unidentified berries and on Siberut Island, a group feasted on ripe figs.

During a 2-year study (1976-1978), Whitten (1982a) found that Kloss gibbons in central Siberut are the least folivorous of the lesser apes that have been studied intensively. A family of 3 individuals used 227 food sources, 96.5% of which provided fruit, in its 31 ha range of hilly lowland evergreen rain forest. They devoted 23% of feeding time to figs, 49% to other fruits, 25% to arthropods, and 2% to the crispy leaf petioles of orchids, *Myrmecodia tuberosa*, and shoots of a leguminous vine (*Bauhinia* sp.). Whitten's (1982a) limited fecal analysis confirmed that Kloss gibbons concentrate on fruit and insects and eliminated tree leaves from their diet. One stool was composed of wood.

The Kloss gibbons generally preferred fruit that were not very ripe. However, they selected ripe arenga palm fruits (*Arenga obtusifolia*) probably because unripe ones contain toxic oxalic acid (Whitten, 1980). Most fruits were small enough (<2 cm in diameter) to be eaten whole. A number of large, dry, fibrous and tough skinned fruit were ignored by the gibbons. The family fed from 9.4 ± 2.2 food sources per day. Whitten (1982a) calculated that the adult male ingested about twice the volume of a gibbon's stomach during a day. Heavy seeds constituted 61% of the ingesta.

On average, Kloss gibbons spent their 10.6-hour (r = 9.2-11.6 hours) activity period as follows: 24% feeding; 19% travelling; 51% resting; and 4% calling (Whitten, 1984b). Although Kloss gibbons foraged infrequently in the early morning hours, over half of their fig quest occurred between 0600 and 0700 h and then steadily decreased during the day. The first foraging peak occurred between 0900 and 1000 h. Their feeding on tree fruit peaked in the early morning and late afternoon, but feeding on the fruit of vines was more evenly distributed through the day. They did not eat termites, leaf parts or palm fruit until afer 1100 h. The median span of 144 feeding bouts was 7 min. The longest bout lasted 47 min (Whitten, 1982a).

In the Ujung Kulon/Gunung Honje Nature Reserve, West Java, groups of moloch gibbons (*Hylobates moloch*) forage quietly with their members between 20 and 100 m (66-330 ft) apart (Kappeler, 1984c). They typically rise at 0530 h and retire at 1630 h, allowing an activity period of 11 hours (n = 18 days).

Five hours (r = 2.5-6.3 hr) were spent foraging, including travel between dispersed resources, and 2.4 hours (r = 1.6-4.3 hr; n = 18 days) were devoted to feeding in 4 or 5 bouts per day. They consumed more fruit in the morning than at other times of the day. On average, they distributed their feeding time as follows: 59% on fruit; 39% on young leaves and shoots; and 1% on blossoms and buds. In addition, caterpillars, stick insects, termites and other arthropods were snatched unimanually, squashed, and eaten, and honey was harvested from deserted combs. When fruit was plentiful, their diet contained 68% fruit, 30% leaves, and 2% flowers (n = 6 days). But when fruit was scarce, their fare consisted of 49% fruit, 50% leaves, and 1% flowers (n = 6 days) (Kappeler, 1984c).

Most (86%) moloch plant foods are the products of trees (108 species) but they also feed on climbers (14 species; 11%), palms (2 species; 2%) and epiphytes (1 species; 1%). They prefer locations at least 10 meters (33 ft) above ground (Kappeler, 1984c).

One moloch group fed from 889 of 975 (i.e. 91% of) trees, which were > 20 meters (66 ft) tall, in their home range. They represented 79% (61 of 77) tree species in the area. Half the moloch's feeding time was spent in only 10 species. And half of these were among the 10 trees with the largest crown volumes in the region (Kappeler, 1984c).

The winter fare of *Hylobates hoolock* in the seasonally cool and dry lowland evergreen rain forest of the Hollongapar Forest Reserve in Sibsāger District, upper Assam, India, consisted of 67% fruit, 32% leaves and 1% insects. Tilson (1979) sensed that fruit was even more frequent in the summer diet but his data were too sparse for quantitative analysis. In winter, the gibbons began feeding 2.4 hours after sunrise (and a long sunbath), continued to feed throughout the day and ceased near sundown. In summer, they began feeding 3 hours after sunrise and spent more time resting and socializing familially. Summer daylight periods are longer in upper Assam. Yet the average daily total winter ($\bar{x} = 5.4$ h; r = 4.3-7.4 h) and summer ($\bar{x} = 4.9$ h; r = 4.2-6.4 h) feeding periods were not significantly different (Tilson, 1979).

Among apes, Peninsular Malaysian siamang have been subjects of the most intensive, long-term noninterventionist feeding studies. Since 1968, Chivers (1971, 1972, 1973, 1974, 1975, 1977b, 1980c) and several colleagues (Chivers et al., 1975; Raemaekers, 1978a,b; MacKinnon and MacKinnon, 1978) have monitored the feeding, ranging, calling and social behavior of siamang at several localities in the lowland and hill dipterocarp forests of central Peninsular Malaysia. Their methods include direct observations of the subjects' behavior, collection of representative foods (Figure 9) and fecal deposits, and recording the phenology of food trees and vines. They have been especially keen to quantify their data and to test the statistical significance of their results. They have not killed siamang in order to inspect the contents of their stomachs or published nutritional analyses of their diets.

The fruit and young leaf shoots of fig vines and trees (*Ficus* spp.) constitute the dietary staple of siamang in Peninsular Malaysia and Sumatra (Chivers, 1971 et seq.; McClure, 1964; Rijksen, 1978; Gittins and Raemaekers, 1980). Siamang devote between 22 and 30% of their feeding time to the produce of figs. While average feeding bouts last 25 minutes, those on figs average about 40 minutes (Chivers et al., 1975).

Although the early study of Chivers (1972, 1973, 1974, 1975) indicated that siamang were predominantly folivorous, his subsequent research (Chivers et al., 1975; Chivers, 1977b,c) indicated that they eat ripe fruits about 40% and new leaves, shoots and leaf stems approximately 50% of the time. The remaining 10% of feeding time was devoted to buds, flowers, termites and caterpillars. Generally, siamang devoted less time to feeding on insects than on figs (Chivers, 1975; Chivers et al., 1975). Gittins and Raemaekers (1980, p. 90; Raemaekers, 1984, p. 210) summarized the distribution of siamang feeding time as follows: young leaves and shoots, 38%; mature leaves, 5%; figs, 22%; other fruits, 14%; flowers, 6%; and animal matter, 15%.

Figure 9: Specimens from siamang food trees: (a) *Antidesma coriaceum*; (b) *Sarcotheca griffithi*; (c) *Bouea oppositifolia*; (d) *Calophyllum* sp.; (e) *Baccaurea motleyana*; (f) *Vitis* sp;   (continued)

Figure 9 (Continued): (h) *Ficus bracteata*; (i) *Ficus annulata*; (j) *Ficus auriantacea*; (k) *Ficus heteropleura*; (l) *Ficus sumatrana*; (m) *Ficus stupenda*. (Courtesy of D.J. Chivers.)

Siamang obtain water mainly from fruits and the moisture adherent to the surfaces of plants. Chivers (1974, 1977c) observed them hand-dipping fluid from tree holes only in the lowland forests.

Siamang spend somewhat more than half of their 10.5 hour average daily activity period feeding. This behavior is rather evenly distributed throughout the day. That is to say, unlike many other catarrhine primates, they do not take a notable midday break when the sun beats hottest. Like sympatric lar gibbons, Peninsular Malaysian siamang characteristically embark upon a morning orgy of fig or other fruit consumption. Thereafter, they concentrate on folivorous foods, sometimes garnished with caterpillars or termites, and then top off the day with a final helping of figs or other fruit. In the quest for food siamang move down from sleeping sites in emergent trees to the middle level of the canopy where they spend most of the day. Further, they move into lower levels of the forest to forage and feed during the hottest period of the day (Chivers, 1974, 1975, 1977c; Raemaekers, 1978a). The peak of winter drinking occurred around noon (Chivers, 1977c).

ORANGUTANS

The basic feeding habits of orangutans are fairly well documented now from field studies of representative populations in most regions where they are extant. The most thorough studies of orangutan behavior (sometimes focussed more on social organization than feeding ecology) have been conducted in Sabah (Davenport, 1967; Horr, 1972, 1975, 1977; MacKinnon, 1971, 1974a); Kalimantan Timur, Indonesia (Rodman, 1973, 1977, 1979, 1984); southwestern Kalimantan Tengah (Galdikas, 1979, 1982c) and northern Sumatra (MacKinnon, 1974a; Rijksen, 1978). Some observers collected and identified food plants and insects and examined fecal deposits. There are no nutritional reports or studies on stomach contents. The feeding behavior of orangutans in Sarawak is still rather poorly known, though the brief contacts of Wallace (1869), Schaller (1961) and Harrisson (1962) indicated that they prefer to eat fruits, especially the large, tough, spiny-skinned durian (Figure 10). They supplement their frugivory with young shoots, leaves, buds and bark (Harrisson, 1962) or the phloem thereof (Schaller, 1961).

In 1964, Davenport (1967) conducted a 7-month study of orangutans in the Sepilok Forest Reserve, Sabah, during a relatively dry period in which little fruit could be found. The orangutans devoted 90% of their feeding time to leaves and shoots. Bouts of feeding were interspersed with rest throughout the day. On average they spent one-third of the waking period feeding.

On the basis of a 2-year study (1967-1969) in the Segaliud-Lokan Reserve, Sabah, and a 2.5-month study (1971) in the Kutai Nature Reserve,

a

b

c

d

Figure 10: Sumatran orangutans (a) eating figs *(Ficus racemosa)* and (b-d) durian fruit *(Durio* sp.). The subject in (b-d) is a rehabilitant. (Courtesy of H.D. Rijksen, 1978.)

Kalimantan Timur, Horr (1975, 1977) concluded that although Bornean orangutans eat a variety of vegetal foods and insects, fruits constitute the critical component of their diet. They ate more than 100 types of fruit in the Lokan area. Commonly the orangutans ate fruit just before it had ripened fully. Fruits in earlier stages of maturation were eaten rarely. Horr (1972, p. 47) noted that the majority of their food was located in the "lower jungle canopy," i.e., between 20 and 60 feet, and that they spent most of their time there. However, they "often" came to the ground for a variety of foods, water, and when they "really wanted to get away from humans." Horr did not report the percent of time that he observed orangutans eating fruit, leaves, flowers, bamboo shoots, phloem, orchids, termites and earth from termite mounds or otherwise quantify the diet of his subjects. Unlike Davenport (1967), Horr (1972) found evidence that orangutan foraging and feeding is distributed bimodally, with notable morning and afternoon activity periods separated by a midday rest.

MacKinnon (1971, 1974a) has provided the fullest report on the feeding ecology of orangutans in Sabah. He studied them for 16 months, between 1968 and 1970, in the primary dipterocarp forests of the Ulu Segama Reserve where a great variety of fruits were available during most of the year (MacKinnon, 1971, p. 153). But preferred items were usually rare, seasonal and widely dispersed in the forest (MacKinnon, 1974a, 1979). Of the 28 major fruit species that were eaten by orangutans in Segama, 18 were rare, 6 occurred occasionally, and only 4 were available frequently. Apparently, 1968 and 1970 were years of low fruit productivity in the Segama area. The orangutans would remain near major fruiting trees for several days until the current crop was exhausted (MacKinnon, 1974a). For much of the year they had to subsist on less nutritional foods (MacKinnon, 1971, p. 163).

MacKinnon (1971) confirmed that orangutans eat some fruits, e.g. durian and rambutans, before they are ripe. In addition to fruits, including the produce of strangling figs, he observed orangutans eating leaves, lianas, epiphytes, wood pith, bark, flowers and shoots, as well as small quantities of ants, bees, wasp galls, honey and apian grubs. He commented (1971, p. 164) that they will try almost anything that looks edible. Some individuals even came to the ground to eat mineral-rich soil (MacKinnon, 1974a). Over 95% of the foods eaten by the orangutans at Segama (and Ranun, northern Sumatra) were from the middle (18-60 feet) and upper (60-180 feet) strata of the forest (MacKinnon, 1974a).

Orangutans pluck small, loosely attached foods, such as berries and leaf shoots, directly with their highly mobile, prehensile lips or with their fingers. Large fruits, epiphytes and ferns are picked and held manually while the orangutan bites them open and scrapes off edible portions with its teeth or fingers. Many foods require manual and oral processing before ingestion. Inedible parts are generally dropped or spat out, though some fruit stones and fibrous wedges are swallowed (MacKinnon, 1974a, pp. 43-44).

MacKinnon (1971, 1974a) never saw wild orangutans drink from streams or eat vertebrate prey. He found no hair, feathers or bones in their feces. Most of their fluid was obtained from succulent vegetation, particularly the tubers of orchids, woody lianas, fruits and shoots. They licked wet vegetation and hair on their forelimbs and sucked water from a puddle in a large crotch. One juvenile manually scooped water from a tree hole to its mouth, which was kept close to the bowl (MacKinnon, 1971, p. 165; 1974a, p. 46).

The orangutans of Segama spent an average of 4.3 hours per day feeding. They exhibited a bimodal activity pattern; feeding and travel peaked before and after a rather long, nested, midday rest. Generally the morning and afternoon feeding bouts were interspersed with additional short bouts of rest. Daily weather conditions sometimes affected their activity pattern. For instance, they arose late on cold wet mornings (MacKinnon, 1971). Further, they tended to feed less and to travel more during periods when fruit was scarce (MacKinnon, 1974a).

The 15-month study of Rodman (1973a,b, 1977, 1979, 1984) and 5 coworkers in the Kutai Reserve of Kalimantan Timur produced observations that are quite consonant with those of Horr and MacKinnon on the feeding behavior of Bornean orangutans in Sabah. The orangutans of Kutai were predominantly frugivorous and spent the majority of their time in the middle canopy (60-90 feet). Their most favored fruits were extremely rare. During the study period, they shifted from one fruit species to another and transferred to a diet composed largely of leaves, bark and the distal twigs of a few strangling figs (*Ficus* spp.) when fruits were meager. They fed from about 80 varieties of trees and vines. Even on days when fruits were plentiful, they intermittently ate some leaves, bark and the succulent bases of leaf stems. They usually swallowed fruit seeds (Rodman, 1977).

Rodman (1977) calculated that the orangutans at Kutai devoted approximately 54% of their feeding time to fruits, 29% to leaves, 14% to bark, 2% to flowers, and less than 1% to insects. Overall, they spent 46% (or about 5 hours) of the mean activity period (11 hr and 16 min) feeding. They had 2 daily peaks of feeding activity (at 0700-0715 and 1615-1630 h) and a notable midday rest period (Rodman, 1977, 1979).

Over a 10-year span, Galdikas and a number of associates clocked more than 15,000 hours observing Bornean orangutans in the Tanjung Puting Reserve (Galdikas and Teleki, 1981). There are at least 600 tree species in the reserve (Galdikas, 1979). They provide the bulk of over 400 food items that are eaten by orangutans. Permanent staples are concentrated in the middle and lower canopies.

The orangutans of Tanjung Puting fed during 60% of the average daily activity period and travelled, mostly in search of food, during an additional

19% of the day (Galdikas and Teleki, 1981). Like other Bornean orangutans, they rely heavily on seasonal fruit and flowers. Figs are scarce at Tanjung Puting (Galdikas, 1979). Unlike the orangutans at Segama, those at Tanjung Puting did not eat remarkable quantities of unripe fruit (Galdikas and Teleki, 1981). They liked their fruit ripe and even ate fallen fruits that had ripened on the forest floor. The leaves and bark of a relatively small number of plant species and perhaps termites are year-round staples (Galdikas, 1979). Galdikas calculated that the orangutans exploited plant products (e.g. fruits and flowers) during 65%, plant parts (e.g. leaves and bark) during 29%, and insects during 4 percent of their feeding time. During the remaining 2 percent of feeding time, observations were inconclusive or they fed on inanimate substances (Galdikas and Teleki, 1981).

The orangutans at Tanjung Puting were not observed to prey on birds and mammals.* Males, in particular, spent notable amounts of time foraging for and eating termites. In 3 instances, males shared isopteran treats with individual females. These are the only observations of food sharing (apart from mothers with dependent young) that have been made at Tanjung Puting between 1971 and 1980 (Galdikas and Teleki, 1981).

Although Carpenter (1938) observed Sumatran orangutans during only one month in 1973, he noted that they ate leaves and bark in addition to fruit and that they had a particularly destructive predilection for durian.

MacKinnon's (1974a) 6-month study of Sumatran orangutans in the Ranun River area in 1971 revealed that the daily activities of Sumatran and Bornean orangutans are basically similar. His Sumatran data were collected during a highly productive fruiting season. The orangutans at Ranun spent 6.3 hours per day feeding. This was 2 hours per day more than the average feeding time of orangutans at Segama. The Ranun orangutans had morning and afternoon feeding periods broken by a long, sometimes nested, midday rest. The vast majority of their activities occurred in the middle and upper canopies of the forest.

MacKinnon (1974a, p. 31) explained that Sumatran orangutans fed on fewer species of fruit than the Sabah orangutans did because they were more abundant than most Bornean food species. But recall that his study in Ranun was 10 months shorter than the latter and most of it was conducted during periods of low fruit productivity. A very acidic fruit (*Dracontomelum edule*), which was eaten sparingly by other animals, was a special favorite of the Ranun orangutans. In 1971, it accounted for 43% of orangutan feeding time in June, 14% in July (when figs accounted for 53% of the diet), and rose to 33% in August (MacKinnon, 1974a, p. 31). The Ranun orangutans ate a variety of other plant produce and insects. There is no evidence that they

---

*Galdikas (1981) mentioned that she received a personal communication (1980) about meat eating by a wild orangutan, but did not give details (Galdikas and Teleki, 1981, p. 317).

ate vertebrates. Like sympatric lar gibbons, the Ranun orangutans ate mineral-rich soil from caves which had been dug by elephants.

Rijksen (1978) and his associates have conducted the most thoroughgoing study of northern Sumatran orangutans. His monograph is based on more than 2,000 hours of observations on wild subjects and 3,500 hours with rehabilitants during a 3-year period (1971-1974) in the Ketambe area of the mountainous Gunung Leuser Reserve. The summary of feeding behavior that I present here is only based on wild subjects.

Preferred, seasonally fruiting trees, like *Durio* spp. and *Ficus* spp., are plentiful in the Ketambe area (Figure 10). The habitat is a mixed rain forest of the upland variety in which there is a low density of dipterocarps; the lower strata form a closed canopy; lianas and climbers are abundant; and the herbaceous ground cover is lush. There are at least 8 species of strangling figs (*Ficus* spp.), which play a major role in the diet of Ketambe orangutans (Rijksen, 1978, p. 46).

They ate 92 different fruits; 13 kinds of leaves; 22 other types of vegetal products and parts, such as shoots of twigs, tubers of orchids, flowers and the growth layers of certain trees; 2 kinds of aerial roots and climbers; at least 2 species of epiphytic fungi; 17 species of insects; avian eggs; cobwebs; leaf galls; small quantities of soil that covered the tunnels of arboreal termites; and perhaps decayed wood. Fifty-three percent of the 114 food plants that Rijksen (1978, p. 54) collected were from trees, 7% from strangling figs, 28% from lianas, rattans and climbers, 8% were epiphytes; and 4% were from herbaceous plants and grasses. Thirty-two percent of them occurred in the highest canopy, 51% in the middle canopy, and 17% in layers close to the ground. Of the 52 food plants that were most important to the orangutans, only 4 species (8%) were common (i.e., 9 or more trees per hectare), 29 species (56%) were represented by fewer than 9 trees per hectare, and 19 species (36%) were quite rare (Rijksen, 1978, p. 65).

The Ketambe orangutans devoted 58% of their feeding time to fruits, 25% to leafy materials; 14% to search for insects; and 3% to bark or chewing wadges. Figs, many of which contained wasps, constituted 54% of the fruits that were eaten by the orangutans (Rijksen, 1978, pp. 52, 54). The majority of ingested fruits were medium sized (1-2 cm long) but 2 of them (*Durio* spp. and *Artocarpus elasticus*) were quite large (15-20 cm long). Most orangutan youngsters are unable to pick and to open durian fruit (Rijksen, 1978, p. 83). One species of durian fruit and 4 species of figs were utilized while unripe. Seeds were swallowed often but they probably had little nutritional value for the orangutans. Forty-four percent of 96 fecal samples contained intact seeds (Rijksen, 1978, pp. 54-55, 97).

Despite apparent preferences for certain species of plants, on average, the Ketambe orangutans included 7 different items in their daily fare. They fed from between 4 and 15 species per day. Even when fruits were plentiful, they regularly ate quantities of leaves, buds and terminal shoots (Rijksen,

1978, pp. 55, 70). Unlike the Segama orangutans, those at Ketambe did not sample everything that looked edible. Rijksen (1978, p. 73) noted that they ignored several species of fruit, including one of *Ficus*, which were also eschewed by other primates in the area.

In addition to fig wasps, ingested with inflorescences, the Ketambe orangutans ate ants, which commonly occurred in epiphytes, 2 species of termites, 2 species of lepidopteran caterpillars, insect eggs and crickets. Most of the insect prey were arboreal. Seventy-nine percent of the fecal samples contained insects. Ants were the insects most often found in fecal deposits. Some stools contained large amounts of fig wasp remains. All fecal samples from adolescents contained insect remains but they were absent from the stools of some adults. Seventy percent of the insect free deposits were from adult males (Rijksen, 1978, pp. 59-60, 93). Rijksen (1978, pp. 52, 93) only twice observed orangutans raiding avian nests. Vertebrate remains were not found in any of the fecal samples (Rijksen, 1978, p. 60).

Young orangutans drank water at least once per day, usually by hand or finger-dipping from tree bowls, which they would make special efforts to reach. They also licked rainwater from leaves and the hair on their forelimbs during showers. They were not seen drinking from streams (Rijksen, 1978, pp. 52-53, 94-95).

The daily activity cycle of Ketambe orangutans was characterized by 2 peaks: one between 2 and 3 hours after leaving the night nest and another around 1500 hr. The midday was spent resting. Feeding was most intensive in the morning and travel was common in the afternoon. Between 0800 and 1100 h, the orangutans generally fed on fruits that were esteemed and plentiful. In the afternoon, they shifted to terminal shoots, insects and less abundant fruits. They usually followed long bouts of frugivory by feeding on leaves from a neighboring liana or tree. Leaves of *Acacia pennata* were eaten before 0800 h or after 1500 h and were ignored at other times (Rijksen, 1978, pp. 75-76).

FOOD COMPETITION AND NICHE DIFFERENTIATION AMONG THE ASIAN APES

Field research on the Asian apes has stimulated interesting hypotheses about possible niche separation between Peninsular Malaysian hylobatid apes, Bornean gibbons and orangutans, and Sumatran gibbons, siamang and orangutans. But our ability to determine the extent to which the Asian apes actually compete with one another for food and the effects that this might have had on their behavior and morphological evolution is greatly limited by ignorance of the comparative physiology of the animals, the nutritional and other chemical contents of ingested items (and potential foods that are avoided), and the long-term and seasonal fluctuations in forest phenology and composition. Further, all potential competitors have not been studied concurrently and with equal intensity and longevity in representative regions (Chivers, 1984).

The most comprehensive comparisons have been conducted on white-handed gibbons (*Hylobates lar lar*), agile gibbons (*Hylobates agilis agilis*), and siamang (*Hylobates [Symphalangus] syndactylus*) in Peninsular Malaysia. Gittins and Raemaekers (1980, p. 99; Raemaekers, 1984) concluded that the feeding and ranging habits of Malaysian white-handed and agile gibbons are so similar that they are ecological equivalents, which could not coexist for lengthy spans.

On the basis of observations over a period of 33 months from a tree platform in the Gombak Valley, Selangor, McClure (1964) concluded that although the 2 species are mainly frugivorous, the white-handed gibbons may eat more leafy food than sympatric siamang do. He also commented that they were less numerous and their wanderings were more restricted than those of the siamang. He did not observe interactions between gibbons and siamang.

Ellefson (1974, pp. 25-26), who also observed the 2 species briefly in the Gombak Valley, speculated that because siamang are heavier than lar gibbons and eat similar foods, they must range over larger territories in order to subsist. He observed that during interspecific interactions, the siamang seemed to displace the gibbons and that gibbons would wait until siamang had finished eating and had moved out of a particular tree before they entered it to eat. Chivers (1971, 1973, 1974), MacKinnon (1977), and Raemaekers (1978b) confirmed the dominance of siamang over lar gibbons at mutually preferred feeding sites. Chivers (1971, p. 84) succinctly summarized the situation thus:

> The most positive response to other primates was directed at the white-handed gibbons, which often tried to feed in the same tree. The males soon chased them out.

But, conversely to the suggestions of McClure and Ellefson, Chivers (1973, 1974) found that siamang occupy smaller territories than lar gibbons do. He also stressed that quantitatively, the diets of Malayan lar gibbons and siamang are quite different. Chivers (1973, p. 115) concluded that lar gibbons must range farther than siamang because their diet is more specialized. His initial studies indicated that siamang were more folivorous and perhaps less insectivorous than the lar gibbons, which Ellefson had studied. The siamang concentrated more on figs than the other Malayan primates did (Chivers, 1974, p. 291). At a given time they had about 2 staples and 3 or 4 supplementary foods, instead of seeking a diversity of fruits, as seemed to be characteristic of lar subsistence behavior (Chivers, 1974, p. 120).

Subsequent research on sympatric lar gibbons and siamang in the Krau Game Reserve, Panang (MacKinnon, 1977; MacKinnon and MacKinnon, 1978, 1980; Raemaekers, 1978a,b), revealed that the dichotomy between the feeding habits of lar gibbons and siamang was somewhat overdrawn by Chivers and that categorizing the former as specialist feeders and the latter as generalized feeders was premature (Tuttle, 1976) and misleading (MacKinnon and MacKinnon, 1978, 1980). The lar gibbons devoted only about 10% more of their feeding time to fruits than the siamang did. A second revision is that the Krau siamang ate more insects than the lar gibbons did (Chivers, 1977b, p. 99). They also ate more flowers and fed on fewer species of fruit than the Krau lar gibbons did (MacKinnon, 1977; MacKinnon and MacKinnon, 1980).

The study of MacKinnon and MacKinnon (1978) showed that among the 5 species of catarrhine primates in the Krau Game Reserve, the potential for food competiton was greatest between the lar gibbons and the siamang. Canopy use by the 2 species was strikingly similar and the correspondence of their diets was almost as great as that expected between 2 populations of the same species. Both species relied heavily on the produce of *Ficus* spp. (MacKinnon and MacKinnon, 1978, 1980).

Further, they were remarkably alike in the way they distributed their feeding on each food class through the day (Raemaekers, 1978a, p. 195). On average, both species arose at 0630 h, after a foodless retirement period of 14 (siamang) or 15 (lar) hours. They initially sought a feeding site which had a large supply of figs or other fruit. The siamang seemed to select more for figs (versus other fruits) than the lar gibbons did. Figs not only provide a source of fluid and quick energy in the form of carbohydrates but also could be an important source of protein, amino acids, and lipids when they are infested with fig wasps and their parasites. Many authors (e.g., Raemaekers, 1978a) have mentioned that insects are an important source of animal protein. But probably of equal or greater importance is the fact that insects, especially their larval morphs, can have quite high lipid contents (Gaulin and Kenner, 1977). Hladik (1977) also noted their potential as a source of free amino acids.

Leaves probably provide nutrients, including amino acids, which the apes cannot obtain from fruits, at least in the amounts that they require. The leaves of some species may be eaten strategically in the afternoon because their sugar content is higher or their toxicity is lower then (Raemaekers, 1978a). Assuming that leaves are digested more slowly than fruits, they would stave off hunger during the long non-feeding period. The final feast of figs or other fruit would also provide energy for the night when temperatures drop. Moreover, the afternoon fruit source usually will be the site of the first feed on the next morning (Raemaekers, 1978a).

Competition between siamang and lar gibbons is greatest at the morning fruit source. Figs accounted for 55% of conflicts between the 2 hylobatid species and only 14% of their feeding visits. These conflicts occurred more frequently than expected in the early morning. In 21 out of 22 conflicts, the first animal to give chase was a siamang. Raemaekers (1978b) concluded that because interspecific confrontations were relatively uncommon during his study period, at least one of the hylobatid species was avoiding the other.

The greater size of siamang and their habit of intensively feeding as a cohesive family group in a single food source enables them to dominate preferred feeding sites vis-à-vis sympatric lar gibbons. They spend about three-fourths of their ranging time in one-fourth of their home range. There was almost perfect correlation between the amounts of feeding and ranging time spent in a given section of their range (MacKinnon, 1977).

Per contra, lar gibbons do not show such close correlations between feeding and ranging time. Their ranges are about twice the size of siamang ranges and they travel twice the daily distance that siamang do (MacKinnon, 1977). Their strategy seems to be one of more extensive travel and ingesting smaller quantities of food from a wider variety of species than is characteristic of the siamang. Thus they are able to avoid frequent disruptions and potentially injurious altercations at food sites.

The extent to which lar gibbons or siamang are better disposed physiologically for one feeding/ranging pattern or the other must be demonstrated by comparative physiologists. Chivers (1972, p. 131) remarked that the relatively long large intestine of the siamang (Kohlbrügge, 1890-91) might be linked to the predominance of young parts of vines in their diet. However, an elaborate quantitative study by Chivers and Hladik (1980) on the gastrointestinal tracts of many primate species and other mammals indicates that *Hylobates lar* may have a more folivorous gut than *Hylobates syndactylus* does.

The dentition seems to offer better support for the idea that siamang are especially adapted for folivory. Kay and Hylander (1978; Kay, 1975, 1981; Hylander, 1975) found that they have relatively well developed molar shearing and crushing mechanisms and narrower upper incisors, as would be expected in a primate that eats quantities of leaves or other vegetation that requires shredding. The relatively wider incisors of common gibbons are considered to be useful for nipping fruits open.

That siamang and lar gibbons may be somewhat different physiologically is also indicated by the fact that in central Malaysia the former are commonly found at higher altitudes than the latter (Medway, 1972; Chivers, 1974, p. 272). Caldecott (1980) proposed that siamang are better adapted than lar gibbons to the increased tree density, smaller trees, lower and more tangled canopy, low floristic diversity, and much reduced availability of food sources that were predicted to be highly prized by hylobatid apes.

Bornean gibbons and orangutans are grossly different in size. They also evince notable differences in the types of food that they eat and have quite dissimilar foraging strategies (MacKinnon, 1977). Rodman (1978) found that in the Kutai Reserve, orangutans were more frugivorous than the gibbons were. He calculated the ratio of leaves to fruit in the diet of Kutai orangutans to be 20 while that of sympatric Müller's gibbons was 75. The two species tended to feed and travel in the same levels of the forest, except the orangutans were much more likely to come to the ground. Rodman did not describe direct interactions between orangutans and gibbons.

In the Ulu Segama Reserve, orangutans ate the same species of fruit that gibbons ate on 35% of fruit feeding records. But the gibbons were not seen eating other fruits, which constituted an additional 30.5% of orangutan fruit feeding records and which were theoretically accessible to them. Although strangling figs were probably the most attractive items on the menus of both apes, the orangutans fed extensively on them only when they were temporarily superabundant for the gibbons. The 11 instances when orangutans and gibbons were seen feeding together occurred in very large food sources, 5 of which were *Ficus* spp. These interspecific feeding aggregations were characteristically peaceful. Twice gibbons threatened and dashed at juvenile orangutans, which seemed to react playfully toward them. Once a gibbon family hurried away from a small fruiting tree as a group of 3 orangutans approached it to feed (MacKinnon, 1977, pp. 751-752).

In comparison with orangutans, the gibbons of Segama were extremely sedentary. They were very selective feeders that could usually maintain a high proportion of fruit in their diet even when supplies seemed to be meager. The Segama orangutans ranged over a much wider area, gorged themselves on luxuriant fruits whenever possible, and otherwise ate quantities of bark, lianas and wood pith (10.5% of their feeding records), which the gibbons were never seen to eat. In short, the gibbons seemed to prefer quality whereas the orangutans pursued quantity (MacKinnon, 1977).

Like bourgeois, orangutans can accumulate large deposits of body fat, which could enable them to endure lean fruiting seasons (MacKinnon, 1974a, 1977). Whereas Segama orangutans ate certain species of fruit unripe, the gibbons, hornbills and squirrels eschewed them until they had ripened fully. Further, their large size and strength enabled the orangutans to process outsized and hard fruits which the gibbons could not manage; 34.5% of all fruit feeding records of the Segama orangutans were comprised of such fruits (MacKinnon, 1977, pp. 750-751).

Aspects of the synecology of northern Sumatran apes have been studied most extensively by Rijksen (1978) and several Indonesian scientists. They found that among the 6 species of catarrhine primates in the Katambe area, the apes were most frequently observed in proximity to one another. In the vicinity of large strangling figs, orangutans and siamang were seen together 47 times. They were not seen together at other localities.

The superfluity of figs seemed to cause less trouble in the forests of

Ketambe than a single fruit did in the Garden of Eden. Interspecific fights were infrequent. During most agonistic bouts, siamang were apparently the major troublemakers. They often actively displaced lar gibbons from certain trees, especially *Ficus* spp. But there was considerable variation in the aggressiveness that individual siamang exhibited toward lar gibbons and orangutans. For instance, one group of siamang sometimes travelled for up to 3 hours with a group of lar gibbons. All grooming and play occurred between the youngsters of the 2 species. The same gibbon group was occasionally harassed by a second group of siamang with which it also shared a common territory (Rijksen, 1978, p.119).

Despite their relatively modest size, siamang attacked young and adult female orangutans on 3 occasions at Ketambe. One male siamang bit 2 youngsters in a group of 4, all of which fled after the attack. Perhaps a siamang bite is worse than eating bark. Rijksen (1978) also reported 4 agonistic encounters between siamang and orangutans in other areas of northern Sumatra.

Orangutans and gibbons seemed to be indifferent to each other during 40 out of the 41 occasions when they were observed together. Once an adolescent male orangutan displaced a male gibbon in a large fig tree and then followed him out of the tree (Rijksen, 1978, p. 117).

Interestingly, the only report of meat-eating by wild *Pongo pygmaeus* is that of a consorting young Ketambe female which totally consumed an infant *Hylobates lar* over a span of 137 minutes. Sugardjito and Nurhuda (1981) assumed that she had scavenged it somewhere in the canopy because its coat was not bloody before she began to eat it. However, she was already eating the infant's head when they first sighted her. Thus it could have been killed by a cranial bite or squeezes from her powerful hands or feet. She did not share the carcass with her adult male consort nor did he beg or otherwise try to acquire it even though he sat 7 m (23 ft) away and watched the proceedings intently for about an hour. She ate the gibbon while suspended from her left fore- and hind limbs. This indicates that she may have been quite selfishly savoring her pre-bedtime treat (Sugardjito and Nurhuda, 1981).

Although the apes of Ketambe spent most of their feeding time in the middle canopy of the forest and shared a predilection for figs, their different ranging patterns, temporal use of common space and harvesting modes tended to minimize direct agonistic encounters (Rijksen, 1978). As in Peninsular Malaysia, the lar gibbons had territories that were about twice the size of siamang territories and each day they travelled about 60% further than the siamang did. Of the 3 species, the lar gibbons were the most selective feeders. They arose one hour earlier and fed on choice fruiting sources before the other 2 species arrived. Rijksen (1978, p. 132) concluded that the orangutan is probably the least selective feeder among the 3 species of Sumatran apes. The siamang is in a difficult intermediate position between the orangutan, which usually travels no further than it does each day and grossly reduces the quantity of preferred fruits in the area, and the lar gibbon which often gets first pick of quality food items that the lesser apes must share (Rijksen, 1978, p. 126).

MacKinnon's (1977) shorter study in the Ranun area and brief reconnaissances elsewhere in northern Sumatra confirmed that siamang are sometimes quite testy when they meet other apes at a fruit source. The Ranun siamang fed on fruit species which accounted for 64% of sympatric orangutan feeding records during MacKinnon's 6-month study. Although the siamang normally fed peacefully in the same tree with orangutans, a family once chased an adolescent male orangutan from a fig tree. On another occasion, the same family attacked an infant orangutan; but they were chased away by its mother.

Ranun lar gibbons were never seen feeding in the same tree with siamang. The territories of 6 lar gibbon groups markedly overlapped those of 6 siamang groups in the Ranun River area. The orangutans congregated along the river when fruit supplies ran low. Like Rijksen, MacKinnon (1977) concluded that the different feeding strategies which the hylobatid apes and the orangutans employ are sufficient to minimize direct competiton between the lesser and the great apes at Runan. His data on lar gibbons and siamang were insufficient to document the nature of niche differentiation between the 2 Sumatran hylobatid species.

COMMON CHIMPANZEES

Nissen (1931) provided the first substantive information on the diet and feeding behavior of wild chimpanzees. In 1930 he spent little more than 75 days during a 3 month period tracking and observing *Pan troglodytes verus* in the forests east of Kindia, Guinea. The study was conducted at the end of the dry season when rainfall was infrequent. His data primarily consisted of unquantified direct observations and examinations of fecal deposits and the leavings of chimpanzees at feeding sites. Eighteen chimpanzees were collected. Because most of them were destined to stock the new Yale Anthropoid Experiment Station in Orange Park, Florida, their abdomens were left intact.

Nissen (1931) found that 28 of the 54 foods eaten by the Guinean chimpanzees were fruits or berries. They also fed on 3 different kinds of stems and stalks, 2 species of flowers and one species of leaves. Most of the foods were seasonal as attested by the fact that they matured and disappeared during Nissen's short visit. The longest diameters of most of the fruits were between one and 3 inches, though one species was 8 inches long. The bulk of food items were bitter, sweet, fruity or, most commonly, astringent to humans. A few were sour, unpleasant or mealy. Ripe, overripe, and underripe fruits were eaten. Sometimes the chimpanzees seemed to wait until the preferred stage of maturation had been achieved before eating a food. They also seemed to avoid certain fruits altogether. Some fruits were swallowed whole. Others were chewed and then certain parts were spat out. Still other fruits were peeled or shelled so that only particular parts would be ingested. The chimpanzees spat out fibrous wedges while feeding on stalks, brown blossoms and one species of fruit. Their feces were usually fibrous and stringy. About 80% of them contained undigested seeds, palm nuts and fruit stones. Nissen (1931, p. 66) noted that during bouts of ravenous feeding, the chimpanzees often swallowed seeds, fruit stems and skins and they were less selective about the degree of ripeness of foods than they were when more nearly satiated.

Nissen (1931) observed chimpanzees digging. But he attributed this behavior to play instead of foraging or feeding. He found no evidence that they ate honey, eggs or animal prey. Twice Nissen (1931, p. 69) saw individual chimpanzees crouch beside a stream and apparently suck up water momentarily. He inferred that wild chimpanzees required and use very little water.

Nissen (1931, p. 53) concluded that because of the low concentrations of calories in most of their foods, chimpanzees must ingest large quantities and spend a fair share of their waking hours feeding. He estimated that they spent between 3 and 6 hours per day feeding. They gorged themselves, with only short breaks, during the first one to 3 hours after dawn. Between 1100 and 1600 h they fed little and were otherwise fairly inactive. The rare instances when they did feed during the middle of the day occurred in shady areas of the forest. They regularly fed vigorously again during the last hour or 2 of daylight.

De Bournonville (1967) published supplementary data on the diets of Guinean chimpanzees, based on a 100-day field survey (1965-1966) in the Fouta Djallon mountain region and reports from the Waters and Forestry Service of Guinea. De Bournonville and his coworkers contacted chimpanzees only 7 times. They saw them eating shaddock, mandarin orange and oil palm trees. On the basis of leavings, they added oranges, fruits of *Ficus ovata*, and berries from a wild vine to their list of chimpanzee foods. The foresters enumerated 40 plant species from which the chimpanzees took fruits predominantly, as well as some seeds, nuts, flowers, leaves, pith and roots. A number of the plants are cultivars. De Bournonville (1967) found that a minimum of 11 species of plants were represented in about 30 stools. Seeds and fruit stones constituted 35% of the dry weight of the fecal sample.

Between 1976 and 1979, members of the Sterling African Primate Project contacted wild chimpanzees (*Pan troglodytes verus*) 367 times and collected fecal samples from nonprovisioned subjects in the vicinity of Mt. Assirik in the Parc National du Niokolo-Koba of southeastern Senegal

(McGrew et al., 1978, 1979a, 1981; McGrew, 1983). This is probably the hottest and the driest site where chimpanzees have been studied over an extended period of time. It is essentially a pristine open grassland and woodland habitat with forest covering less than 3% of the area. The chimpanzees appear to be well adapted to the less than lush conditions (McGrew et al., 1981).

Their reports on feeding habits have stressed meat eating, based on examinations of 380 fecal specimens and direct observations (McGrew et al., 1978, 1979a; McGrew, 1983). One stool contained pieces of a sloughed cobra (*Naja* sp.) skin. And 10 (2.6%) of the scats contained hair and/or bits of mammalian skin, muscle, bones and teeth. The catch included one bushbaby (*Galago senegalensis*), 3 *Galago* sp., and 2 pottos (*Perodicticus potto*). Two prosimian limbs had passed through the predator's alimentary canal with much of the flesh undigested. An adult female rehabilitant to the area was stripping bark from a dead tree when she found 2 bushbabies in a hollow. She promptly dispatched them and, during the next hour, dined on them with wadges of leaves (McGrew et al., 1978).

The chimpanzees of Mt. Assirik mainly fed on 26 fruits, 7 seeds and pods, 5 leaves and shoots, 4 flowers and inflorescences, one stem and one bark from 35 genera of plants (McGrew et al., 1982). They concentrated their termite fishing during a 4-month span between May and August in ecotonal regions between the plateau and woodland areas. However, the major foods eaten in May and June were located chiefly in the forest and woodlands; 81% of fecal samples evinced fruit of *Saba senegalensis*, 40% had fruit of *Cola cordifolia*, 33% had fig (*Ficus* spp.) seeds, and 28% had stones and fruit from *Lannea* spp. Between August and January the chimpanzees of Mt. Assirik fed more intensively on vegetal species in the ecotonal region, particularly *Hexalobus monopetalus* (in 60% of stools—August and September), *Grewia lasidoscus* (in 57% of stools—October and November), *Tamarindus indica* (in 49% of stools—November and December) and *Zizyphus* sp. (in 9% of stools between November and January) (McBeath and McGrew, ;1982).

During the 77-day visit to the Tai Forest in southwestern Ivory Coast, Boesch (1978) collected some data on the feeding behavior of *Pan troglodytes verus*. He concluded that they are basically frugivorous. They chewed the pulp and skins of drupes of *Sacoglottis gabonensis* and then spat them out. They squeezed the pericarps of drupes of *Chrysophyllum taiense*, sucked out their interiors, and spat out the stones. They also ate fruits of *Uapaca heudelotii* and *Pycnocoma* sp., and the kernals of *Coula edulis*, *Sacoglottis gabonensis*, *Detarium senegalense*, *Panda oleosa*, and *Parinari excelsa*, which they extracted with crude tools (Boesch and Boesch, 1981, 1982, 1984a,b,c; Chapter 5; Figure 15a). Boesch (1978) also saw an arboreal adult male holding a dead young *Colobus badius* while other chimpanzees seemed to solicit for it.

Information on the naturalistic diets and feeding behavior of *Pan troglodytes troglodytes* is based on studies by Jones and Sabater Pi (1971) and Sabater Pi (1977a, 1979) in Equatorial Guinea and Hladik (1973, 1974, 1975, 1977), McGrew and Rogers (1983), Tutin and Fernandez (1985) and Tutin et al. (1984) in Gabon. Sabater Pi (1979) and his coworkers clocked 119 hours of contact with chimpanzees in the Okorobiko and Matama mountain areas during 2 study periods between 1963 and 1969. They timed feeding activities, examined leavings and stools, and collected plants for taxonomic identification and tasting.

The Okorobiko-Matama chimpanzees fed on 43 species of plants and 2 species of termites (*Macrotermes*). Fifty-four percent of their food was found in dense and secondary forests, where they climbed as high as 150 ft (50 m) to feed. The remainder of their food was taken from plantations and areas of vegetational regeneration. Twenty-eight percent of their feeding bouts occurred on the ground. Terrestrial feeding was especially common during months when fruits were sparse in the forests. Then the chimpanzees raided plantations and fed on *Aframomum* spp. and other low plants. Three species of *Aframomum* may have constituted as much as 15% of their annual diet (Sabater Pi, 1979).

Overall, the diet of the chimpanzees consisted of approximately 45% fruits, including pomes, berries, drupes and arils; 32% leaves, sprouts and pith; 8% seeds; and 4% termites. They also ate bark and honey (Sabater Pi, 1979).* Approximately 35% of their foods were orange (23%) or red (12%), Sabater Pi (1979) classified 58% of them as sweet, 25% as mealy, and 3% as bitter.

Sabater Pi (1979) noted that several chimpanzees smelled and visually inspected foods before the group commenced feeding. Fruits were usually picked individually by hand and then placed in the mouth. Fruit skins were spat out. Clusters of small fruit, leaves and sprouts were either removed by hand or directly by mouth. Sugar cane was carried from fields into the undergrowth where it was eaten. Husks of *Brachystegia mildbraedii* were usually removed manually before the seeds were ingested. Fibrous wadges were spat out after chewing the pith out of sugar cane and *Musa* spp. (Jones and Sabater Pi, 1971; Sabater Pi, 1979). Modified sticks were apparently used to open terrestrial termite nests (Jones and Sabater Pi, 1969; Sabater Pi, 1974a).

The daily activity pattern of chimpanzees at Okorobiko-Matama was characterized by morning and afternoon peaks of feeding and a midday period of reduced activity. Maximum feeding occurred between 0800 and 1100 h and a second peak of feeding fell between 1600 and 1700 h. In the

---

*Previously, Sabater Pi (1977a) had stated that their diet consisted of approximately 60% fruits, 30% leaves, sprouts and pith, 5% seeds, and the remainder of insects and other invertebrates.

morning, the peak of travel (0600-0800 h) occurred before the peak of feeding whereas in the afternoon they coincided (Sabater Pí, 1979).

Hladik (1973, 1974, 1975, 1977) and his laboratory collaborators (Hladik and Gueguen, 1974; Hladik and Viroben, 1974) have executed the most thorough nutritional study of free-ranging chimpanzees. During one year (1971-1972) and 3-month (1975) expeditions, Hladik followed 8 well habituated subadult chimpanzees that had been released onto "l'ile aux Singes" in the Ivindo River near Makokou, Gabon. He timed their feeding activities, collected food items, and estimated the quantities of each food that they ate. He clocked about 451 hours of contact with them. The Ipassa area, where the island is located, is characterized by rain forest from which more than 900 plant species have been identified. Fruiting lianas are common among them. Most of the chimpanzees had been captured in the Ipassa area and kept for laboratory studies prior to their release onto the island. They quickly adapted to the local conditions. Thirty percent of their annual diet consisted of provisioned bananas (Hladik, 1977).

Hladik (1977) found that there was poor correlation between calculations of the percentage of major food categories in the chimpanzees' diets based on time spent feeding versus the amounts of food ingested. For instance, one day the chimpanzees devoted 49% of their feeding time to leaves and bark, 14% to fruits and 37% to insects. But these items constituted 33, 63, and 4% respectively, of the food ingested that day. Therefore, Hladik used estimates of the food actually ingested to calculate the annual diet of Ipassa chimpanzees. It consisted of 68% fruits, 28% leaves and stems, 4% animal food, and very small amounts of earth (Hladik, 1973 et seq.). The daily portions of fruit ranged between 40 and 90% with most records showing between 55 and 80% of the total food intake. Animal food, consisting chiefly of ants, termites, avian eggs and fledglings, varied between 2.5 and 6% of the daily diet (Hladik, 1977). Seasonality was marked. Although primarily frugivorous, the Ipassa chimpanzees increased the amount of shoots and stems in their daily diets up to 50% during seasons when indigenous fruits were sparse.

The Ipassa chimpanzees foraged more often in the middle forest canopy than on the ground. They ate greater than 20 different food items each day. Their daily food intake varied between 1200 and 2500 gm (Hladik, 1973). Hladik (1977) identified 141 plant foods, 33 animal foods and 5 mineral foods. The animal foods included avian and arthropod eggs, fledglings, a murid rodent, perhaps an Allen's galago, spiders, scorpions, cicadas, caterpillars, crickets, grasshoppers, mantises, beetles, grubs and cochineal insects (Hladik, 1973). The mineral foods consisted mostly of clay and other phyllitous soil that had been processed by insects. An additional 144 foods of the Ipassa chimpanzees were unidentified specifically. No single species accounted for a large proportion of the annual diet. The Ipassa chimpanzees fed more on fruits in the first half of the day and more on leaves in the afternoon. During rainy seasons, a marked peak of leaf-eating characterized the last 2 hours of daily activity (Hladik, 1977).

The Ipassa chimpanzees used stems, which they had stripped of leaves, to fish for ants. They whacked scorpions manually before eating them. Bark from different trees and lianas was chewed with scorpions, insects, fledglings, eggs and other animal foods. Fibrous wadges of *Hypselodelphis violecea* were spat out after the stems had been masticated. Sometimes wadges were used to extract water from tree holes (Hladik, 1973, 1974, 1977).

On the basis of chemical analyses of nutriments and alkaloids in representative foods eaten by the Ipassa chimpanzees, Hladik (1977) concluded that they select items largely for specific nutrients that they contain instead of basing their selection on the avoidance of small quantities of alkaloids. In order to achieve nutritional balance, they must eat a notable variety of foods that are widely dispersed in space and time (Hladik, 1977, p. 500). They obtained energy from the glucids in fruits and the lipids in certain seeds and arils. Because the protein component of fruit pulp is very low and seeds and arils are often scarce and rather low in protein, the chimpanzees mainly relied upon young leaves and shoots and invertebrates for protein (Hladik, 1977). The leaves and shoots were particularly important when invertebrates were scarce. Leaves provided calcium and animal foods are a good source of potassium. Indeed some plant foods, like the stems of *Hypeselodelphis violacea* and the petioles of *Musanga cecropioides*, seem to be eaten more for their minerals than for other nutriments. Arthropods, for which the chimpanzees spent 30 to 50% of their time foraging, were important sources of certain essential amino acids that are low in plants. Some arthropods, e.g. scorpions, also provided glycogen and phospholipids (Hladik and Viroben, 1974; Hladik, 1977). The small quantities (10-20 gm) of soil that were eaten once or twice a day probably were not an important source of minerals or other nutriments for the Ipassa chimpanzees. Hladik and Gueguen (1974; Hladik, 1977) suggested instead that the soil may absorb the tannins that are contained in certain plant foods.

Tutin and Fernandez (1985) examined 25 stools that were left by wild chimpanzees and the leavings at 35 nest and feeding sites in a pristine primary rain forest around Belinga in northeastern Gabon. Belinga is approximately 100 km (60 miles) northeast of Makokou (McGrew and Rogers, 1983). Tutin and Fernandez (1985) documented 46 plant foods from 43 species and 4 animal foods in the Belinga chimpanzee diet. Fruits constituted 78% (39 species) of the total foods eaten. In addition, they ate 3 species of stems and pith, 2 of leaves, 2 of bark, and 4 of invertebrates, including a snail (*Limicolaria* sp.), the honey and grubs of *Trigona* sp., and termites (*Macrotermes nobilis*). The Belinga chimpanzees made and used probes for termite fishing (McGrew and Rogers, 1983; Chapter 5). Remains of *Macrotermes* were found in 24% of the chimpanzee stools. The snail and an unidentified insect were each found in one stool. The chimpanzees collected honey by indexically probing the nest and licking the extract from their fingers (Tutin and Fernandez, 1985).

Although eastern chimpanzees (*Pan troglodytes schweinfurthi*) have been more extensively and intensively studied than the other subspecies of common chimpanzee, knowledge of their naturalistic diets falls somewhat short of that provided by Hladik on the Ipassa chimpanzees. The acquisition of detailed information on the feeding ecology of the best known populations has been limited by difficult observational conditions, the provisioning that was implemented in order to overcome this impediment, an emphasis on social versus maintenance behavior, or various combinations of these factors.

Rahm (1967) reported that 14 of the 25 food plants of chimpanzees that they collected near the western shore of Lake Kivu in Zaire were from abandoned banana plantations. The chimpanzees ate fruits, buds, leaves, shoots, pith and petioles from a variety of trees, shrubs, ferns and herbs. They moved between areas of vegetational regeneration and lowland rain forest according to the availability of preferred foods. In the forest, they fed mainly on the fruits of trees. But the produce of certain herbs and shrubs constituted the bulk of their diet (Rahm, 1967, p. 198).

Beginning in 1962, chimpanzees of the Budongo Forest in Bunyoro District, western Uganda, were studied by Reynolds and Reynolds (1965; Reynolds, 1965), Sugiyama (1968, 1969) and Suzuki (1971, 1975). Reynolds and Reynolds (1965) observed them for 300 hours during an 8-month study, which was concentrated in an area of colonizing, mixed and swamp forest that had been logged selectively a quarter of a century earlier. The canopy height was approximately 140 feet (47 m) in the mixed forest.

Reynolds and Reynolds (1965) identified 35 plant species, nearly one-third of which are *Ficus* spp., from which the chimpanzees ate fruits, leaves, bark and pith. They estimated that the Budongo chimpanzee's diet consisted of 90% fruits, 5% leaves, 4% bark and stems and only 1% insects. They found no evidence for avian eggs, termites or vertebrates among the chimpanzee's foods or that they drank from terrestrial water sources. They observed an adult male hand-dipping and drinking water from a tree bowl.

Reynolds and Reynolds (1965) estimated that Budongo chimpanzees fed and foraged between 6 and 7 hours of the 12 hour diurnal period. Their daily ranging behavior was affected markedly by the seasonal ripening of fruit trees and vines. During their study, a succession of a few particular fruit species seemed to constitute the staple foods (e.g. 50% or more of the estimated daily fare) for periods of 2 months or less.

Following Langdale-Brown et al. (1964), Sugiyama (1968) characterized the vegetation of the Budongo Central Forest Reserve as medium altitude moist semi-deciduous forest, in which there are 3 or 4 strata. He clocked

360 hours of contact with chimpanzees over a 6-month period in 1966 and 1967. Sugiyama (1968) habituated some chimpanzees to his presence and attempted to provision them only a few times. He listed 14 important plant food species, 6 of which were *Ficus* spp. The chimpanzees variously ate the fruits and seeds from all of the species except *Celtis* spp., from which they foraged young leaves. He noted that they generally chose only fruits that were fully ripe and which did not taste bitter. Sugiyama (1968) concluded that figs were the most important food of the Budongo chimpanzees. However, during the dry season, when figs were no longer abundant, they fed intensively on the seeds of *Cynometra alexandrii*. Sugiyama did not observe meat-eating or deliberate captures of arthropods, though the Budongo chimpanzees did ingest small insects that infested the figs.

Suzuki (1975), who succeeded Sugiyama, found that during the period between 1967 and 1973, the Budongo chimpanzees fed from about 50 plant species. He confirmed observations by Reynolds and Sugiyama that each season they concentrated on a few species of foods for as long as they were abundant. When fruits were meager they ate more leaves than when fruits were plentiful. At the beginning of the dry season they fed intensively on the seeds of *Cynometra alexandrii* (Suzuki, 1975). He estimated that the social unit of 70-80 individuals ranged over 15-20 km.^{2}

Unlike his predecessors, Suzuki (1971, 1975) observed meat-eating by the Budongo chimpanzees. It appears that they can be as calloused as their rumps. Initially, Suzuki (1971) encountered an adult male manipulating and eating a mutilated newborn baby chimpanzee that remained alive during part of the lengthy consumption. Subsequently, Suzuki (1975) observed them eating 3 colobus monkeys (*Colobus guereza*), a blue monkey (*Cercopithecus mitis*) and a blue duiker (*Cephalophus monticola*). Bouts of meat-eating were characterized by the interspersion of leaves, grass and fruit.

Suzuki (1975) also added tree gum from *Khaya anthotheca* and avian eggs to the food list of Budongo chimpanzees.

Ghiglieri (1984a,b) conducted a systematic 16-month ecological study of wild, unprovisioned chimpanzees in the Ngogo Nature Reserve, which is part of the Kibale Forest Reserve, in the Toro District, western Uganda. He chiefly contacted the shy subjects (406 hrs) by staying all day near large fruiting trees. In addition, he monitored the phenology of 107 specimens (mainly *Ficus* spp.), representing 14 species, of prospective food sources and grossly calculated the crown sizes of fruit trees that were used by the Ngogo chimpanzees.

Kibale chimpanzees are basically frugivorous. They spent 78% of feeding time eating fruit pulp, often swallowing pips therewith. Ripe fruits of *Ficus* spp. were prized probably because they are easily harvested, have low toxin levels, and contain substantial protein (Janzen, 1979). The remainder of their feeding time was distributed as follows: 9.2%, young leaves, particularly those low in phenols; 4.0%, mature leaves; 4.2%, blossoms; 1.5%, leaf buds; 1.0%, seeds (apart from fruit pips); and 2.2%, cambium, bark and wood. This produce came from 51 plant types and species. In addition, they probably preyed on a red colobus monkey and ate termites from a dead tree. They sucked water directly from creeks and dipped it from holes in boles (Ghiglieri, 1984a,b).

Ghiglieri (1984a, p. 86) concluded that chimpanzees are specialist feeders which preferentially exploit rare food types of rare species. Most of the trees which produced the bulk of fruits eaten by the Kibale chimpanzees were relatively rare and dispersed in the Ngogo Forest. Over a 17-month span there were 2 fruiting peaks but individual trees seemed to be unpredictable phenologically (Ghiglieri, 1984b).

On average, 3.6 subjects (r = 1-24; n = 687 feeding aggregations) assembled at choice feeding sites. The sizes of feeding parties were correlated directly with crown sizes of the fruiting trees. The chimpanzees seemed to travel in smaller bands during lean fruiting periods. The arrival of later parties at an occupied food site was significantly associated with the utterance of food calls (i.e., pant-hoots; Chapter 7) by early arrivals. Although the male callers must share the fare, they may be groomed by newcomers, enjoy the protection of a larger party and breed with responsive estrous females (Ghiglieri, 1984b).

At Kibale, male chimpanzees spent more time foraging (62.1 versus 52.4%) and travelling (12.1 versus 10.0%) and less time resting (25.8 versus 37.6%) than females did. Both sexes exhibited morning and late afternoon peaks of foraging with a midday peak of resting. But the males foraged predominantly during all their waking hours, while the females rested more than half the time between 1000 h and 1400 h (Ghiglieri, 1984a,b).

During the initial 2.5 years (1960-1963) of study in the Gombe National Park, northwestern Tanzania, Goodall (1963a, 1965) conducted noninterventionistic observations on the feeding behavior and natural foods of eastern chimpanzees. She continued to add new foods to the list on the basis of observations away from her camps and daily examinations of more than 800 stools, which the chimpanzees deposited at an artificial feeding site, especially during the great banana bonanza between 1965 and 1969 (Lawick-Goodall, 1968, 1971, 1975).

The Gombe National Park occupies a narrow mountainous area along the eastern shore of Lake Tanganyika. The valleys and lower slopes are covered with dense gallery rain forest in which trees reach heights of up to 24 m (80 ft). Deciduous woodlands, with trees seldom more than 12 m (40 ft) high, are characteristic of the upper slopes. Many of the peaks and ridges sport areas of grass, which grows seasonally as high as 4.2 m (14 ft) (Goodall, 1965).

Lawick-Goodall (1968, 1971, 1975) established that the diets of Gombe chimpanzees contain 48 kinds of fruit, 28 leaves and shoots, 6 blossoms, 10 seeds, 7 piths, 2 resins and 2 bark or wood items from 80 species of plants. Most of their plant foods were found in trees. Sometimes they ate ripe fallen fruits and the leaves of low plants (Goodall, 1963).

Fruit eating techniques were quite individualistic. Some chimpanzees popped whole fruits into their mouths while others first peeled them with their lips and teeth or nipped off pieces of fruit, chewed them briefly, and spat out the peeling before taking another bite. When they were very hungry, they tended to gulp down unripe, as well as ripe, fruits—peels, pips and all. As their appetites were appeased, the pace of eating slowed and they became more selective (Goodall, 1963). Small fruits and fruit stones often passed virtually unchanged through their alimentary canals (Lawick-Goodall, 1968). Large fruits were picked manually. Ones with hard rinds, like *Strychnos*, were banged against tree trunks or rocks. Berries, buds and blossoms were commonly plucked labially. The chimpanzees tore bark off trees with their hands and teeth and stripped off the phloem with their incisors (Goodall, 1963, 1964). They ripped open large seed pods with their hands and teeth and processed small pods with their teeth, lips and tongues (Lawick-Goodall, 1968).

Gombe chimpanzees consumed the larvae of 2 species of gall fly, 2 species of termites, 3 species of ants, large beetle grubs, moth caterpillars, ant and mantid eggs, bee larvae, and honey. Bees occurred in some stools. The grubs were chewed with a garnish of leaves. Grass stalks, sticks and twigs were used to extract insects from their nests (Lawick-Goodall, 1968). The chimpanzees also licked insects directly off environmental substrates and their bodies (Goodall, 1963).

McGrew (1974) concluded that termites and driver ants constitute a significant source of animal food for Gombe chimpanzees (Figure 14). He conservatively estimated that the average 15 minute episode of ant dipping yielded 17.6 gm of ingested prey. Remains of driver ants were found in the feces of Gombe chimpanzees during 8 months of the year (Lawick-Goodall, 1968, p. 187).

Like the Ketambe orangutans, Gombe chimpanzees exhibit sexual differences in the ingestion of insects. Fecal analyses revealed that 56% (45 of 81) of female stools contained insect remains while only 27% (31 of 113) male stools did (McGrew, 1979, p. 448). An even more striking sexual difference appeared in the occurrence of vertebrate remains in the 194 stools. Whereas 13 stools from male chimpanzees contained bits of birds and mammals, only one from a female did (McGrew, 1979, pp. 448-449).

The vertebrate foods of Gombe chimpanzees include avian eggs, fledglings, red colobus monkeys (*Colobus badius*), young baboons (*Papio doguera*), young bushbuck (*Tragelaphus scriptus*), young bushpigs (*Potamochoerus porcus*), a blue monkey (*Cercopithecus mitis*) and a redtail monkey (*Cercopithecus ascanius*). Before 1968, observers at Gombe National Park saw chimpanzees eat 20 mammals and they found the remains of 36 additional prey in chimpanzee stools. Primates, including 2 human infants

(Goodall, 1977, p. 278), are the most common order of mammalian prey (Lawick-Goodall, 1968; Goodall, 1981). But the figures probably do not represent naturalistic predation accurately because the 9 young baboons were killed and consumed at the feeding camp during the peak period of banana provisioning (Wrangham, 1974a). The prey were captured manually, dismembered with hands and teeth, and, like grubs, consumed with helpings of leaves (Lawick-Goodall, 1968).

On 5 occasions, Goodall (1963) saw chimpanzees manually scrape soil from a cliff face and eat it. The soil contained small quantities of sodium chloride. When crossing perennial streams, the Gombe chimpanzees commonly crouched momentarily to suck up water with their lips. They extracted water from tree bowls with sops of chewed leaves and licked rain drops from plants (Lawick-Goodall, 1968).

The Gombe chimpanzees fed between 6 and 8 hours per day. Generally they spent an hour or 2 at each food source and travelled between one and 10 miles per day (Lawick-Goodall, 1968). Goodall (1965) estimated the annual range of Gombe chimpanzees, except mothers with dependent young, to be between 67 and 78 km^2 (25-30 mi^2). They might feed at any time of the day but certain periods were favored over others. The most prominent peak of feeding occurred between 1530 and 1830 h. They also fed more often between 0700 h and 0900 h than at midday (Lawick-Goodall, 1968).

All members of a foraging party usually fed on the same food source. Sometimes food laden branches were carried for short distances during leisurely feeding and departures from food trees. During most months the chimpanzees balanced their diets between at least 2 types of important foods, most of which, including many insect species, were only seasonally available (Lawick-Goodall, 1968).

During 12 months in 1972 and 1973, Wrangham (1977) conducted the most intensive study on the feeding ecology of Gombe chimpanzees. He followed 36 independent and 19 dependent young subjects that were well habituated to humans and recorded the feeding behavior of 14 focal adult males and the chimpanzees that were with them.

Wrangham (1977) added 77 new foods to the Gombe chimpanzee menu. As of 1983 (Nishida et al.), 203 plant foods are known to have been eaten at least once by Gombe chimpanzees. Still there are no feeding records for more than 300 plant species in Gombe National Park and there is no evidence that the chimpanzees prey upon amphibians, reptiles or fish (Nishida et al., 1983). Although the Gombe chimpanzees ate about 60 types of food per month, they tended to concentrate on a few items over periods as long as 2 months. Their daily feeding round included visits to about 14 major food sources. About 5 species of fruits and 5 species of leaves were the most important foods each day. Animals constituted a small portion of the daily food intake.

When they arrived at a new food source, they inspected it carefully before feeding. If one or more adults were already eating, members of a new foraging party would join them without prolonged inspection of the food source. But if youngsters or baboons were the only individuals feeding, new foragers inspected the source themselves. Tall trees that would require long vertical climbs were particularly subject to thorough examinations (Wrangham, 1977, p. 510). Large foods were inspected before ingestion and many were rejected. Pods from several trees were sampled before the chimpanzees fed on seeds intensively in one of them for periods up to 4 hours. Leaves were eaten more directly than seeds and fruits. The selectivity of chimpanzee feeding is indicated also by the fact that 15 of the 27 foods that were recorded only once in their diets were commonly available in Gombe National Park (Wrangham, 1977, p. 517).

During the wet season, Gombe chimpanzees spent 88% of their feeding time in trees. During the dry seasons they spent as little as 57% of their feeding time in trees. Apparently, they do not like to sit on wet ground. There might have been more fruit and other preferred foods on shrubs and other low plants during dry seasons (Wrangham, 1977).

During a somewhat shorter study (786 hours), Teleki (1981) found that 31 Gombe chimpanzees spent 42.8% of the average 15-hr diurnal period feeding, 13.4% travelling, 18.9% resting and 24.9% interacting socially. Most travel was related to the food quest. He noted no modality in their feeding activity. Instead they fed at a steady rate from sunrise to sunset. Their feeding budget was apportioned as follows: fauna, 20.4% (of the 15-hr span); plant parts (mainly foliage), 32.0%; plant products (fruits, nuts, etc.), 47.1%; and minerals, including water, resin, cocoons, etc., 0.5%. Teleki's subjects focussed on plant products in the morning and on plant parts in the evening. Fauna, consisting chiefly of insects, was sought in the late morning and midday hours.

The Gombe chimpanzees spent more time feeding in the dry seasons (57 and 60% of activity budget) than during the wet season (46% of activity budget). Wrangham (1977) noted 3 peaks of feeding, separated generally by periods of grooming. They occurred between 0730 h and 0830 h and around 1300 h and 1700 h. Most of the daily feeding time occurred in a few uninterrupted bouts, 37% of which lasted at least 5 minutes and accounted for 88% of total feeding time. In general, fruits were eaten earlier in the day than leaves were (medians 1133 h and 1422 h, respectively). However, the young leaves of *Aspilia pluriseta* and *A. rudis* were always selected carefully and eaten slowly before 0800 h (Wrangham, 1977, p. 520; Wrangham and Nishida, 1983). Animals were preyed upon throughout the day. The average day ranges of adult males (3.3-6.4 km) were larger than those of adult females (2.0-3.9 km). Adult males annually ranged over 12.5 km². The annual range of anoestrus Gombe females was smaller (Wrangham, 1977).

Since 1961, members of the Kyoto University Africa Primatological Expedition have studied the distribution, ecology and social organization of

Pan troglodytes schweinfurthi at 5 localities south of Gombe National park in the Kigoma and Tabora regions of Western Tanzania. They have identified many of the foods eaten by chimpanzees on the basis of direct observations and the examination of fecal samples and have stressed the manner in which chimpanzees exploit the diverse microhabitats of the region.

The vegetation is a mosaic of deciduous woodlands, savanna that is characterized by the same species of trees as the woodlands, grass patches, bamboo bush, swamps, and lowland and mountain evergreen and semi-deciduous riverine forests. Although woodlands of Caesalpiniaceae are predominant (63%) in the region, the chimpanzees are heavily dependent upon the riverine forests that cover only about 5% of the total study area. The chimpanzees prefer the hilly and mountainous segments of the region where they range widely in the narrow strips of forest (often only 5-20 m wide), neighboring *Brachystegia* woodlands, and ecotones between the forests and woodlands (Kano, 1971b, 1972).

Trees in the woodlands and savannas are between 5 and 20 m (16.5-66 ft) high and those of the forest canopies are between 7 and 40 m (23-132 ft) high. The largest (and presumably the most productive vis-à-vis the diet of chimpanzees) riverine forests are of a mixed type (Richards, 1952, pp. 248-254), in which there are many tree species, climbers and thickets. Kano (1972) doubted that the chimpanzees fed much in the grassy areas, savannas, swamps and bamboo bush. Whereas they use most types of forest intensively, only one of the 15 types of woodland (*Brachystegia bussei*) provided them with a major seasonal source of food (Kano, 1972).

Initial studies by Azuma and Toyoshima (1961-1962) in the Kabogo Point (= Kagogo Head) region established that the chimpanzees fed from at least 38 plant species, including trees, woody climbers (*Saba* sp.), herbs (particularly the fruits, leaves and stalks of *Aframomum* sp.), and shrubs (*Parinari* sp.). They mostly ate fruits (31 types), but also fed on 8 kinds of leaves and one type each of flowers and stems. Azuma and Toyoshima (1961-1962) witnessed only one instance of chimpanzees ingesting animal food, viz. termites or beetles from rotten wood.

In 1963, the Kyoto University team moved inland to the Kasakati Basin where perennial riverine forests were denser and more common than at Gombe National Park (60 mi [76 km] northwest) and Kaboga Point (about 12 mi [19 km] southwest). During 4 months of the rainy season in 1963-64, Izawa and Itani (1966) recorded 43 food items in the chimpanzees' diet, including 15 kinds of fruit, 12 stalks from herbs, grasses and shrubs, 6 kinds of grass blades and leaves of herbs, 4 species of seeds, 4 kinds of bark from trees and vines, one species of flowers, and honey. Although fruits seemed to be preferred foods, they constituted only 36% of the food items. The Kasakati chimpanzees changed their diet 3 times during the 4 month study period. They concentrated on specific abundant fruits until the supplies were depleted and supplemented them with a wide variety of food items. Most fruits were between one and 3 cm in diameter, though one species of fibrous, aggregate fruit was about 25 cm in diameter. Izawa and Itani (1966) observed no chimpanzees eating insects, vertebrates, avian eggs, soil or tree leaves and found no trace of insects in the 14 stools that they inspected.

The feeding methods that Izawa and Itani (1966) described are similar to those reported by Nissen, Goodall and others. The Kasakati chimpanzees often swallowed fruit seeds, which passed unaltered through their alimentary canals, and spat out fruit skins and wadges of bark. They commonly tossed aside partially eaten fruits. Although *Aframomum* spp. were relatively rare in the Kasakati region, the chimpanzees ate their fruit and pith, but not their leaves or subterranean parts. One female extracted honey from a tree hollow with a small twig.

Izawa and Itani (1966) noted that the Kasakati chimpanzees fed chiefly between 0830 h and 1100 h and between 1500 h and 1700 h. They arose between 0630 h and 0730 h and retired between 1800 h and 1900 h each day.

Kawabe (1966) and Suzuki (1966) recorded additional examples of animal foods in the diet of Kasakati chimpanzees. Kawabe (1966) witnessed a group of 6 individuals chasing, capturing and killing an adult redtail monkey. Only 14 out of 174 fecal samples that Suzuki (1966) collected during a 15-month period in 1964 and 1965 contained traces of insects and other animal foods. Nine of them had remains of 2 species of ants, 4 contained termites and one included a bit of mammalian cartilage or hoof. All 14 of these stools were collected during the rainy season. Suzuki (1966) also found remains of ants in 2 of 21 fecal samples that were collected near the Kasakati Basin area. In 1965, he witnessed termite fishing by chimpanzees with tools that they had made from twigs and bark. He added elephant beetles to their diet on the basis of discarded wings that evinced the teeth marks of chimpanzees.

Suzuki (1969, 1975) further expanded the list of plant species eaten by Kasakati chimpanzees to 78, from which they took 61 fruits, 16 seeds, 4 flowers, 4 leaves, 2 stalks and 1 kind of shoot. Twenty-seven of the fruits were eaten when juicy, 33 were dried and one was fibrous. The Kasakati chimpanzees concentrated on a few major foods but fed from an average of 14 species per month. The major food or foods changed every one to 3 months. Most of them were found in the forests. Three noteworthy foods were foraged from the deciduous woodlands. In August, 1964, hard seeds of *Brachystegia* and *Julbernardia* constituted more than 80% of fecal weights. Like soybeans, they have high nutritive values. The unripe produce is eaten pods and all. Ripe pods are cracked with the incisors and only the seeds are eaten. Naturally exposed seeds are gathered from the ground (Suzuki, 1969). Fruits from 2 species of *Strychnos* were obtained in the dry woodlands, though they did not seem to be numerous in the annual diet. The sweet, fleshy, half-dried fruits of *Uapaca* spp., *Vitex* spp., *Parinari* sp. and *Garcinia* sp. provided about 6% of the annual diet as determined by their representation in fecal samples.

Suzuki (1969) concluded that during the dry season, the characteristic foods of Kasakati chimpanzees are hard seeds of Caesalpiniaceae and Papilionaceae, the half-dried fruits of several species, the fruits of *Strychnos*, and the grain heads of *Brachiaria brizontha*, a grass that grows at the edges of riverine forests. Because half-dried fruits constituted 40.5% of the food items of Kasakati chimpanzees and hard seeds made up an additional 10% of their food list, Suzuki (1969) concluded that dry foods were predominant in their diets. It is remarkable that drinking behavior was not mentioned in his report or in the early reports of other researchers in the region.

Suzuki (1969) stated that the nomadic range of the Kasakati chimpanzees was greater than 200 km². However, Izawa (1960) determined that the ranges of 2 Kasakati groups (designated Z and L) were 122 km² and 124 km², respectively.

During 11 months in 1965 and 1966, Kano (1971a, 1972) studied chimpanzees in the Filabanga area, which is 20 km (12 mi) east of the Kyoto University Kasakati base camp. The climate was drier and the vegetation was more open at Filabanga than at Kasakati, Kabogo and Kasoje. Only about 3% of the area was forested. Chimpanzees were relatively scarce (Kano, 1971a). Because he could not approach the subjects closely without spooking them, Kano recorded very few of their foods. However, he noted that between October, 1965, and January, 1966, they fed almost exclusively on the fruits of *Landolphia* spp. in the forests. They appear not to have used available fruits of *Parinari curatellifolia* and *Aframomum* spp. to the extent that the Kasakati chimpanzees did. They spent much of the intense dry season foraging in the *Brachystegia bussei* woodlands. Kano (1971a) estimated the range of the Filabanga unit-groups to be 450 km².

Itani and Kano surveyed the even more sparsely forested Ugalla area, east of Filabanga, during 7 days in the dry season of 1966. Three fecal samples and 3 food remnants indicated that the Ugalla chimpanzees ate grass and the fruits of *Strychnos* sp. and *Aframomum* sp. (Itani, 1979).

Although the researchers from Kyoto University attempted to habituate chimpanzees at several Tanzanian localities to their presence, they succeeded best at Kasoje (= Kasoge) in the Mahale Mountains area, 138 km (83 mi) south of Kigoma near the shore of Lake Tanganyika. Beginning in 1965, Nishida and several other Japanese scientists and their Tanzanian associates followed chimpanzees in various parts of the region. They moderately provisioned chimpanzees of the K-group from 1966 onwards and the M-group from 1973 onwards, both at a stationary feeding area in the Kasoje camp and sometimes when they encountered them away from camp. The study area is considerably more heavily forested than the other Tanzanian study sites. Eighty percent of the Mahale region is covered with semi-deciduous forests and most of the remaining area is characterized by deciduous woodlands (Nishida et al., 1979).

The Mahale chimpanzees naturalistically ate 328 different items from 198 species of plants, including 6 cultivars (Nishida and Uehara, 1983; Nishida et al., 1983; Takasaki, 1983a). The 192 wild species include 83 trees (43.2%), 52 lianas (27.1%), 18 shrubs (9.4%), 29 herbs (15.1%), and an assortment of other plant types and composites (5.2%). Eighty (41.7%) of them are forest species, 73 (38%) are openland species, 10 (5.2%) are lakeshore species and the rest are in ecotones between these primary communities (Nishida and Uehara, 1983).

The food items include at least 61 species of leaves (Nishida, 1972b) and bark from 21 species of trees, shrubs and woody vines (Nishida, 1976). There are greater than 472 plant species in the Mahale area (Nishida et al., 1983). Like the Asian apes, chimpanzees are probably important agents for the dispersal of seeds (Takasaki, 1983b; Takasaki and Uehara, 1984).

Nishida's (1976) long-term studies revealed that the diet of the K-group included 6 species of major foods (i.e., staple items in many years) and 13 species of important foods (i.e. staple or supplementary items in some years). Fruits were often meager during the first half of the dry season and at mid rainy season. Before human provisioning became effective, the Mahale chimpanzees seemed to depend upon the bark of woodland trees during the latter period, when the phloem might have been most nutritious, providing carbohydrates and protein (Nishida, 1976). The Mahale chimpanzees devoted one-third (12 out of 36) of their bark-eating episodes to the phloem of *Brachystegia bussei*. However, they also ate bark from 9 species of forest plants and 4 species that were found both in forests and woodlands (Nishida, 1976). They held large strips of bark from trunks or boughs bimanually and scraped off the phloem with their incisors. They directly gnawed smaller branches. In both instances, they spat out fibrous material after chewing the ingesta awhile (Nishida, 1976).

The Kasoje chimpanzees ate the leaves, pith, shoots, flowers, bark and wood from 15 species of plants that were located mainly along lakeside beaches, in *Cyperus* marshes, or in both areas. They also spent notable amounts of time licking and nibbling rocks along the lake shore and adjacent rock faces (Nishida, 1980).

The Mahale chimpanzees include more than 25 species of insects in their diet. For example, they eat eggs, larvae, pupae and adults of at least 5 species of *Crematogaster*; adults of 4 species of *Camponotus, Oecophylla longinoda*, and at least 8 other hymenopterous species; honey; 3 isopterous species; hemipterous larvae; and coleopterous, orthopterous and lepidopterous imagoes. Three genera of ants (*Crematogaster*, 50%; *Camponotus*, 25%; *Oecophylla*, 5%) were the main insect prey of the Mahale K-group (Nishida and Hiraiwa, 1982). Termites (*Pseudacanthotermes* spp.) were found in only 2.3% (35/1,543) of fecal samples from the K-group (Uehara,

1982). Although they ate clods of soil from mounds of *Pseudacanthotermis spiniger* year-round, they ate their inhabitants almost exclusively during the wet season. Then termites are accessible near the surface and while they swarm on nuptial flights.

The northern neighboring B-group exhibited a higher frequency (5.2%) of termites in their stools and their predation was concentrated early in the wet season while that of K-group was most intensive in the latter half of the wet season. Uehara (1982) reasonably attributed this temporal difference to the fact that whereas K-group preyed upon *Pseudacanthotermes* spp., B-group concentrated on *Macrotermes* sp. This coincides with unequal availability of their mounds in the ranges of the 2 groups. K-group has not been seen to eat *Odontotermes patruus* even though there are at least 2 mounds in their range (Uehara, 1982).

Ant fishing was primarily an arboreal activity (up to 15 meters above ground), the majority (56%) of which occurred in *Brachystegia* woodlands (Figure 14b). Mahale chimpanzees also fished for ants in the forests and forest fringe areas. The 27 episodes that Nishida (1973) witnessed occurred in only 10 different trees. If a tree hole is large enough, the chimpanzees hand-dip for ants. If the hole is too small for hand-dipping, they employ a variety of probes, some of which they manufacture.

Nishida (1973) concluded that although ant fishing might be very important for the chimpanzees nutritionally, they engaged in the behavior almost as a pastime. Or they may simply like the spicy taste of formic acid (Nishida and Hiraiwa, 1982). The chimpanzees visited the Kasoje feeding site mostly between 0800 h and 1100 h and between 1500 h and 1900 h. But tool assisted insect eating was only observed between 1000 h and 1700 h. It was most intensive between 1100 h and 1600 h. While some chimpanzees fished for ants, their companions groomed or rested (Nishida, 1973, p. 361).

Nishida et al. (1979) reported 20 instances of successful and attempted predation on mammals and adult birds by the Mahale chimpanzees based on direct observations, fecal examinations and circumstantial evidence. The chimpanzees were seen capturing, eating, carrying and/or holding 5 young blue duikers (*Cephalephus monticola* [Kawanaka, 1982a, p. 372]), a 3-year old chimpanzee (Figure 11), one greater galago (*Galago crassicaudatus*), one vervet monkey (*Cercopithecus aethiops*), one giant rat (*Cricetomys eminii*), and probably a giant forest squirrel (*Protoxerus stangeri*). Four fecal samples contained remains of 3 redtail monkeys (*Cercopithecus ascanius*) and a white-tailed mongoose (*Ischneumia albicauda*). Except for the vervet monkey, all of these prey species are primarily forest dwellers in the Mahale region (Nishida et al., 1979). Nishida (1980) and his associates later reported that members of the K-group caught and ate yellow spotted hyraxes (*Heterohyrax brucei*), which live in the rocky zone by the shore of Lake Tanganyika. There is no evidence that they ate the crabs that also inhabit the lakeshore (Nishida, 1980).

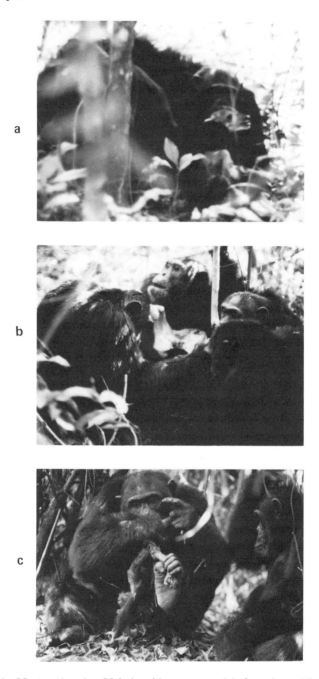

Figure 11: Meat-eating by Mahale chimpanzees. (a) 3 males with a recently captured bushbuck calf. (b) 4 adult males feed on the bushbuck prey. (c) an adult male eating an infant chimpanzee while another adult male watches intently. (Photos courtesy of Kenji Kawanaka.)

Over a 15-month span in 1978 and 1979, the sizeable M-group of Mahale chimpanzees engaged in 15 predatory episodes, directed at small mammals and the young of medium-sized ones. Kawanaka (1982a) and his associates observed 5 kills (4 bushbuck and 1 bushpig) and 6 unsuccessful attempts to capture prey (3 blue duikers, a red colobus monkey, a redtail monkey, and an elephant shrew [*Rhynchocyon cirnei*]). Fecal analyses revealed remains of a blue duiker, a red colobus monkey, a squirrel and an unidentified species of mammal (Kawanaka, 1982a).

During the successive 34-month span from mid-1979 to mid-1982, Mahale chimpanzees were documented to engage in 54 predatory episodes (Takahata et al., 1984). Six of them were only attempts, but most of the remainder were probably actual, as attested by observed kills (3), meat-eating (33), carcass carrying (6) and fecal analyses (6). It is also possible that Mahale chimpanzees acquired some of the meat by scavenging dead mammals (Hasegawa et al., 1983; Takahata et al., 1984). Most of the victims were blue duikers (24%), bushbuck (19%), red colobus monkeys (16%), or redtail monkeys (13%). The majority of prey species inhabit gallery or primary forests though the Mahale chimpanzees now also prey on warthogs (*Phacochoerus aethiopicus*), an openland species.

Takahata et al. (1984) documented that predation by Mahale chimpanzees is seasonal, with the bulk of episodes occurring in the dry season, when favored fruits are abundant and proteinaceous young leaves are scarce. For instance, gross examinations of stools from M-group, collected during the dry (n = 2,837) and rainy (n = 1,380) seasons evidenced vertebrate remains in 1.4% of the former and 0.5% of the latter (Takahata et al., 1984). Hopefully, future fecal samples will be tested chemically for hematin, which would reveal more cryptic vertebrate remains (Spencer et al., 1982).

The Mahale chimpanzees occasionally ate fledgling weaver birds (*Melanopteryx nigerrimus*) and unsuccessfully tried to capture an adult scaly francolin (*Francolinus squamatus*) and something in the nest of a crowned hawk eagle (*Stephanoetus coronatus*) (Nishida et al., Takahata et al., 1984). Further, Norikoshi (1983) listed the white-browed coucal (*Centropus superciliosus*), a weaver bird (*Ploceus* sp.), and domestic chickens (*Gallus gallus*) as prey of the Mahale M- and K-groups. He also mentioned domestic geese that were kept at the study camp as victims of chimpanzee predation.

Because most of the vertebrate prey were medium to small and were commonly youngsters, Nishida et al. (1979) concluded that the Mahale chimpanzees probably do not capture animals heavier than 7 kg (15.4 lbs). Further the predation by chimpanzees seemed to be infrequent in the Mahale area. Nishida et al. (1979) recorded only 18 incidents in a 10-year period. And they found the remains of only one mammal in 989 stools that they had collected systematically during one year (1975-1976). Takahata et al. (1984) noted an increase of predatory episodes, perhaps due to increased observations and the faunal growth that followed the departure of indigenous people from the study area in 1974.

The Mahale chimpanzees masticated meat with wadges of leaves. But usually they did not ingest leaves when they chewed skin. The leaves included not only common food species but also mature ones that were ingested rarely or only with meat (Nishida et al., 1979; Kawanaka, 1982a).

Nishida (1980) provided the principal report on the drinking behavior of chimpanzees living south of the Gombe National Park. The Mahale chimpanzees licked rain drops from their bodies and drank from streams and Lake Tanganyika. Sopping water from tree holes was not observed at Kasoje. More than 85% of the drinking bouts were recorded in the late afternoon. Chimpanzees were seen drinking twice in the early morning and twice in the evening (after 1900 h). Members of the K-group tended to drink most frequently during the dry season. They crouched and sucked water labially during bouts that lasted between 8 sec and 10 min ($\bar{x} = 2.5$ min). Nishida (1980) observed chimpanzees use the lake during 3 of 29 episodes. His assistants saw them drink from the lake 10 times and from streams only 4 times.

Nishida (1979) determined that the home range of K-group was 10.4 km² and that of M-group was more than 13.4 km².

BONOBOS

The feeding habits and ecology of bonobos are known only sketchily. Available accounts contain some tantalizing preliminary observations, circumstantial evidence and hearsay, which, if confirmed by ongoing detailed studies, would set *Pan paniscus* somewhat apart from *Pan troglodytes* as reported hereabove.

Nishida (1972a) related Zairois hunters' reports that bonobos in the secondary swamp forests southwest of Lake Tumba, Bandundu Region, ate fruits, stems, stalks, leaves, and certain subterranean parts of 12 wild plant species and 9 cultivars, a variety of insects and their larvae, honey, and mud fish that they collected from pools on the forest floor.

Horn (1980) observed feeding bonobos for 6 hours during a 2-year study (1972-1974) of black mangabeys (*Cercocebus aterrimus*) in a secondary forest on the west side of Lake Tumba. He listed 12 plant species from which bonobos were seen to eat or were assumed to have eaten fruit, nuts, stems, shoots or seeds. Seven of the plants were foraged on the ground, 2 were fed upon arboreally and 3 were utilized arboreally and terrestrially. Horn concluded that bonobos had ravaged 57 mounds of two species of termites (*Trinervitermes* sp. and *Microtermes* sp.) for their edible inhabitants. Once he saw a terrestrial bonobo ingest a grub which it had picked off a log. Bonobos sometimes carried fallen fruits up to 40 m from a tree before eating them. Horn (1980) watched a terrestrial adult male bonobo tearing apart branches of a small palm tree (*Ancistrophyllum secundiflorum*) and eating their pith. He inferred that bonobos waded into the shallow swamps to feed on the stems and shoots of *Haumania librechtsiana* and *Renealmia africana*.

In 1973, Kano (1979) inaugurated a series of studies on bonobos east of Lake Tumba in the Equateur Region of Zaire, between the Lopori and Tshuapa Rivers. From 1974 onwards, the research was focussed in the vicinity of Wamba village (Kuroda, 1979, 1984; Kano and Mulavwa, 1984; Kitamura, 1983). The vegetation of the region is a mosaic of semi-deciduous lowland rain forest, swamp forest, and swidden fields interspersed with secondary forests (Kano, 1979; Kuroda, 1979; Kano and Mulavwa, 1984). The bonobos utilized all types of forest and forest fringe areas. Kuroda (1979), who focussed on the B-group (Chapter 8), stated that they were most frequently located in the secondary and semi-deciduous forests. But Kano and Mulavwa (1984), who focussed on E-group, concluded that they most frequently exploited the upper crowns and shrub stratum of the dry primary forest ($\bar{x} = 94\%$ of days; $r = 78$-100%; $n = 99$ days).

Kano (1979) inferred from his preliminary survey that juicy and pulpy fruits and fibrous foods (shoots, stalks and leaves) are more important in the bonobo diet than hard items, like beans, are. He accepted Zairois hearsay and circumstantial evidence that the bonobos dug holes (up to 40 cm deep and 50 cm in diameter) in order to obtain subterranean mushrooms. He was somewhat skeptical of reports that they preyed upon fish and he gathered no reliable evidence that they ate other animals.

On the basis of an 8.5-month stay (1974-1975), Kuroda (1979) concluded that during the relatively dry season between July and September, bonobos fed intensively on fruits while at other times they relied more upon fibrous foods, supplemented with available fruits. He listed 19 plant species from which the bonobos utilized 15 different fruits, 3 leaves, pith from 3 stalks, 2 shoots, and 2 seeds. He estimated the range of one group (B) to be between 30 and 40 km².

Kano and Mulavwa (1984) provided the most detailed account of the feeding and ranging behavior of Wamba bonobos based primarily on a 4-month study of the E-group ($n = 65$) in 1981-1982. Bands of the E-group covered 58 km², 66% (38 km²) of which overlapped the ranges of other unit-groups (Chapter 8). Despite the fact that many members of E-group were well-habituated to human tagalongs, observation conditions were problematic due to the density of the forest canopy and the tendency for individuals to spread out amid herbs and shrubs on the ground (Kano and Mulavwa, 1984).

Direct observations and fecal analyses revealed that the average diet of Wamba bonobos consisted of approximately 80% fruit pulp, 15% fibrous foods (e.g. leaves and shoots), and 5% seeds. Animal foods constituted a minute part of their fare. They ate at least 133 parts and products from 114 vegetal species, including 67 fruits, 24 leaves, 12 seeds, 8 shoots, 5 petioles, 4 stems, 4 piths, 3 mushrooms, 2 pods, and 1 each of bark and flowers. Most vegetal foods were collected from tall (> 25 m) trees (44 spp.) and woody vines (23 spp.), followed by medium (15-25 m) and low (5-15 m) trees (22 spp.), herbs and herbaceous vines (12 spp.), shrubs (2 spp.) and fungi (3 spp.) (Kano and Mulavwa, 1984).

Although a variety of cultivars are available at Wamba, the bonobos only ate sugar cane and pineapple pulp outside the provisioning area, where they also were tempted to try other domestic foods. In addition to pineapples and sugar cane, they accepted banana pith (but not fruit) and chicken eggs. Naturalistic animal foods, which were found in about one percent (n = 8) of fecal samples (n = 1,001), included a flying squirrel (*Anomalurus erythronotus*) and earthworms. In addition, the Wamba bonobos ate lepidopteran larvae, honey and the nests of 2 species of stingless bees, and the earth of termite (*Cubitermes* sp.) nests. They were seen drinking from standing water only once so it is assumed that they satisfy their hydric needs chiefly with juicy foods (Kano and Mulavwa, 1984).

Because of large annual and seasonal fluctuations in the availability of preferred foods, many of which are caused by asynchronous fruiting cycles, Wamba bonobos range widely and eat a notable variety of foods each month. Fruits of *Dialium* spp. were the most common staple but they too were seasonal. Many potential food (e.g. *Aframomum* and *Palisota*) were not consumed in proportion to their abundance. Kano and Mulavwa (1984) calculated that the bonobos ate about 10 food types per day and 40 per month and that, on average, 50% of the total menu comprised 93% of the overall diet. In addition to harvesting windfalls of preferred fruits and caterpillars while the quantities lasted, they would spend long periods foraging for rarer foods, e.g. digging holes and wading shallow streams to probe earthworms from their beds. The sizes of foraging bands and clumping of individuals within them are correlated directly with the abundance and density, respectively, of food items at their sources (Kano and Mulavwa, 1984; Kitamura, 1983).

Wamba bonobos spent about 30% of an average activity period feeding arboreally and terrestrially. They rose and began to forage between 0500 and 0600 h and lodged for the night between 1700 and 1900 h. An average daily activity round included the following: arboreal feeding (18%), terrestrial behavior, including feeding (20%), travel (13%) and rest, often with day nesting (43%). Arboreal feeding peaked in the early morning and late afternoon; terrestrial activities peaked after arboreal feeding; and resting occurred largely at midday (Kano and Mulavwa, 1984).

Badrian and Badrian (1977) initiated the second major field study of bonobos. They were joined briefly by MacKinnon (1976) during their initial 7-month study (1974-1975) in the Lomako Forest, Equateur Region, north of the Wamba locality between the Lomako and Yekokora rivers. The area is characterized by primary rain forest, with trees up to 150 feet (50 m) high, patches of secondary forest at the sites of abandoned villages, and extensive swamp forest, with trees over 100 feet (33 m) high, bordering the Lomako River (Badrian and Badrian, 1977; MacKinnon, 1976; Badrian et al., 1981).

Badrian et al. (1981) observed feeding activities during 56% of their 106 contacts (98 hr) with the bonobos. They also collected and analyzed approximately 50 fecal samples.

Lomako bonobos are mainly vegetarians that feed primarily on fruits. Badrian et al. (1981, p. 175) stated that they ate 47 plant foods from 39 species, including 23 (49%) fruits, 10 (21%) young and mature leaves, 7 (15%) stems and pith, 4 (9%) seeds and 3 (6%) flowers.*

On the basis of a fuller study in 1980-1982, Badrian and Malenky (1984) provided a more detailed report on the feeding ecology of Lomako bonobos, which are unprovisioned, unhabituated, and therefore very difficult to locate, follow and observe extensively. They documented 113 plant foods from 81 species, with fruit as the most prominent item in the bonobo diet. It constituted 54.9% of the 113 plant foods. In addition, their menu included 21.2% leaves, 7.1% seeds, 6.2% petioles, 6.2% pith, and 4.4% flowers. Ten species of fruit accounted for 70% of feeding bouts. Many of the most popular fruits were in large trees that grow up to 60 m (198 ft) tall (*Dialium* spp., *Ficus* spp., and *Antiaris toxicara*) or lower trees with large crowns (*Dialium pachyphyllum* and *Uapaca guineensis*), which can accommodate up to 17 hungry bonobos per tree.

At Lomako, bonobos sometimes spent 5-7 hours per day feeding and resting in trees of *Dialium* and *Ficus*. They also terrestrially collected fallen ripe fruits, e.g. large compound fruit of *Anonidium mannii* (up to 50 cm long) and seedy fruits of *Treculia africana*. Although plentiful, leaves generally were eaten by smaller foraging parties. And when seasonal fruit sources could accommodate fewer bonobos, they tended to roam in smaller bands (Badrian and Malenky, 1984).

Like common chimpanzees, bonobos sample fruits orally when they arrive at a prospective feeding site. They seem to prefer ripe fruits over unripe ones. After feeding for awhile, they leisurely chew wadges of fruit skins, leaves and seeds (Badrian et al., 1981).

Bonobos pick large fruits manually and smaller fruits, flowers and leaves either orally or manually. Sometimes they bite open fruits of *Scotellia conensis* and *Dioga zenkeri* and labially ingest the seeds while at other times they pop the fruits into their mouths whole and later spit out the pericarps. They incisally and ungually scrape pulp off the stones of *Pycnanthus*

---

*But in table 1 Badrian et al. (1981, p. 176) list only 36 species and 44 plant foods: 22 (50%) fruits; 10 (23%) leaves; 2 (4%) flowers; 4 (9%) seeds; and 6 (14%) pith.

angolensis after opening the fruits manually. They also swallow many of the fruit stones. The Lomako bonobos nibbled and sucked the pulp from large fruits of *Anonidium mannii* that had fallen to the ground. They cracked open large seed pods of *Anthonotha fragrans* and *Anthonotha macrophylla* with their anterior teeth and removed the seeds digitally and labially. They commonly snapped off food laden branches and carried them to more secure and comfortable feeding sites. (Badrian et al., 1981).

In addition to fallen fruits, the Lomako bonobos fed terrestrially on a variety of herbaceous plants, including the shoots of *Megaphrynium macrostachum* and *Sarcophrynium schweinfurthii*, which were pulled up manually and opened incisally and digitally. They used several stripping and breaking methods to extract the pith from tough, cane-like stalks of *Haumania librechtsiana* and stems of *Palisota ambigua* (Badrian et al., 1981). Canes, shoots, and herbs were carried from their sources and strewn along terrestrial trails by the bonobos (MacKinnon, 1976; Susman, 1980).

The Lomako bonobos incorporated a noteworthy variety of insects and snails in their diets as evidenced by fecal analyses and direct observations. The invertebrate species include the larvae or adults or both morphs of 3 coleopterans, 3 hymenopterans, one bee, 2 isopterans, 3 lepidopterans, 2 orthopterans, 1 giant millipede, 1 land snail and 2 aquatic snails (Badrian and Malenky, 1984; Badrian et al., 1981). Millipedes were most concentrated in their feces between March and May, 1975, when edible fruits seemed to be scarce (Badrian et al., 1981). In January and February, 1981, many bonobo stools contained lepidopteran caterpillars, and in January, 1982, they were seen eating caterpillars with leaves of the infested trees. They also collected caterpillars from the ground (Badrian and Malenky, 1984).

Badrian et al. (1981) saw bonobos stripping leaflets off stems of *Barteria fistulosa*, knocking off the stinging adult ants (*Pachysemia aethiops*) that swarmed over them, and then nibbling and picking out larvae from the lumens of the stems. They also observed an adult male bonobo munching termites which he had plucked from a mound that had been carried 2 m from its base and broken over an exposed tree root. Badrian et al. (1981) "frequently" encountered ruptured mounds and sometimes found sticks and stems nearby with termite soil adhering to them.

Lomako bonobos occasionally foraged in shallow streams, where, in addition to plants, they may have collected small fish or shrimps (Badrian and Malenky, 1984; Badrian and Badrian, 1984a, p. 43).

One bonobo stool included hair, the mandible, and other bones from a shrew. Two other samples contained hair and bones of unidentified small mammals. And a fourth stool included dorsal and ventral scales from a small snake (Badrian et al., 1981). Further, the Lomako bonobos captured 3 infant (<5 kg) duikers (*Cephalophus dorsalis* and *Cephalophus nigrifrons*) which they either ate or abandoned due to disturbance by human observers (Badrian and Malenky, 1984).

GORILLAS

Feeding ecology has been very much at the forefront of the topics upon which students of gorillas have focused during the past quarter of a century. Because of the rapacious decimation of gorilla populations and their habitats by humans and the presumed status of *Pan gorilla* as the largest extant primate herbivore, a number of conservational scientists have exerted herculean efforts to discern the links between their feeding habits, ranging behavior, and sociality. Following the pioneer study of Bingham (1932), the most intensive field studies of gorillas have been conducted on eastern gorillas in Rwanda, eastern Zaire, and southwestern Uganda. However, between 1956 and 1969, Sabater Pi (1960, 1966, 1974, 1976, 1977b, 1979) and several associates (Sabater Pi and Lassaletta, 1958; Sabater Pi and Groves, 1972; Jones and Sabater Pi, 1971) conducted an informative series of studies on the feeding ecology of western gorillas in Equatorial Guinea. They systematically recorded the species and the colors and tastes of plant parts eaten by gorillas. The foods were identified by direct observations of subjects eating them, remnants at feeding sites, and examinations of stools and the contents of the digestive tracts of individuals that were killed for their meat by local hunters (Sabater Pi, 1966).

The diet of gorillas in Equatorial Guinea included various parts of 92 plant species, 74% of which were foraged chiefly in areas where the vegetation was regenerating, following human destruction. Three species of herbs (*Aframomum*) constituted approximately 47% of their diet.* Their fruits were eaten between January and May and their pith, leaves and sprouts were consumed year-round (Sabeter Pi, 1977). Another herb (*Costus lucanusianus*), a shrub (*Trema orientalis*), a vine (*Quisqualis latialata*) and 3 tree species (*Musanga cecropioides, Ficus exasperata,* and *Vernonia conferta*) were also important sources of food for the gorillas, which spent most of their time in the regenerating secondary forests where they grew (Sabater Pi, 1977b; Jones and Sabater Pi, 1971).

The gorillas obtained only 13.5% of their food from dense forests and 12.5% of it from plantations, especially when the fruits of forest species and *Aframomum* were meager (Sabater Pi, 1977b). In both areas they fed mainly on the produce and parts of herbs and vines on or near the ground.

Leaves, pith, and shoots constituted 55% and bark, tubers, flowers and a few other rare items made up an additional 5% of the gorilla's diet. Fruits were a surprisingly common (40%) dietary item, especially vis-à-vis the generally held assumption that gorillas are markedly herbivorous.

Sabater Pi (1977b) concluded that western gorillas are quite partial to fruit but that their bulk prevents them from harvesting the great quantities

*This figure is a notable revision of the estimate (80-90%) provided by Jones and Sabater Pi (1971).

that grow high (20-40m) in the canopy between September and December. Naturally fallen fruits are generally eschewed because they are already tainted by fermentation. However, he observed that young gorillas sometimes climbed into trees to feed and cast down branches laden with fruits, which were consumed by older individuals. Fruits of *Gambeya lacourtiana*, which are the size of oranges, were most commonly eaten by the gorillas in the dense forests of Equatorial Guinea. They constituted 80% of the stomach contents from 5 bagged individuals (Sabater Pí, 1977b).

The western gorillas are selective feeders. They ate the red fruits from plants of *Aframomum* and *Costus* and generally left their green fruits intact. The pericarps of *Aframomum, Costus,* and *Sarcophrynum* and the outer coverings of buds of *Musanga* were usually not ingested but bits of the fruit rinds of *Theobrama* apparently were consumed. The tubers and basal parts of *Manihot* (manioc) were pulled out of the ground and eaten (Jones and Sabater Pí, 1971).

Sabater Pí (1960) observed one young male gorilla chewing wax from a nest of subterranean bees (*Trigona* sp.). Apart from unverified hearsay that the gorillas also ate invertebrates, Sabater Pí (1977b) obtained no evidence that they utilized other non-vegetal foods.

The majority (53%) of the western gorilla's foods were hues of green and yellow. Thirty-eight percent of them ranged from deep red to pale orange and 8% of them (viz. certain roots and berries) were brown. Sixty percent of the foods had some sourness, 16% of them tasted bitter, and 19% of them were insipid. Many of the foods were judged by human tasters to be both sweet and sour (Sabater Pí, 1977b).

The daily feeding activities of western gorillas peaked around 0800 h and 1600 h. The majority of their feeding occurred between 0630 h and 1000 h and between 1400 h and 1730 h. They fed very little at midday (Sabater Pí, 1977b). They commonly fed intensively near their nests and took shorter midday rests during wet weather than during dry weather (Jones and Sabater Pí, 1971).

Bützler (1980) inferred from feeding sites in diverse forest and forest fringe areas of 8 regions that Cameroonian gorillas ate stalks from 2 species of *Aframomum* and *Haumania dankelmaniana*; flowers, branches and stalks of *Tabernathe iboga*; buds of *Musanga cecropioides*; pith from the trunks of *Musa paradisiaca*; berries of *Maesobotrya dusenii*; seeds of *Berlinia* sp.; and fruits of *Gambeya lacourtiana*.

On the basis of a 15-month study (1976-1977) at Campo, Calvert (1985) determined that the vegetal diet of coastal Cameroonian gorillas included 69 items from 50 species. Thirty-eight percent are from herbs, 24% from saplings, 23% from vines, and 15% from trees. Five of the 13 major food species were of *Aframomum*. There was evidence for feeding on *Aframomum* at three-quarters of the feeding sites that she scrutinized.

Fecal analyses (n = 280) show that stems and shoots consitute more than half of the Campo gorilla diet. Half the stools contained remains of fruits. In addition, they consumed leaves, bark, and roots. Calvert (1985) found no evidence in feces or stomach contents (n = 1), at feeding sites (n = 1,400), or via direct observations that Campo gorillas ingested animal matter.

Calvert (1985) also conducted chemical analyses on 36 samples from 27 species of gorilla food plants, with special emphasis on nutriments, digestibility, and potentially toxic phenolic compounds that they contain. She concluded that the most significant factors in food selection by Campo gorillas are lignin, digestibility and crude protein. They prefer plant products and parts that are proteinaceous, fatty, watery and digestible and low in lignin, a structural fiber which is nondegradable by gut microbes and which can retard the digestion of other constituents (Waterman et al., 1980).

The Campo shoots were high in protein, digestibility, and water content, low in lignin, high in some other fibers, and lowest of the major food items in total phenols. Indeed tannins were virtually absent from all of Calvert's (1985) samples of shoots.

Campo stems were high in water and nonstructural carbohydrates and very low in protein, gross energy, lignin and phenols. Of the major food types, stems and shoots were most similar to each other and distinct from leaves (Calvert, 1985).

Campo leaves were relatively high in protein, gross energy, fat, nonstructural carbohydrates and phenols. They were distinguished from fruits primarily by their higher protein content (Calvert, 1985).

Campo fruits were high in fat and gross energy, very high in lignin, and low in water and protein. They had measurable phenolic compounds, including condensed tannins. Calvert (1985) noted that contrary to what one might expect intuitively, the most significant distinctions among the plant parts and products that are eaten by Campo gorillas are between shoots, on the one hand, and fruits and leaves, on the other, instead of fruit versus leaves, stems and shoots. The facts than Campo fruits are woody and, like leaves, contain notable quantities of digestion-inhibiting tannins may explain why fruits are not more prominent in the diet of Cameroonian gorillas.

Largely on the basis of fecal samples (n = 246) from primary forests around Belinga, Gabon, Tutin and Fernandez (1985) contested the classification of western gorillas as folivores. On average, 2.5 species of fruit were identified per gorilla stool. Only 2.4% of the samples lacked fruit remnants.

Tutin and Fernandez (1985) recorded 104 plant foods, representing 89 species, that were used by Belinga gorillas. Fruit was the most numerous item (67%) on the total food list (n = 107). Stems and pith (17%) were second most common and leaves (6.5%) were a distant third. Bark (2%), tubers and rhizomes (2%), flowers (1%), wood (1%) and mushrooms (1%) were sparse on the list.

The Belinga gorillas also ate 3 kinds of insects. Remains of a small species of termite (*Cubitermes sulcifrons*) occurred in 30.5% of the gorilla stools. Further, Tutin and Fernandez (1983) found evidence that they had eaten termites from 54 mounds on 23 of 165 trails that included food litter. Apparently, the gorillas broke off chunks of the mounds manually in order to gain access to their inhabitants. Most mounds were totally destroyed by the giant bug-eaters. Belinga gorillas also occasionally ate caterpillars and grubs and larvae from dead wood (Tutin and Fernandez, 1985).

Although eastern gorillas have been studied much more intensively than the western gorillas, relatively few of the approximately 60 isolated and semi-isolated populations (Emlen and Schaller, 1960a) have been the foci of long-term research. Emlen and Schaller (1960a,b; Schaller, 1963; Schaller and Emlen, 1963) surveyed much of the range of Grauer's gorillas (*Pan gorilla graueri*) and noted the types of habitats that they preferred and the foods that they ate.

They trailed gorillas for 10 days in the Mwenga-Fizi region, west of the northern end Lake Tanganyika, Kivu Region, Zaire. Fizi constituted the southernmost extremity of the gorilla distribution. The Mwenga-Fizi gorillas fed variously on the stems, pith, shoots, bark, petioles, leaves and fruits from 2 species of grass, 3 ferns, 5 herbs, 6 vines, 2 shrubs, 6 forest trees and cultivated banana plants (Schaller, 1963, p. 366). Most of their food was foraged in secondary forests and stands of bamboo (Schaller, 1963, pp. 44-45).

Emlen and Schaller tracked gorillas for 2 weeks in the lowland rain forests of the Utu region, west of Lake Kivu, between the Lowa and Lungulu Rivers. Schaller (1963) returned to the region for 9 days in 1960. The Utu gorillas mainly foraged in young secondary forests and cultivated fields, where banana plants, *Ficus* spp. and *Aframomum* spp. were abundant. They fed most heavily on pith from banana stems (*Musa* sp.) and the pith and fruit of *Aframomum* (Schaller, 1963, p. 151). Schaller (1963, p. 367) listed 2 species of ferns, 5 herbs, 7 forest trees, and 3 cultivars, including the tubers of manioc (*Manihot* sp.), as sources of their food.

During a 16-day reconnaissance by Schaller and Emlen in the Mt. Tschiaberimu region at the northern end of Lake Edward, Kivu Region, Zaire, the gorillas fed primarily on bamboo shoots (*Arundinaria alpina*). Schaller (1963, p. 364) compiled a food list containing 2 species of grass, 5 herbs, 2 vines, 3 shrubs, 5 trees and 2 cultivars, from which the gorillas selectively ate stems, shoots, leaves, bark, twigs and pith.

The feeding ecology of Grauer's gorillas (Groves and Stott, 1979) is best known from studies in the Kahuzi-Biega National Park, which lies west of the southern end of Lake Kivu in Kivu Region, Zaire (Casimir, 1975; Goodall, 1977). Casimir and Butenandt (1973; Casimir, 1975) followed a group of approximately 20 gorillas during 15 months in 1971 and 1972 with the primary goal of documenting their diets and ranging behavior. Because they avoided visual contact with the animals, most of their information on the foods of Kahuzi gorillas is based on leavings and fecal samples. The study area of approximately 70 km² consisted of 37% secondary forest, 25% primary montane forest, 17% *Cyperus* marsh, 15% bamboo (*Arundinaria alpina*) forest, and 7% human artifacts. The gorillas spent the greater part of an annual activity cycle in the secondary forests where the most (30 out of 56) food plants were available. They migrated into the bamboo forest in October and November, when young shoots constituted approximately 90% of their diet (Casimir and Butenandt, 1973).

From the 56 documented food plants, the Kahuzi gorillas selectively ate only the leaves of 27 of them, only bark from 15 species, only the pith from 3 and only the fruits from one. They ate both leaves and bark from 6 species and the bark, leaf tips and fruit from the giant yellow mulberry tree (*Myrianthus holstii*).* They ate only the leaf bases from *Cyperus latifolius*, the internodal segments from stems of *Pennisetum purpureum* (elephant grass), and the bulbs of one species of orchid.

The Kahuzi gorillas dug holes 20 cm in diameter and 20 cm deep in order to retrieve young bamboo shoots. The tenderness of the shoots was apparently matched by that of their digestive tracts, as they suffered severe diarrhea during the first 2 or 3 weeks of the bamboo bonanza (Casimir, 1975).

The selectivity of feeding by Kahuzi gorillas was at least partly determined by secondary compounds in certain plant parts. For instance, they ate the leaves and discarded the closely juxtaposed fruits of *Basella alba*, only the latter of which contained highly toxic saponin (Casimir, 1975).

The Kahuzi gorillas waded knee deep into marshes to forage *Cyperus latifolius*. None of the 43 fecal samples, which were collected at fairly regular intervals over a one year period, contained remains of vertebrates or invertebrates. The gorillas did not distrub active bird and honeybee nests that were clearly visible near their own nests. Neither did they invade subterranean bee nests as the Kuhuzi chimpanzees did (Casimir, 1975).

Casimir (1975) reported the protein content of parts eaten from 14 plants and of parts not eaten from 9 of them. The protein content in the leaves of *Rumex abyssinicus* (42.5% of dry weight), *Basella alba* (35%), and *Piper capense* (27%) and the shoots of *Arundinaria alpina* (19%) was relatively high. In 3 species (*Carapa grandiflora, Galiniera coffeoides* and *Urera hypselendron*), the bark, which the gorillas chewed, contained less protein than the leaves, which the gorillas eschewed. Essential amino acids were usually more concentrated in the parts that the gorillas ate, though they would have to eat a variety of plants in order to obtain a balance among

---

*Casimir (1975) referred to *M. holstii* in the text and a figure legend but he did not list it in his tables of food plants (pp. 100 and 103).

the 20 amino acids that Casimir (1975) assayed. The mineral contents (Ca, K, Mg, Na) were generally higher in the nondietary parts than in the dietary parts of 5 plant foods that Casimir (1975) analyzed. The dietary plant parts that were sampled had a mean water content of 81.5%. Casimir and Butenandt (1973) calculated the annual home range of their study group to be 31 km² and speculated that over a period of years it would be between 40 and 50 km².

Goodall (1977) conducted an intensive 7-month study on the feeding ecology and ranging behavior of Kahuzi gorillas. His main study group of 20 gorillas were neighbors of the groups that Casimir and Butenandt tracked. His study ran concurrently with theirs in 1972. He clocked 273 hours of direct contact with the subjects, identified food plants, collected leavings and fecal samples, and had nutritional analyses run on representative food items.

Goodall (1977) reported that 104 food species provided foods for Kahuzi gorillas. The plants included 33 vines, 25 herbs, 20 trees, 13 shrubs, 6 grasses, 4 ferns and 3 semi-parasites. His main study group ate 160 items from 78 species. Goodall (1977) confirmed that Kahuzi gorillas were generally highly selective about the items that they would ingest. Only a few herbs (*Rubia cordifolia*), 3 species of epiphytic ferns, and perhaps a few vines (*Galium simense*) were wholly eaten. Three of 4 parts were consumed from some of their main food species. But the majority of plants provided fewer food items for the gorillas (Goodall, 1977).

Analyses of 39 foods and 28 nondietary items (some of which were eaten by sympatric guenons and chimpanzees) revealed no major differences in their potential value to gorillas as sources of energy and protein. The crude fiber content of 22 foods ranged between 7.4% of dry weight for shoots of *Arundinaria alpina* and 43% for leaves of *Shefflera kraussiana* (Goodall, 1977). The low fiber content of bamboo shoots might have contributed to the diarrhea of the Kahuzi gorillas when they initially gorged them.

Goodall (1977) estimated that silverbacked males and adult females respectively consumed 30 kg (66 lbs) and 18 kg (40 lbs) of food per day. The largest silverback often fed on the ground while other individuals foraged above ground, especially when they were mixed secondary forests. Youngsters usually moved highest into the trees. But all individuals climbed as high as possible for the leaves of *Loranthus* sp. and the seasonal fruits of *Myrianthus holstii* and *Syzygium guinense*.

Goodall (1977) calculated that they could easily obtain enough water from their food to obviate the need to drink it. Their succulent foods included herbs, vines and bamboo shoots during rainy seasons, fruits of *Myrianthus holstii* and the bases of *Cyperus latifolius* in the dry seasons, and the pith of wild banana plants (*Ensete* sp.) year-round. Goodall (1977) obtained no direct evidence that they drank from streams or pools, though he twice saw individuals lick dew from leaves.

Like Casimir, Goodall (1977) noted that the Kahuzi gorillas ignored eggs and fledglings and did not invade the nests of bees. None of his "many" fecal samples contained animal remnants. Goodall (1977, p. 474) concluded that the Kahuzi gorillas could obtain sufficient energy and protein from their vegetal diet.

Goodall (1977) determined the 7-month range of his main study group to be 22.6 km². It included 69% primary vegetation types (i.e., high montane, bamboo, and swamp forests and *Cyperus* swamps) and 31% secondary vegetation (i.e. mixed secondary montane forests, open vegetation dominated by *Hagenia abyssinica*, meadows, and herbaceous patches on recent clearings). He concluded that at Kahuzi, the distribution of food was more determinant of gorilla movements than the presence of other groups was and that the gorillas seemed to choose foods primarily because of their nutritive value.

Bingham's (1932) 7-week study in 1929 in the Parc des Virungas (formerly Albert National Park), Kivu Region, Zaire, produced scant information about the foods of mountain gorillas (*Pan gorilla beringei*). On the basis of remnants and feces, he concluded that mountain gorillas subsist mostly on succulent, fibrous plants, like wild celery, lobelia stalks, and young bamboo shoots. These plants were their major source of water also. He found no evidence that they dug in the ground, entered trees to feed, were carnivorous, or foraged for honey. Once Bingham (1932) found circumstantial evidence that the gorillas had scraped an opaque white substance, in which winged arthropods were embedded, from a decaying hollow log.

Thirty years later, Schaller (1963, 1965a) conducted the most detailed study on Zairean Virunga mountain gorillas. The Virunga Mountains are located some 100 km (60 mi) northeast of the Kahuzi area. In addition to identifying food remnants, tasting food plants and examining fecal deposits, Schaller clocked 466 hours of direct observations on 10 groups in the Kabara study area over a 13-month period in 1959 and 1960. The Kabara site included 65 km² of *Hagenia* woodland on the steep slopes of Mts. Mikeno and Karisimbi and the saddle between them (Schaller, 1963, 1965a).

The range of habitats utilized by Virunga gorillas was less variable than that of the Kahuzi gorillas (Goodall, 1977, p. 450). Only 4, altitudinally zoned, vegetational types (*Hagenia* and *Hypericum* woodlands, giant senacio, and bamboo) were used commonly by the Kabara gorillas (Schaller, 1963, pp. 60-62). The woodlands in which *Hagenia abyssinica* was predominant were the most favored habitat of gorillas (Schaller, 1963, p. 63). Because the canopy covered only about half the area, *Hagenia* woodland resembled open parkland, with a very dense layer of herbs, growing 5-8 ft (1.5-2.4 m) high. The canopy height rarely exceeded 70 ft (21 m). *Hagenia* woodland was prevalent between 9,200 and 11,000 ft (2,760-

3,300 m). Above 11,000 ft *Hagenia* trees were fewer and the woodlands were characterized by *Hypericum lanceolatum* (the second most common tree in *Hagenia* woodland). At about 11,400 ft (3,420 m) the zone of giant groundsels (*Senacio* spp.) and outsized arborescent herbs (*Lobelia* spp.) began. They are between 10 and 20 ft (3-6 m) tall. Grasses were also common in the giant senacio zone, which the Kabara gorillas visited only sporadically and for a few days at a time.

At Kabara, patches of scrubby, vine entangled bamboo, 10-15 ft (3-4.5 m) high, and interspersed with small marshes, herbaceous meadows, and buffalo and elephant trails, occurred between 7,300 and 9,200 ft (2,190-2,760 m). The gorillas visited them sporadically and during the wetter months when young shoots were available (Schaller, 1963, pp. 63, 150, 357).

The Kabara gorillas variously ate the stems, shoots, petioles, pith, roots, leaves, flowers, fruit, twigs, bark and rotten wood from 29 plant species, including one each of grass (*Arundinaria alpina*), sedge (*Carex petitiana*) and fern (*Polypodium* sp.), 11 herbs, 5 vines, 3 shrubs and 7 trees. Schaller (1963, p. 150) estimated that in the *Hagenia* woodland *Galium simense* (a vine eaten wholly) and the stems and sometimes other parts of 3 herbs (*Peucedanum linderi, Carduus afromontanus, Laportea alatipes*) constitute 80% of their fare. The fruits of *Pygeum africanum* (tree) were consumed extensively when they ripened in April and the fruits of *Rubus runssorensis* (shrub) were eaten year-round. In the giant senacio zone, the Kabara gorillas mainly ate the pith and bases of leaf clusters from *Senacio* spp. and various items from 3 species of herbs. Schaller (1963, p. 150) was puzzled by the fact that the gorillas commonly ingested dry slabs of bark from *Hagenia* and gnawed the bark and rotten logs of *Hypericum*. Goodall (1977, p. 473) suggested that bark provided bulk and may stimulate the motility of intestinal contents in gorillas.

Most (59%) of the 27 food items from 22 plant species that Schaller (1963 pp. 165, 372) tasted were bitter and/or astringent. Many fewer were insipid or palatable (22%), sour (7%), sweet (7%) or mealy (4%).

Schaller (1963, p. 156) noted that food collection by Kabara gorillas was almost entirely manual, though sometimes they used hands and mouth conjointly to remove branches. If plant parts required further processing, e.g. the branches of the shrub *Vernonia adolphi-frederici*, they employed the hands and mouth about equally to tear and shred them. They dug up very young bamboo shoots, ripped them apart or peeled them, and left the outer husks in piles. They snapped the tender green stems of more mature plants at their internodes and chewed only the lower portions thereof. Some small ferns (*Polypodium* sp.) were plucked slowly one-by-one from the branches of *Hagenia* with thumb-to-index finger grips. They detached stalks of wild celery (*Peucedanum* spp.) near the ground, removed the upper leafy portions manually or orally, peeled and bit off the bark, and consumed only the tender centers of the stems. Chemical analysis of *P. linderi* revealed that the discarded leaves are probably more nutritious than the ingested stems (Schaller, 1963, pp. 153, 368). Some gorillas drew the stalks of virulent stinging nettles (*Laportea alatipes*) and thistles (*Carduus afromontanus*) between the thumb and index finger to strip off clumps of leaves which were chewed with aplomb. Other individuals plucked and ate the leaves singly (Schaller, 1963, pp. 158-160). The taproots of *Cynoglossum* spp. were pulled from the ground and eaten. *Galium* was carefully wadded before ingestion, probably to minimize the effects of hooks on its leaves. Blackberries (*Rubus*) were plucked gingerly to avoid prickles from adjacent thorns. They passed intact through the alimentary canal. Schaller (1963, p. 163) remarked that the Kabara gorillas climbed high into the branches of *Pygeum africanum* to obtain twigs from which they ate the bark. A silverbacked male fed from a branch that had been dropped inadvertently from a tree by a feeding female.

Schaller (1963, pp. 166, 371) found 2 places on Mt. Mikeno where gorillas had scraped incisally and eaten a chalky tasting volcanic soil, samples of which tested high in potassium and sodium. He found no evidence that they raided apian nests, which were common at Kabara, ingested animal foods, or drank water. There were no permanent sources of standing water at Kabara (Schaller, 1963, p. 169).

The daily activity cycles of Kabara gorillas were punctuated by two periods of intensive feeding and movement between feeding sites. Feeding usually began by 0600 h and peaked between 0700 h and 0800 h. The groups usually ceased feeding and rested between 1000 h and 1400 h. Feeding and movement resumed between 1500 h and 1700 h. Then they retired (Schaller, 1963, p. 147).

Schaller (1963, p. 127) estimated that each group of gorillas at Kabara had an annual home range of between 26 and 39 km^2. Schaller (1963, p. 130) detected no seasonality in their patterns of movement within home ranges, except for visits to the bamboo zone in certain months.

Osborn (1963) and Donisthorpe (1958) studied mountain gorillas consecutively for 4 months and 8 months, respectively, in 1956 and 1957, on the northern slopes of Mts. Muhavura and Gahinga in the 23.4 km^2 (9 mi^2) gorilla sanctuary near Kisoro, Kigezi District, southwestern Uganda.

Donisthorpe (1958) compiled a list of at least 6 staple foods, including parts of giant celery (*Peucedanum* spp.), bamboo (*Arundinaria alpina*), 2 genera of shrubs (*Vernonia* spp. and *Pychnostachys goetzenii*), a climber (*Basella alba*), and a large herb (*Cynoglossum ampifolium*), 9 supplementary (i.e., seasonal and snack) foods, and 6-8 foods that were eaten only occasionally. Osborn (1963) added 2 additional foods to the list. They obtained no direct evidence for animal foods in the gorillas' diets. Further, with one doubtful exception, no fruits were identified in their diets (Donisthorpe, 1958, p. 205). Instead they consumed pith, stems, leaves, petioles, bark, shoots, roots and nettle tops. Donisthorpe (1958) questioned the local belief that the movements of gorillas among the various vegetational zones were determined by their nutritional requirements. Many foods were to be found on the upper and lower slopes, though the latter had the greatest variety of plants. Both authors were struck by the fact that the gorillas placed the leavings from some foods in small piles.

Schaller (1963) and Emlen stayed in the saddle between Mts. Muhavura and Gahinga for 17 days in March and April, 1959, and tracked gorillas in the surrounding area of the sanctuary. Schaller (1963, pp. 51-53) described 4 types of vegetation that were utilized by the Kisoro gorillas. The dense mountain woodlands were characterized by a fairly continuous canopy, between 12 and 18 m (40-60 ft) high, of *Hagenia abyssinica, Hypericum lanceolatum*, and other tree species, underlain by saplings and shrubs up to 6 m (20 ft) and overgrown by a variety of vines. The ground was covered densely by herbs up to 1.5 m (5 ft) tall. Patches of bamboo (*Arundinaria alpina*) occurred throughout the mountain woodland. Mt. Gahinga also had extensive stands of bamboo, much of which was overgrown by vines. At about 3,000 m (10,000 ft) the forests were dominated by trees of *Hypericum lanceolatum* up to 9 m (30 ft) high. These woodlands were broken by small meadows, patches of scraggly bamboo and clearings covered with grass and herbs that grew up to 2.4 m (8 ft) high. Giant senacio forest replaced the woodlands at altitudes between 3,300 and 3,450 m.

The Kisoro gorillas moved between the 4 vegetation types in search of herbaceous forage. Schaller concluded that because of the large variety and abundance of food plants in the mountain woodland and the time that the Kisoro gorillas spent there, it was probably their preferred habitat. Schaller (1963, p. 362) listed 3 species of grasses and sedges, 12 herbs, 5 vines, 3 shrubs and 4 trees which were food sources for the Kisoro gorillas. Only 5 of them occurred in the giant senacio zone, which was visited by the gorillas only sporadically.

For 64 days (May-August) in 1959, Kawai and Mizuhara (1959-1960) also studied gorillas on Mts. Muhavura and Gahinga as part of the Japan Monkey Centre's Second Gorilla Expedition. They once saw gorillas climb into bamboo but never a tree. They identified 25 plant species that were eaten by the Kisoro gorillas and which were mostly the same as those listed by Donisthorpe and Schaller. They confirmed that 5 items—bamboo shoots, celery, the stems from 2 genera of shrubs, and the roots of *Cynoglossum amplifolium*—were staple foods. Like Donisthorpe (1958, p. 210), Kawai and Mizuhara (1959-1960) noted that when the season for bamboo shoots (June-July) ended, the gorillas shifted to celery as their staple. Schaller (1963, pp. 54, 345) and Emlen observed that the Kisoro gorillas continued to eat celery during the bamboo season. The food list of Kawai and Mizuhara included only one fruit while that of Schaller included 3 fruits. They found no evidence for animal foods in the gorillas' fare. The

Kisoro gorillas dug holes 10 cm in diameter and 5-10 cm deep, in order to extract the roots of *Cynoglossum*. Kawai and Mizuhara (1959-1960) expressed surprise that the gorillas ate so few foods and had a smaller daily intake than they had expected them to have. They found that 8 out of 17 gorilla foods, including 4 of the 5 staples, tasted bitter.

The longest, most continuous study of gorillas has been conducted by Fossey and a number of associates on the southern slopes of Mt. Visoke in the Parc National des Volcans, Rwanda. Fossey began her research in 1967 and was later joined by Harcourt, Stewart, Vedder, and Watts, who focussed on the feeding ecology and ranging behavior of several habituated groups in the main study area of 25 km². Although Fossey and her co-workers encountered 8 groups, totaling 65 animals, they concentrated on 2 groups, designated 4 and 5, which were permanent residents in the study area (Harcourt and Stewart, 1984; Watts, 1984, 1985a,b; Vedder, 1984). During the initial 7 years, feeding behavior was less systematically recorded than social behavior and observations were not evenly distributed throughout the day (Fossey and Harcourt, 1977). Fossey succeeded brilliantly in habituating group 4 to the presence of researchers and without the biasing effects of provisioning.

The Parc National des Volcans extends over 120 km² and spans altitudes between 2,100 and 4,434 m (7,000-14,780 ft) (Harcourt, 1977; Harcourt and Curry-Lindahl, 1978). It is characterized by 6 vegetational zones (Spinage, 1969) that are similar to those described by Schaller (1963) for the Zairean and Ugandan Virungas. However, not all vegetational types were represented in the ranges of each group of gorillas. For example, zones of bamboo (*Arundinaria alpina*) and *Pygeum africanum* did not occur in the range of the best known Visoke group (Fossey, 1974).

Fossey (1974) concluded that the home range of group 4 included the following 6 vegetational types: saddle (40.8%), nettles (11.5%), brush ridge (14.6%), vernonia (3.8%), herbaceous slopes (26.2%) and sub-alpine and alpine (11.5%).[*]

The saddle zone lay adjacent to Mt. Visoke. It was covered heavily with trees of *Hagenia abyssinica* and *Hypericum revolutum*, upon which edible vines, ferns and a fungus (*Gaaoderma applanatum*) grew, and sported a rich herbaceous stratum, characterized by celery (*Peucedanum linderi*), thistles (*Carduus afromontanus*) and dock (*Rumex ruwenzoriensis*) and tangles of *Galium spurium*. Near the base of the mountain was a dense zone of tall nettles (*Laportea alatipes*). The brush ridge vegetation was a combination of shrubs (e.g. *Rubus runssorensis*) and *Pygeum africanum* along the edges of

---

[*]These percentages were computed on the basis of 130 total grid squares (Fossey, 1974, p. 569). Adding the percentages, one obtains a total of 108.4%. The grand total of grid squares listed in the first column of Fossey's Table II is 142 because some squares contained more than one vegetation type.

ravines that were scattered throughout the study area. *Vernonia adolphi-frederici* occurred in small clusters on and adjacent to the slopes of Mt. Visoke. *Hypericum revolutum* were common on the first few hundred meters of the slopes. But the slopes were mainly covered with fairly low, dense herbs (Fossey, 1974).

The Visoke gorillas fed selectively from 42 species of wild plants and ate no crops. Further, they did not eat some common wild plant species, e.g. balsam (*Impatiens*), which were eaten by other herbivores at Visoke. Three plants (*Galium, Carduus,* and *Peucedanum*) accounted for 60% of the gorillas' feeding records and 70% of their feeding time. *Vernonia* (flowers, pith and rotten wood), 2 species of nettle (*Urtica, Laportea*), and 2 species of blackberry (*Rubus*) accounted for an additional 20% of the feeding records for group 4. Fossey and Harcourt (1977, p. 426) estimated that the Visoke gorillas probably spent 95% of their feeding time on only 9 (21%) of the 42 plant food species. The bulk (86%) of their diet consisted of young and mature leaves, shoots, and stems. The remainder of the diet consisted of wood (7%), roots (3%), flowers (2%) and fruits (2%). Fossey and Harcourt (1977, p. 424) reported that the gorillas sometimes ingested free water by bending down and sucking it up.

Like other gorillas, those at Visoke usually gathered and ingested only certain parts of most plants. Given a choice between a commonly available food plant species and a less common one, they usually ate from the latter. They seemed to prefer thistles and celery over *Galium* and usually relished the leaves of thistles more than their stems (Fossey and Harcourt, 1977, p. 427). Their feeding techniques were basically like those described by Schaller (1963). Young adults and youngsters spent more time feeding in trees than silverbacks did, especially when they ate *Galium* (Figure 8). Adults climbed trees most often for fruits, which were rare in the area. Youngsters generally fed less on the most common food species than the adults did, but this may be due to vertical stratification during some feeding bouts and the limited capacities for youngsters to uproot celery (Fossey and Harcourt, 1977, pp. 432-434).

They peeled away slabs of bark, and digitally or labially plucked off slugs, grubs and insects and ate them (Fossey, 1974; Fossey and Harcourt, 1977). Larvae were also obtained deliberately from split stems and the surfaces of leaves and were ingested inadvertently with other vegetal foods. Fossey and Harcourt (1977, p. 444) concluded that the Visoke gorillas probably obtain sufficient quantities of arthropods to meet their requirements for vitamin B_{12}.

Fossey (1981, 1983, 1984) found fragments of bone, teeth and hair from an infant gorilla in the dung of 2 female gorillas. This constitutes a singular case of vertebrate ingestion by mountain gorillas.

Though they lacked grins to betray it, Visoke gorillas also occasionally engaged in coprophagy. Harcourt and Stewart (1978a) reported 25 coprophagous instances on 21 separate days, many of which were rainy. In most cases, only small amounts of feces were ingested. In 13 of 17 cases in which the depositor was identified, it ate its own feces. In the remaining 4 instances, youngsters ate feces from adults. Commonly, the consumer defecated into its own hand and immediately carried the deposit to mouth. Harcourt and Stewart (1978a) pooh-poohed the idea that copraphagy supplied an important source of vitamin B_{12} for Visoke gorillas. Instead they suggested that it might serve other nutritional needs or simply be born of boredom or the desire to ingest something warm after long, cold, wet foodless periods.

Fossey and Harcourt (1977) estimated that Visoke gorillas spent about 25% of the daily activity period feeding. They generally arose between 0600 h and 0645 h and retired between 1800 h and 1845 h. Groups 4 and 5 often took rests at midday while a third group (8) continued to feed (Fossey and Harcourt, 1977, pp. 424-425). In 2 consecutive years, groups 4 and 5 ranged over 4.0 and 4.9 km² and 8.1 and 7.7 km², respectively (Fossey and Harcourt, 1977, p. 434). The only evidence that the movements of groups between vegetational zones were affected by food preferences were 2 shifts of group 5 to the bamboo zone between September, 1972, and September, 1974, when new shoots were abundant there (Fossey and Harcourt, 1977, p. 439). Groups of gorillas usually did not exhaust the food supply at a given locality before moving on (Fossey and Harcourt, 1977, p. 427). However, lone silverbacks may over-use an area (Fossey, 1974, p. 578).

Based on a later study, Harcourt and Stewart (1984) suggested that Fossey and Harcourt (1977) had underestimated the feeding times of Visoke gorillas. During a 2-year span in 1981-1983, the median adult in groups 4 and 5 spent 45.5% of its day feeding. Youngsters increased their feeding time on solid food until their fifth year, after which they fed as long per day as adults did. A lone adult male at Visoke spent 65% of his day feeding (Harcourt and Stewart, 1984).

Watts (1984, 1985a,b) and Vedder (1984) performed model studies on the feeding ecology of Visoke groups 4 and 5. They not only collected quantitative data on the diets, habitat structures and ranging behavior of their focal groups but also sent samples of representative foods for chemical assays to identify their nutrients, digestibility, and secondary compounds which could inhibit their digestion (Waterman et al., 1983; Waterman, 1984).

Watts (1984, 1985a,b) focussed on group 4 (874 h over a 16-month span) and also observed group 5 (164 hr) and a third (Nunkie's) group (92 hr). Vedder (1984) concentrated on group 5 during 18 months in 1978-1979, most of which were concurrent with Watts' tenure at Visoke. The chief difference between the habitats of group 4 (n = 12) and group 5 (n = 13) was the availability of bamboo (*Arundinaria alpina*) in the range of the latter (Watts, 1984).

Members of Visoke group 4 ate 75 foods from 38 plant species (Watts,

1984). The bulk of their food consisted of fresh leaves (67.7% of ingesta) and stems (25%) from ground level herbs, vines and shrubs (Chapter 2). They also consumed much smaller quantities of pith (2.4%), epithelium from excavated roots (1.4%), bark (1.4%) and flowers (1.1%) and roots, fruit, fresh and rotten wood, fungus, dead leaves, galls, gorilla dung, bryophytes, twigs, seeds and petioles (each accounting for < 0.5% of the overall ingesta). Cocoons from dead stems were the only animal matter that they purposefully ingested. The leaves and stems of three species (*Galium ruwenzoriense*, 35.7%, *Carduus nyassanus*, 31.4 percent; and *Peucedanum linderi*, 9.0%) accounted for more than three-fourths of their overall food intake (Watts, 1984). They are all succulent, digestible, and free of condensed tannins. Further, the thistle leaves (*Carduus*) and vine of *Galium* are quite proteinaceous (Waterman et al., 1983, p. 245).

Watts (1984) concluded that the Virunga gorilla forage is low in easily assimilatable nutrients in comparison with fruit, flowers and animal matter. However, in undisturbed areas of the Virungas food is relatively abundant for the gorillas because the most common items in their diet are available perennially. Still they must range notable distances, sampling a large number of sources, in order to provide variety for nutritional balance and sufficient amounts of staples for all group members. Watts (1984) estimated that silverbacks ate 30 kg (66 lb) per day, mature blackbacked males consumed 22 kg (48 lb)/day, adult females ate 18 kg (44 lb)/day, and 4-5 years old ate 12 kg (26 lb)/day.

Members of Visoke group 4 foraged cohesively in each vegetational zone. Overall dietary overlap between individuals varied between 72 and 90% ($\bar{x}$ = 83.9 ± 4.8%). They did not share food, except for mothers with their infants; and, in contrast with chimpanzees and orangutans, even this was not actively practiced (Watts, 1984, 1985a,b). Indeed infants probably learned what to eat by watching their mothers and sampling bits of her meals or plants like them. Although they also experimented with other items, by the time they are 3 years old, young gorillas have diets that are basically like those of their elders (Watts, 1985a).

The basic diet of Visoke group 5 is quite similar to that of group 4. They had 53 food items in common. In addition, members of group 5 ate small amounts of 15 items that were not included in the diet of group 4 (Watts, 1984). The major difference lay in the much greater amount of seasonally harvested bamboo shoots in the diet of group 5 (Watts, 1984; Vedder, 1984).

As measured by percentage of feeding time, 4 foods constituted three-quarters of the diet for group 5: the vine, *Galium ruwenzoriense* (36.5%); celery stalks (*Peucedanum linderi*, 24.1%); thistle leaves (*Carduus nyassanus*, 9.4%); and bamboo shoots (*Arundinaria alpina*, 5.1%). The latter are relatively high in protein and the gorillas prefer them to other foods during the 2-3 months when they sprout (Vedder, 1984). Based on estimated wet weights of ingested items, Watts (1984) confirmed that the same 4 foods make up three-fourths of the diet of group 5. The diet of Nunkie's group completely overlapped those of groups 4 and 5 (Watts, 1984).

Like Watts, Vedder (1984) concluded that mountain gorilla ranging patterns are at least partly determined by the availability of food. Group 5 spent the most time in areas with relatively rich food supplies, that is to say, where they could obtain a variety of items, many of which are highly proteinaceous. Sheer abundance of a potential food is not sufficient to keep them in an area.

The Kayonza Forest (aka Bwindi Forest Reserve and Impenetrable Central Forest Reserve) in western Kigezi District, Uganda, is the eastern-most extremity of the distribution of mountain gorillas (Groves and Stott, 1979; Chapter 1). Pitman (1935) obtained little first-hand knowledge of them during 2 very brief visits in 1933 and 1934. In 1959, Emlen and Schaller (1960a) spent 11 days tracking Kayonza gorillas and Schaller returned to the Kayonza for a 12-day study in 1960.

The Impenetrable Central Forest Reserve encompassed about 250 km² of mountain forest, in which logging was (and is) practiced. Altitudes of the valleys and ridges range between 1,650 and 2,100 m (5,500-7,000 ft). In undisturbed areas, particularly the ridges, trees reached heights between 36 and 42 m (120-140 ft) and ground cover was sparse. On the lower slopes and in the valleys, large trees were scattered or absent. Instead there were dense stands of tall (6 m high) tree ferns (*Cyathea deckenii*), shrubs, herbs and vines, which justify the former official name of the reserve (Schaller, 1963, pp. 46-48).

Schaller (1963, p. 48) concluded that the Kayonza gorillas foraged mostly on the lower slopes where herbs and vines were abundant. He documented that they variously ate shoots, pith, leaves, bark and fruits from 23 plant species, including 3 ferns, 4 herbs, 9 vines, 3 shrubs, 3 trees and one cultivar (the pith of banana stems (*Musa* sp.), which they obtained by raiding plantations). Assistants added a sedge, a vine and a tree to Schaller's (1963, p. 363) list. Schaller (1963, pp. 150-151) noted that the fronds of *Cyathea deckenii* and the bark and leaves of 5 vines (*Urera hypselendron, Piper capense, Basella alba, Mikania cordata*, and especially *Momordica foetida*) were important foods of the Kayonza gorillas. Out of a total of 57 food species which Schaller (1963, p. 153) collected in the Kayonza Forest and the Virunga Volcanoes, only 9 (16%) of them were shared by gorillas in the 2 areas. He attributed this largely to the different vegetations of the two regions.

Kawai and Mizuhara (1959-1960) tracked gorillas in the Kayonza Forest for one week in 1959. They noted that the pith from fronds of tree ferns was a staple food. The remarked (p. 33) that comparisons between the feeding habits of Kayonza versus Ugandan Virunga gorillas were impeded by the extreme differences in the vegetation of the two regions.

COMPARISONS OF THE FEEDING ECOLOGIES OF THE AFRICAN APES

In a table titled "synoptic comparison of anthropoid apes," Yerkes and Yerkes (1929, p. 556) listed the naturalistic diets of chimpanzees (and the Asian apes) as "vegetarian" and that of gorillas as "herbivorous or frugivorous." Schaller (1965b, p. 477) compiled a summary table in which he listed the diet of the chimpanzee as "primarily frugivorous, also herbivorous, also eats insects, antelopes, monkeys" and that of the gorilla as "primarily herbivorous, also frugivorous; feeding on animal matter not reliably reported." He premised these generalizations on contemporary field data collected by Goodall and the Reynolds on chimpanzees and by himself on gorillas. Because gorillas inhabiting rain forests in the Zaire basin and the Kayonza Forest were primarily herbivorous and the Budongo chimpanzees were chiefly frugivorous, Schaller (1965b, p. 481) inferred that when the 2 species inhabited the same forests there probably would be little competition between them for food.

Modern field studies support the following generalizations about the feeding habits of the African apes: Chimpanzees are heavily frugivorous, with a special predilection for figs in regions where they are available. Chimpanzees are much more euryphagous than gorillas are. In addition to fruits, eggs, honey, and a wide variety of arthropods and vertebrates, chimpanzees eat many leafy, pithy and fibrous foods, though not to the extent that gorillas do. Chimpanzees can subsist chiefly on quantities of rather hard seeds and desiccated fruits over notable periods of time.

Gorillas are predominantly herbivorous, though they too have a taste for fruit and will forage them if they are accessible. Some gorillas actively prey on arthropods in addition to ingesting them inadvertently with plant foods. But they have not been observed to kill and eat vertebrates or avian eggs. Chimpanzees and gorillas eat soil and more rarely their feces. Drinking behavior is relatively infrequent. It is assumed that chimpanzees obtain most of their fluids from juicy fruits and other foods and that the hydric needs of gorillas are met mostly by ingesting herbage.

Too little is known about the nutritional needs and digestive physiology of the African apes to permit firm inferences about the extent to which specific kinds of foods, let alone individual plant and animal species, determine the ranging patterns and behavior of chimpanzees and gorillas in different regions of Africa. Further, exact data on annual and longer term ranges of African pongid groups are rare in the literature, especially for chimpanzees. When thorough studies were conducted, generally ranges were found to be smaller than the estimates of earlier observers. For instance, Goodall (1965) estimated that adult Gombe chimpanzees (without dependent infants) had home ranges between 67 and 78 km². But Wrangham (1977) discovered that adult male Gombe chimpanzees had ranges of only 12.5 km² and that anestrous females had even smaller ranges. Wrangham's figures are remarkably close to the average range (11.9 km²) for the two intensively studied groups of Mahale chimpanzees (Nishida, 1979a). Similarly, whereas Schaller (1963) estimated that the Kabara mountain gorillas had ranges of 26-39 km², the much more intensive studies of Fossey and Harcourt (1977) revealed that the home ranges of Visoke mountain gorillas averaged only about 6.2 km². This is quite close to the average range (6.3 km²) for 6 groups of western gorillas that Jones and Sabater Pí (1971) studied in the Mt. Alen and Abuminzok-Aninzok regions of Equatorial Guinea.

Close, carefully controlled comparisons of the feeding ecologies of sympatric chimpanzees and gorillas have not been conducted. However, the study of Jones and Sabater Pí (1971) provided some suggestive information about potential niche differentiation of the 2 species in Equatorial Guinea. Jones and Sabater Pí (1971, p. 82) rarely detected chimpanzees and gorillas near one another. Twice, noisy chimpanzees were apparently ignored by gorillas. They found that during the dry season, both species increased their forays on human plantations. But during the wet season, gorillas spent even more time in relatively open areas while the chimpanzees were observed chiefly in the upper strata of the forests. This may be attributed to the location of their preferred foods (Jones and Sabater Pí, 1971, p. 38). Differences in their daily activity patterns probably also kept the 2 species separated. Although they all left their nests shortly after dawn, the chimpanzees usually moved to another area to feed while the gorillas foraged at the sleeping site. The chimpanzees tended to feed more intensively than gorillas did in the afternoon. Whereas the gorillas fed and moved about in their sleeping sites, the chimpanzees settled down rather promptly. Both species were more active on wet days than on dry days (Jones and Sabater Pí, 1971, p. 56).

The Belinga area and Lope Reserve (Tutin et al , 1984) of Gabon appear to be excellent locations for future studies on the feeding ecology of sympatric chimpanzees and gorillas. Preliminary data indicate that there is considerable overlap in the diets of the 2 species, especially in the consumption of fruit. Tutin and Fernandez (1985) found that although stems, pith and leaves were more often eaten by the Belinga gorillas, 60% of their food species were utilized also by sympatric chimpanzees. Seventy percent (n = 35) of fruit species that are eaten by western gorillas are also eaten by western chimpanzees. Further, the Belinga gorillas apparently ate 15 species that were not evidenced in the feces or leavings or by direct observations of the chimpanzees (Tutin and Fernandez, 1985).

In the Lope Reserve, 60% of the foods that were eaten by gorillas were also eaten by sympatric chimpanzees. Lope gorillas ate fruit regularly as evidenced by its presence in 98% of fecal samples. Lope chimpanzees also ate fruit (100% of stools) and consumed leaves, stems and pith in lesser amounts than gorillas did. Tutin et al. (1984) concluded that the dietary overlap is sufficient to indicate potential ecological competition between the 2 species.

Gross comparisons between the diets of gorillas, common chimpanzees and bonobos are likely to be confounded by the notable interpopulational variability in the diets of each species. For instance, of the 286 potential food types that are common to the Mahale region and Gombe National Park, the chimpanzees are known to share only 104 (36.4%) of them in their diets (Nishida et al., 1983).

MEAT-EATING, HUNTING, FOOD-SHARING AND CANNIBALISM

Contemporary field studies have provided a sizeable corpus of data pertaining to the Yerkeses' question about the naturalistic carnivorousness of apes. It is now securely established that the hylobatid apes and orangutans ingest notable amounts of animal protein, mostly in the form of arthropods. Gorillas appear to ingest much less arthropodan protein, especially in proportion to their great size. Among the apes, chimpanzees, including bonobos, are the most dramatic meat-eaters (Butynski, 1982). They not only consume invertebrates regularly, but also prey upon a remarkable variety of small and young, medium-sized mammals and fledgling birds. Like insects, these warm-blooded prey are sometimes actively sought, as well as being collected during chance encounters.

The Gombe chimpanzees are the most thoroughly studied pongid population regarding all aspects of vertebrate capture and consumption. Goodall (1963) was the first field behavioralist to document that chimpanzees eat freshly killed mammals. During the initial 15 months of her study, she saw arboreal chimpanzees eating a decapitated and disembowelled infant bushpig, a bushbuck and an unidentified small mammal. She also found part of a monkey's foot in feces from a chimpanzee. However, she did not observe them capturing and killing the 4 prey.

Later, Lawick-Goodall (1965, p. 823) reported that Gombe chimpanzees had unsuccessfully stalked a young baboon that had strayed from its group and had eaten flesh from the skull of an adolescent baboon. She first observed the arboreal hunting of an adult red colobus monkey by 2 adolescent male chimpanzees. One of them climbed into a neighboring tree and sat still. While the quarry watched him, his companion climbed into its tree, ran along the roost, and pounced on the victim. It apparently died quickly during the manual capture. Six members of the chimpanzee group, including the first adolescent, entered the death tree and most of them obtained bits of the carcass (Goodall, 1965, p. 445).

After the banana provisioning area was set up, Lawick-Goodall (1968) and her associates obtained more detailed observations on how chimpanzees capture and consume young baboons. Teleki (1973a) summarized the first 10 years of data on the predatory behavior of Gombe chimpanzees, with special emphasis on 30 predatory episodes that occurred between March 1968, and March, 1969. Twelve of the 30 attempts were successful. They confirmed that, like certain carnivoran predators, chimpanzees acquire vertebrate prey by a variety of methods, ranging from sudden snatches of individuals that happen to be nearby to careful and sometimes circuitous stalks of more distant quarry. The hunters may act singly or in company with other chimpanzees (Lawick-Goodall, 1967, 1968, 1971, 1973). For instance, during a common display, an adult male suddenly grabbed an infant baboon from its mother's lap and flailed it about, like a branch. The victim was carried off by another male chimpanzee after the captor had dropped it while beset by adult baboons. On another occasion, 4 male chimpanzees rose abruptly and moved quietly toward a small group of baboons. When Lawick-Goodall (1967, pp. 64-66; 1968, p. 190; 1971, pp. 199-201) saw them next, one had caught a juvenile baboon which he held by a hind limb and swung overhead then downward so that its head bashed the ground.

Field records between 1960 and 1970 on a population of approximately 50 chimpanzees around the Kakombe field station revealed that they had killed and eaten at least 95 mammals and that 37 others had escaped during their predatory attempts. Gombe researchers observed 46 kills; 38 others were documented by fecal analysis; and chimpanzees carried 11 dead mammals into the banana feeding area (Teleki, 1973a, p. 53; 1973b). Fifty-six specimens of mammalian prey were identified. Sixty-five percent of them were primates (21 *Papio doguera*, 14 *Colobus badius*, 1 *Cercopithecus mitis* and 1 *Cercopithecus ascanius*) and 35% were ungulates (10 *Potamochoerus porcus* and 9 *Tragelaphus scriptus*). All prey weighed less than 20 lb (9 kg). Teleki (1973b) predominantly witnessed predation on baboons, which constituted 10 of 12 kills by the Kakombe chimpanzees and all 18 escapees during his one-year field study.

Teleki (1973a, p. 56; 1973b) observed that Kakombe hunts were performed primarily by 13 adult males, though adult females and adolescents had also been seen hunting and capturing mammals when no adults males were nearby.

Kakombe chimpanzee hunts took the form of (1) opportunistic lunges and grabs for unwary or poorly protected victims, (2) chases, and (3) stalks. The first method struck observers as being explosive and instantaneous. Chases are usually of short duration. But the hunters sometimes ran tenaciously over several hundred yards and attained speeds near 30 miles per hour (48 km/hr) (Teleki, 1973a, p. 130). Stalks were the most time-consuming mode of mammalian predation ($\bar{x}$ duration = 28 min). Seizures and short chases were generally performed by individuals. Longer chases and stalks were commonly joined by other chimpanzees (Teleki, 1973a).

Whether this qualifies as cooperative hunting is moot. Busse (1978) argued persuasively that Gombe chimpanzees do not hunt cooperatively. Takahashi et al. (1984) noted that this is probably also the case at Mahale.

During a 30-month span (March, 1968 - August, 1970) the Kakombe chimpanzees made 21 kills during 44 attempts (Teleki, 1973a, p. 53), a success rate of 48 percent. Teleki classified 22 pursuits as seizures, chases or stalks. Three of 7 (43%) seizures were successful and 6 of 11 (55%) chases culminated in captures. The 4 stalks fizzled (Teleki, 1973b).

In addition to battering captives against the ground and standing objects, single captors commonly attempt to dispatch them with bites on the nape of the neck or spine or by wringing their necks. Sometimes the prey remains alive during part of its consumption. If several individuals converge on the quarry, it expires more quickly from fragmentation. Unlike certain large felid carnivores and human murderers, chimpanzees have not been seen to throttle prey orally or manually.

Pandemonium ensues immediately after quarry is caught in the company of other chimpanzees. This contrasts sharply with their stony silence during the pursuit (Teleki, 1973a,b). Unless the captor is inaccessible to other chimpanzees, they may boldly snatch away portions of the prey during the first moments after its capture. Thereafter, they generally calm down and resort to beggary, taking pieces from the possessor's hand or mouth, and grubbing for morsels that have been dropped or abandoned.

Beggars sit very close to the consumer and stare at its face or the coveted prize (Figure 11). The salivating suppliant may reach forward with a supinated hand and sometimes gently touch the diner's chin, mouth, or meat. These gestures may be rewarded with a juicy morsel, a token of skin or bone, or a much masticated wadge of leaves. Alternatively, the proprietor may turn its back, move away, or evince irritation by vocalizing, pushing or striking at the nuisance (Teleki, 1973a,b; Nissen and Crawford, 1936; Norikoshi, 1982; Kawanaka, 1982a; Takahata et al., 1984). Teleki (1973a,b) observed that 29% (114/395) of solicitations were rewarded. He never witnessed a melee over meat. Consumption of a carcass takes between 1.5 and 9 hr ($\bar{x}$ = 3.5 hr). Unlike canids, chimpanzees do not bolt chunks of flesh. Instead they ingest small pieces with leaves and spend respectable spans sucking and chewing on the wadges.

Surprisingly large numbers of individuals ($\bar{x}$ = 8; r = 4-15) acquire bits of the carcass, even though it is not large (Teleki, 1973a). Only the brain is eaten selfishly by the head keeper. The extensive flesh-sharing of chimpanzees has fueled suggestions that their hunting is motivated by undefined social factors, as well as being of some unmeasured nutritional importance (Teleki, 1973a,b; Kortlandt, Beck and Vogel in Tuttle, 1975, pp. 301-304). The Kakombe chimpanzees often took a cercopithecoid savory after they had pigged-out with bananas (Teleki, 1973a, p. 109). In 1968, 8 of 9 infant baboon prey were killed after the chimpanzees had fed heavily (Wrangham,

1974a). Thus, general hunger pangs probably should not be advocated as the primary motivation for all chimpanzee predatory incidents. Pelham and Burton (1976, 1977) suggested that high intakes of bananas may have caused chemical imbalances which the chimpanzees could correct by ingesting animal protein. This idea was roundly rejected by Teleki (1977). It remains untested experimentally.

Teleki (1973a; 1981) elected to minimize distortions of chimpanzee predatory behavior that could have been induced by the banana bonanza at Kakombe. Artificial feeding was reduced during Teleki's study. His account differs from that of Lawick-Goodall (1968, p. 166), who stated that from mid-1967 they fed the chimpanzees at irregular intervals, only once or twice per week. According to Teleki (1973a, p. 37), artificial feeding occurred almost daily during 1967, was reduced to a random 7 days per fortnight early in 1968, and was further curtailed to once or twice weekly (and sometimes less) beginning in June, 1968. Littery banana peels, which acted as baboon bait, were collected promptly by camp personnel. By early 1969, singletons and members of small aggregations of chimpanzees were fed a few bananas each, once every 2 weeks via a system of mechanized semisubterranean bunkers.

Baboons constituted 83% of prey (12 kills) between March, 1968, and March, 1969, and only 22% of prey (9 kills) during the succeeding 18 months of 1969 and 1970. Teleki (1973a, pp. 54-55 and 108) concluded that the banana factor probably affected the type of prey captured more than the yearly rate of predation.

From data collected during 10,000 hours of observation on Gombe chimpanzees, Wrangham (1974b) concluded that their predation did not vary seasonally in a regular cycle and that killing baboons was rare away from the banana bunkers. However, in the bush, chimpanzees commandeered the kills of baboons (Wrangham, 1974b; Morris and Goodall, 1977).

Wrangham (1974b) observed that red colobus monkeys of any age or sex, but especially lone adults, were the most common mammalian prey of chimpanzees away from camp. They were likely to become victims when the chimpanzees caught them in broken canopy. Busse (1977) estimated that chimpanzees killed between 8 and 13% of the Gombe red colobus population during 1973 and 1974. Nineteen of 29 identified colobine quarry were subadults. Wrangham straightforwardly concluded that chimpanzee hunting is a foraging activity instead of a symbolic demonstration of the chimpanzees' power to kill.

The banana factor probably also affected the intensity of flesh-sharing by camp chimpanzees. Wrangham (1974a, p. 87) found that on banana days, the sizes of aggregations declined as the probability that fruit was still available decreased. Further, camp aggregations were much smaller or nil on banana-free days. The chimpanzees generally came to camp between 0630 h and 0930 h and left if it was fruitless.

The factors that determine which chimanzees will receive complimentary morsels from a possessor are poorly understood. Teleki (1973a, p. 146) noted no vocal or gestural signals that are specific to flesh-eating. Access to meat does not conform neatly to a predictable social pattern and it is not determined by overt competition. For instance, possessors sometimes reward subadult and adult female beggars with bits of flesh and refuse higher ranking males without incurring recognizable penalties.

Teleki (1973a, pp. 155-157) stated that bananas were more selfishly guarded than meat was and competition for them depended more upon brawn and social status. However, McGrew (1975) reported that 27% (124/457) of bouts with banana beggary were successful. This is very close to the frequency for flesh. Nissen and Crawford (1936) found that in 56% (149/266) of instances captive cadgers obtained food or tokens from possessors, which were more reluctant to give up food than tokens.

Teleki (1973a, pp. 158-160) observed that bits of flesh are rarely passed to tertiary individuals except when mothers share them with their dependent youngsters. The extent to which one prolific female, Flo, shared meat with her 5 progeny decreased from the youngest to the oldest. Plant foods are also most commonly shared between mothers and their subadult offspring. During a 21-month period in 1972 and 1973, McGrew (1975) and other Gombe researchers recorded 457 transfers of bananas from capitalists to cadgers. Approximately half the 124 incidents of banana sharing were between mother-offspring pairs. Bananas passed from mothers to offspring in 92% of these episodes and from youngsters to their mothers in the remaining 8% of cases. The greatest number of non-familial sharings (73%) consisted of adult males giving fruit to "unrelated" adult females. Adult males rarely begged for bananas and some of their supplications were fruitless.

McGrew (1975) maintained that observations in the bush confirm the conclusion that most food sharing occurs between mother chimpanzees and their dependent youngsters. When the former open tough fruit (*Strychnos*) and sticky pods (*Diplorhyncus*), they often give some to their infants. Silk (1978) confirmed that the natural foods most frequently and successfully solicited by Gombe infants were generally difficult for them to procure or process.

In the Kibale Forest, Uganda, Ghiglieri (1985a, p. 75) saw an infant chimpanzee pull the fig laden hand of its mother to mouth and ingest part of the fruit. Another Kibale chimpanzee mother picked a fig and placed it in her infant's mouth.

In the Mahale K-group, during 32 begging bouts, half the cadgers were handed vegetal food by possessors. Most (87.5%) of the interactions were between adult or adolescent female cadgers and adult or adolescent male capitalists (Nishida, 1970).

In the Tai forest, Ivory Coast, mother chimpanzees did not refuse to share kernels, which they had extracted from nuts, with their begging infants and juveniles (n = 42 observations) (Boesch and Boesch, 1984b). Youngsters cannot open the nuts (Chapter 5). One adult female cracker allowed a male to take panda kernels that she had extracted; presumed siblings shared nut meats on 6 occasions (Boesch and Boesch, 1984b).

Teleki (1973a, pp. 162-163) also believed that estrus affects the likelihood that females will acquire meat when they cadge from adult males. Although he described no instance wherein a florid female obtrusively put out for protein, his figures are suggestive. The success rate of the estrous females was 69% (91/132 instances) while that of anestrous females was 44% (42/104 instances). Teleki commented that estrous females remain closer to the predatory adult males and are more persistent beggars than anestrous females are.

The modes of acquisition, killing, division, and consumption of mammalian prey by chimpanzees outside Gombe National Park are basically similar to those reported by Gombe researchers. However, there are a few differences, the significance of which cannot be weighed now because observations away from provisioning areas are fragmentary, few, and episodic.

Observations by Boesch (1978), Sugiyama (1981), and Anderson et al. (1983) in West Africa and Suzuki (1971, 1975) and Ghiglieri (1984a) in Uganda clearly established that predation on vertebrates and meat-eating are widespread practices among chimpanzees, and that, as at Gombe and other localities in western Tanzania (Kawabe, 1966; Nishida et al., 1979; Kawanaka, 1981, 1982a; Norikoshi, 1982, 1983; Itani, 1982; McGrew, 1983; Takahata, 1984), they occur readily away from provisioned areas.

Because baboons have not been reported as quarry outside Gombe (McGrew, 1983) we might attribute their initial victimization to the banana bonanza. There are no reports that female chimpanzees have secured baboon bonbons. Physically they should be able to do so. Perhaps they are intimidated by adult male baboons as well as by mature male chimpanzees.

Kawabe's (1966) account of 6 Kasakati chimpanzees preying upon redtails is outstanding for the full participation of 5 adult and subadult females in the chase and in the vigorous 10-minute struggle over a carcass. Suzuki (1975, p. 262) also noted several furious struggles among 4 adult Budongo males over parts of a blue duiker that they had killed. His observations spanned 2.5 hr. Like Kawabe, Nishida et al. (1979) and Takahata et al. (1984) stressed that females play a substantial role in the predation of Mahale chimpanzees on mammals and birds. They tend to grab young ungulates while the males are more disposed to chase monkeys (Takahata et al., 1984).

During meat consumption, the Mahale chimpanzees seem to interact more aggressively over carcasses than Gombe chimpanzees do (Kawanaka, 1982a; Norikoshi, 1982; Takahata et al., 1984). On average, 5.1 ± 3.4 chimpanzees (r = 1-15 chimpanzees, n = 36 bouts) shared the victims of predatory episodes at Mahale. Dominant males often got the lion's share regardless whether they were the original captors (Takahata et al., 1984). Thirty-one percent (26/83) of manually gesticulating Mahale beggars were rewarded with a bit of prey. This is quite close to the Gombe success rate (29%). Further, some cadgers were allowed to eat from carcasses while they were being held, sometimes pedally, by the possessors (Norikoshi, 1982; Kawanaka, 1981; Takahata et al., 1984). This was especially common in adult male dyads. Kawanaka (1982a) and Takahata et al. (1984) noted that the distribution of meat among males reflected their special social relationships (Chapter 8). Apparently chimpanzees that prey together stay together.

Grooming the possessor did not particularly increase one's chance for being handed meat. But it got the groomer closer to littery tidbits. Per contra, copulation did seem to do the trick. Five of 7 females which copulated with possessors received meat between 4 and 20 min after their services were rendered (Takahata et al., 1984). Kawanaka (1982a) commented that sometimes whether a female receives meat or not may be more dependent upon her relationship with the male possessor than upon her estrous condition.

At Bossou, Guinea, Sugiyama (1981) saw an arboreal adult female or juvenile male capture a tree pangolin (*Manis tricuspis*) by the tail. The captor carried it out of view and was followed by the rest of the chimpanzee group. When Sugiyama caught up with them he witnessed a Neronic scene. An arboreal adult female was eating leafy garnishes and flesh from the tailless caudal half of the carcass while others rested, wrestled, or engaged in a ménage à trois nearby. Later Sugiyama found the pangolin's tail but no other parts of it. No pangolin pieces were found in feces from 5 chimpanzees during the next 2 days and no mammalian bits were in > 200 stools examined during Sugiyama's studies.

Chimpanzee cannibalism was first observed by Suzuki (1971) in the Budongo Forest. He happened upon a familiar large arboreal male which held a live, partly eaten, newborn baby. Other males sat close to the possessor, groomed him, and reached for the victim. It was passed back and forth between the initial possessor and his groomer. The possessor ate from its lower body with garnishes of leaves, allowed a few cadging males to touch the prey, and, after two hours, descended to the ground with the baby in his mouth, and disappeared into the bush, followed by the entourage.

Bygott (1972) observed infanticidal cannibalism by 5 adult males of a temporary aggregation, which he had seen feeding and resting during a 3-hour span before they suddenly encountered 2 adult females. They vigorously attacked the eldest, which Bygott had not seen in the area before. After a 2-minute tumultuary hiatus in Bygott's observations, the female had disappeared, the males were clustered in a tree, and one of them held a live 1.5 year old infant, with which Bygott was unfamiliar. The possessor held it by the hind limbs, battered its head against a branch, and bit flesh from its thighs. A second male rent away a foot and ate it. During 1.5 hours, the possessor intermittently bit, sniffed, poked, groomed, punched, flailed, and otherwise toyed with and abused the carcass. After he had abandoned it, the foot-eater dined on the body for 1.5 hours. The first possessor returned and flailed it. Then more males snacked on it. However, they mainly ingested other foods while holding the body. When they abandoned it 6 hours after the capture, one hand, the hind limbs and the perineum had been consumed. Bygott (1972) remarked that the infant was less appetizing and more of a curiosity than the other mammalian prey of Gombe chimpanzees.

While there is a reasonable chance that Suzuki, Bygott, and other Gombe researchers observed exocannibalism by male chimpanzees (Goodall, 1977), the next chronicle is clearly about endocannibalism.

In 1975, Gombe researchers saw an adult female, Passion, attack a female associate, kill her 3-week old infant, and methodically eat it in company with her own adolescent daughter, Pom, and infant son. In 1976, Pom grabbed and dispatched the recently born infant of the same female and picnicked on it with her mother and brother. One month later, Pom struck again, killing the new baby of another mother in the group (Goodall, 1977). These events prompted Goodall (1979) to examine the demographic records for the Gombe study population during a 3-year span between 1974 and 1976. To her horror, she found that in the shadow of Moloch only one infant had survived longer than a month.

No further cannibalistic killings were reported, though Goodall (1979, p. 619) physically intervened during a subsequent infanticidal effort by Pom. The presence of communal adult males appeared to be even more effective than Goodall's cudgel in curbing the infanticidal urges of Pom and Passion.

Data are so insufficient that attempts to explain these events, either nutritionally or socially, would be premature (Polis et al., 1984). Why Gombe female chimpanzees might be more inclined toward endocannibalism while the males are disposed to exocannibalism is now as indeterminate as it is intriguing. The females devour their quarry (n = 3) straightforwardly as they would eat cercopithecoid and ungulate prey (Goodall, 1979). None of the episodes of cannibalism by males (n = 4) was seen to end with consummate ingestion of the carcass and they tended to brutalize it (Goodall, 1977).

The only reported case of cannibalism in gorillas involved a mother and her young adult daughter, which shared the firstling of a female in their group. The male infant was probably the product of father/daughter incest. The presumed sire was the silverback of Visoke group 5 and the elder cannibal was the dominant female therein (Fossey, 1981, 1983, 1984).

Observations on infanticide (n = 4) in the Mahale Mountains have muddied the bloody issues of infanticide and cannibalism in African apes

(Kawanaka, 1981; Norikoshi, 1982; Nishida et al., 1979; Nishida and Kawanaka, 1985). Kawanaka (1981) concluded that some episodes cannot be categorized neatly as exocannibalism versus endocannibalism. However, all 4 cases of infanticide at Mahale occurred in or near the area of overlap between the ranges of M-group and K-group and the latter was being absorbed by the former at the time (Chapter 8).

One 1.5 year old victim of male predators at Mahale was entirely consumed over a 3-hr span, though an adult female kept the carcass awhile, and the males did not manhandle it anymore than was necessary to consume it (Kawanaka, 1981). Likewise, a 2.5 month old victim of endocannibalism was systematically consumed without tenderization by M-group males (Norikoshi, 1982).

Despite the fact that adult chimpanzees are sometimes brutally killed during intergroup altercations (Chapter 8), there are no reports of pongid cannibalism wherein a victim is older than 3 years (Kawanaka, 1981). The mother from K-group that had lost her 1.5 year old son to M-group cannibals was viciously attacked with her second son when he was 40 months old and 49 months old. The muggers concentrated mostly on her but the youngster was also bloodied. Nishida and Hiraiwa-Hasegawa (1985) noted behaviors, like grooming, embracing and kissing the targets, which suggested ambivalence on the parts of the aggressors. If cannibalism was the prime motive for their actions, perhaps the age of the intended victim caused them some confusion. We will never know because the drama was terminated both times by human intervention (Nishida and Hiraiwa-Hasegawa, 1985).

Insofar as their habits are known, bonobos appear to be relatively genteel in comparison with their cannibalistic, infanticidal and selfish congeners. They rarely eat vertebrates and are not reported to kill and eat other bonobos. They seem to avoid sharing meat (Badrian and Malenky, 1984). But their spookiness has prevented sustained observations on them following a vertebrate kill.

Bonobos are the apes most likely to share vegetal food across a wide range of individuals in addition to mother/infant dyads (Chapter 8). Kano (1980) recorded 261 bouts of food sharing in the provisioned Wamba bonobos. Items passed both ways between members of all age-sex classes except from infants to other group members. Among adults (n = 89 bouts), most (61%) sharing occurred between females and much less (9%) between males. A fourth of the adult sharing consisted of males giving food to females; the reverse was rare (6% of bouts). Females commonly begged for food while males cadged relatively rarely (Kano, 1980).

Cadgers used a variety of gestures to signal their desires to the generally unenthusiastic possessors, including cautiously, but straightforwardly, reaching for an item while grinning; more or less patiently waiting for a bit to be proffered; mouthing the food in a possessor's hand or its mouth;

touching the mouth of a possessor immediately after it had ingested a morsel; and probing the muncher's oral cavity and perhaps extracting a tidbit. Rejection was usually signalled by the possessor turning its back on the beggar or moving away from the nuisance. However, the outstretched hands of two suppliants were bitten (Kano, 1980; Kuroda, 1980, 1984).

Large, rare fruits of *Anonidium mannii* and *Treculia africana* were shared mostly among adult bonobos, sometimes by being passed around the begging cluster (Kano, 1980; Kuroda, 1980). Young cadgers received sugar cane chiefly from adults (typically their mothers), which were better able to peel off the tough epidermis. Eight species of small and medium-sized fruits, which were easy to collect and to process, were shared almost exclusively by adults with youngsters (Kano, 1980).

Kuroda (1984) quantified and augmented descriptions on the basic interactions of Wamba bonobos over sugar cane and pineapples that humans supplied at a provisioning area. The subjects shared food in 53.7% of 947 transfers. Agonism, in the form of threats or struggles over food, was infrequent (10.9% of interactions), mostly occurred between males (especially over pineapples), and accounted for only 5.2% (28/537) of food transfers. Physical attacks were mild and rare (n = 15). Only 2 of them, both between females, involved biting. Sometimes (n = 11) subordinates secured food from threatening dominant ones and at times the latter failed in attempts to grab food from their subordinates (n = 14).

Youngsters and older offspring most often begged for sugar cane from their mothers. The former were rewarded with morsels more often than the latter were. Females seemed to discourage beggars by giving them well chewed wadges and scraps. Young females commonly begged for sugar cane from dominant males even when it was readily available to them directly on the provisioning ground (Kuroda, 1984).

Subordinate possessors avoided nuisances by carrying their treasures away from the provisioning arena. But high-ranking males stayed and ultimately shared prizes like pineapples with others. Estrous (n = 24) and anestrous (n = 1) females copulated with males before taking some of their sweets. One young female placed a piece of pineapple in the palm of a male that had just had his way with her (Kuroda, 1984). Is it possible that in pongids, as in people, "prostitution" can work both ways?

4

Lodge Sites and Nesting

"One wonders whether the art of nest building is acquired by the individual through imitation of its elders or is instead a natural ability which appears at an appropriate stage of growth and development; whether it is carried from generation to generation by racial tradition, by heredity, or both. The question has not yet been satisfactorily answered." (Yerkes, 1943, p. 16).

If they could enunciate English, parental apes would directly inform their progeny that "It's a jungle out there." Secure sleeping sites are a premier ecological need of primates that are potential victims of nocturnal predators, parasites, pests, and inclement weather. Even the most terrestrial species seek shelter in trees or on rocky ledges that can bear their weight and numbers. Like cercopithecoid monkeys, the hylobatid apes roost arboreally with the aid of ischial callosities ("sitting pads") whereas the pongid apes support their greater bulk by building nests in which to recline (Anderson, 1984a).

The following survey indicates that apes spend a lot of time at their lodge sites, presumably sleeping. We really do not know why, though it must conserve energy. There is much room for further research which would employ sophisticated physiological apparatus (Campbell and Tobler, 1984).

HYLOBATID APES

Carpenter (1940) did not specify the stratum of the northwestern Thai forests in which his subjects (*Hylobates lar carpenteri*) slept. He reported that they selected trees with dense tops toward the centers of their ranges

126

and that they rarely went to the very tops of trees. He suggested that protection from strong winds was a prime characteristic of their lodge trees. He speculated that they changed sleeping localities because of "increasing odor" (presumably their own), shifting winds, disturbances by humans and perhaps predators, pressure from neighboring groups and fluctuations in food availability.)

Ellefson (1974) observed that the gibbons (*Hylobates lar lar*) at Tanjong Triang, Peninsular Malaysia, slept high (above 90 ft) and (except for infants) in separate trees that were centrally located in their territories. They did not reuse sleeping trees on consecutive nights and apparently did not sleep in fruiting trees. He did not discuss the factors that determined their selection of sleeping sites.

In the Krau Game Reserve, Peninsular Malaysia, *Hylobates lar lar* slept in the central areas of their ranges (MacKinnon and MacKinnon, 1980). According to Gittins and Raemaekers (1980), they roosted between 20 m and 40 m (66-132 ft) approximately 75% of the time and above 40 m approximately 25% of the time. MacKinnon (1977, p. 764) reported that their night sleeping was "confined almost wholly to the upper storey of emergent crowns.")

Agile gibbons (*Hylobates agilis agilis*) at Sungai Dal in the Gunong Bubu Forest Reserve, Peninsular Malaysia, slept high in the canopy [55% of the time between 25 and 35 m (82.5-111.5 ft); 40% of the time above 35 m; and 5% of the time between 15 m (49.5 ft) and 25 m (Gittins and Raemaekers, 1980)]. Group members often roosted "tens of metres" apart (Gittins, 1982). But the group concentrated on sleeping trees in two areas (89% of nights) which were on high ground and often some distance from their food trees. Gittins (1982) suggested that morning calls from lodge trees on ridges could be heard over greater distances than calls from the valley could and that the gibbons also avoided nocturnal inversions of temperature that cool the valleys more than the ridges.

Rijksen (1978) reported that in the Ketambe area of Sumatra *Hylobates lar vestitus* slept on consecutive nights in different trees that were close to the centers of their territories and near a fruit tree.

Thai pileated gibbons (*Hylobates pileatus*) usually slept in emergents (> 25 m [82.5 ft]) or trees of the upper canopy. They preferred dipterocarps. The lodge trees are widespread in their range. On average, one group of 6 individuals spent 15.8 hours/day in lodge trees. They spread out to sleep in different trees 10-15 m (33-49.5 ft) apart. The infant slept with its mother in the center of the group. A young juvenile female slept near the mother/infant pair (Srikosamatara, 1984).

Initially, Tenaza (1975) reported that, unlike the gibbons noted hereabove, Kloss gibbons (*Hylobates klossii*) in the Saibi Ulu region of eastern Siberut island, slept near the perimeters of their territories because the lianaless emergent trees, which protect them from nocturnal predation (by unnamed agents), only grow on narrow hill crests. Subsequently, Tilson and Tenaza (1982) stated that lodge trees are centrally located in Kloss gibbon territories. They were 35-55 m (115.5-181.5 ft) tall and projected 5-25 m (16.5-82.5 ft) above the middle canopy. Each family used between 2 and 5 trees ($\bar{x} = 3$), one of which was preferred over the others, on average being occupied 84% of nights. All members of a family, except fully grown subadults, lodged together in one tree per night. Tenaza and Tilson (1985) concluded that in central Siberut, Kloss gibbons prefer lianaless emergents (91% of lodge trees) because local hunters climb vines in order to shoot primates with arrows (Figure 12). Pythons might also get to them via lianas.

Whitten (1982b) found that nearly all lodge trees of Kloss gibbons in the Paitan area of central Siberut also were centrally located in their home ranges. His subjects covered a high proportion of their day range during the last 2 hours of the activity period, probably in an effort to locate an appropriate lodge tree. An adult male that Whitten (1982b) studied intensively slept in emergent trees of the upper canopy more than two-thirds of the time and in the middle and lower canopies on the remaining nights. The Kloss gibbons of Paitan selected lodge trees that bore few lianas and no epiphytes of *Myrmecodia tuberosa*, which shelter pestiferous biting ants (Whitten, 1982c,d).

Peninsular Malaysian siamang (*Hylobates syndactylus continentis*) spend the night in towering emergent trees (Chivers, 1973; Chivers and Raemaekers, 1980; MacKinnon, 1977). Indeed siamang roost in trees that are taller than 40 m (132 ft) high more often than the other 2 species of Malayan lesser apes do. Gittins and Raemaekers (1980) found that they lodged above 40 m about 85% of the time and between 20 m and 40 m about 15% of the time.

Like Kloss gibbons, siamang group members usually sleep in the terminal branches of one tree or closely neighboring trees (Chivers, 1972, 1974). Only 5 trees (of 2 species) among the 34 sleeping sites used by the main study group at Kuala Lompat provided them food sometime during the study (Chivers, 1974). They spent 74% of nights in only 44% (15/34) of the sleeping sites. At Ulu Sempam, the siamang tended to sleep away from the areas of daily activity, whereas they more commonly slept centrally in their territories at Kuala Lompat. They used the same trees consecutively on less than 6% of nights (Chivers, 1974).

At Ketambe, each group of Sumatran siamang (*Hylobates syndactylus syndactylus*) had between 2 and 5 regular sleeping sites. They consisted of emergents or trees on vantage points and often were located near the periphery of the territory (Rijksen, 1978).

ORANGUTANS

Sumatran orangutans (*Pongo pygmaeus abelii*) at Ketambe (Figure 13a) generally built night nests 13-15 m (43-49.5 ft) above the ground though the

Figure 12: Lodge trees of Kloss gibbons (B) and Mentawi langurs (A). In (B) the arrow points to a liana-free tree 54 m tall. In (A) the arrow points to one of several woody vines that are sturdy enough for human hunters to climb. The tree in (A) is 45 m tall. (© Richard Tenaza 1984)

Figure 13: (a) A Sumatran orangutan night nest with rain cover. The adult male in the nest is holding the cover open in order to look at the photographer. (Photo courtesy of H.D. Rijksen, 1978.) (b) Terrestrial nest of a mountain gorilla that is soiled with feces. (Courtesy of David P. Watts.)

forest structure at some sites caused them to choose higher and lower supports (Rijksen, 1978). They preferred to nest at vantage points, like slopes on the edges of gaps in the foliage, from which they could see a large section of the forest. They usually changed locales on consecutive nights though individuals occasionally reused nests near attractive food sources after refreshing the linings (Rijksen, 1978). The Ketambe orangutans usually reoccupied nesting sites and refurbished nests after spans of between 2 and 8 months.

Rijksen (1978) noted that they never slept at night in food trees though they sometimes built day nests in them. Indeed they rarely lodged at night near trees with ripe fruit. Carpenter (1932) saw nests in groves of durian trees at one locality in Aceh, Sumatra, but he did not say when they were used.

Sugardjito (1983) studied 172 lodge sites of 10 habituated Ketambe orangutans with particular attention to their heights above ground and locations vis-à-vis the last food trees of the day. He found that the subjects which were most vulnerable to nocturnal, arboreal predation (presumably by clouded leopards, *Neofelis nebulosa*) built their nests higher in trees and further from the last food tree of the day. Thus adolescents (n = 5) and females with infants (n = 4) usually selected lodge trees that were fruitless and more cryptic and inaccessible than an adult male and 2 solitary adult females[*] did. Indeed these bolder adults often nested in or close to their last feeding sites of the day (Sugardjito, 1983).

MacKinnon (1974) did not report differences in the nesting practices between Sumatran and Bornean (*Pongo pygmaeus pygmaeus*) orangutans. His account appears to be based chiefly on observations in the Ulu Segama Reserve, Sabah. He reported that they frequently nested in or near the last food tree and fed in it the following morning. They usually built a new nest each night but 3 adult males spent 2 consecutive nights in their nests after feeding late in the lodge trees. Nests were clumped on westward slopes, which presumably receive warmth from the afternoon sun and shelter from evening winds and provide an extensive view of the region. MacKinnon (1974) inferred that some orangutans travelled long distances to nest in particular lodge trees. In low flat tracts of the Segama forest they built nests in trees that were beside or overhung streams and in sheltered areas beneath hillsides (MacKinnon, 1971).

The Segama orangutans nested between 3 m (10 ft) and more than 27 m (90 ft) above the ground. The majority of nests were built between 12 m (40 ft) and 27 m. The mean height is 19.5 m (65 ft) (MacKinnon, 1971). The heights of nests seemed to be determined by branch structure. Large nests

---

[*]Both these females gave birth during Sugardjito's study. Hence the discrepancy hereabove between 10 actual subjects and a total of 12 if one adds n in the 4 age-sex classes.

were located at lower levels than small ones. MacKinnon (1974) concluded that some trees were not used for night lodgings because they were too tall (e.g. *Dipterocarpus* spp. and *Dryobanus* spp.) or lacked appropriate nesting materials.

Yoshiba (1964) located 490 orangutan nests during a 2-month reconnaissance in eastern Sabah. Most of them were in the middle story of the forest: 59.2% were 16-25 m (58-82.5 ft) above ground; 21% were at 6-15 m (20-49.5 ft); and 19.4% were from 26 m (86 ft) to more than 36 m (119 ft) above ground. Only 2 nests were at 5 m (16.5 ft) or lower. Yoshiba (1964) could not confirm local reports that old orangutans sleep on the ground. Some nests were overhung by branches while others were exposed to the sky. Most nests were isolated but some of them were paired or treble. Yoshiba (1964) could not discern whether they had been used contemporaneously.

Davenport (1967) observed 18 instances of orangutans settling for the night in Sabah. They all built new nests, often in trees in which they had been eating leaves. And occasionally they snacked on leaves before dropping off to sleep. Horr (1977) reported that near the Lokan River orangutans preferred belian trees (*Eusideroxylon* sp.) for lodging because their branches facilitate the construction of secure nests. They made nests each night (Horr, 1972).

Schaller (1961) observed 228 nests in mixed swamp, lowland and hill forests of Sarawak. But he did not indicate whether they were used as nocturnal or diurnal rests. Eighty-two percent of them were between 9.3 m and 21 m (31-70 ft) above ground, i.e. in the middle of the lower story of the forest. A few nests occurred in the tops of emergent trees (up to 130 ft) and in very low trees (11-20 ft).

Rodman (1979) reported that habituated orangutans in the Kutai Nature Reserve of East Kalimantan built nests on 75% of 105 evenings when he saw them tucked for the night. Two adult females built new nests on most nights but a juvenile female often used an old nest and an adult male slept nestless in the crotch of a large tree.

COMMON CHIMPANZEES

Nissen (1931) provided a thorough early account on the dry season nesting habits of Guinean chimpanzees (*Pan troglodytes verus*) even though his direct encounters with them were limited. Nests were located in dense valley forests and in trees of the more sparsely wooded hills and highlands. He detected no preference for tree types or vegetational zones. Nests were in leafy trees of many species, including fruiting ones, but palms were nestless. The chimpanzees did not nest in bushes or on the ground. Nissen (1931) concluded that the selection of lodge trees was largely random. He suggested that old nests were sometimes used after being relined with fresh leafy sprigs. As many as 10 chimpanzees built nests together in a single tree. However, on a particular night a group's nests were generally distributed over 2 or more trees that were no more than 60 m (200 ft) apart. The chimpanzees built nests between 4 m (13 ft) and 31.5 m (105 ft) above ground. The average height of 100 nests was 11.5 m (38.4 ft). Six nests were lower than 6 m (20 ft) and 7 were higher than 22.5 m (75 ft). Nests were higher in valley forests than in the highlands because appropriate lodge trees were taller in the former areas. Nissen (1931) emphasized predator avoidance as a cause of nesting by Guinean chimpanzees. He noted that they were most watchful at nightfall and that they removed branches immediately above their nests, perhaps to keep "snakes and panthers" from dropping in for supper. In highland trees, nests sometimes allowed their occupants a clear view over adjacent grassy areas.

Bournonville (1967) studied 184 chimpanzee nests during a dry season in Guinea. They were found in more than 21 species of trees, including a palm (*Elaeis guineensis*). More than two-thirds of them were in only 6 species of trees. They were generally away from food trees. In some areas, the chimpanzees appeared to prefer particular species as lodge trees. Some of them had lush foliage while others were rather thinly leafed. The latter occurred in areas where leafy trees were relatively rare.

A vast majority (94.5%) of nests were between 4 m (13 ft) and 16 m (53 ft) above ground. The lowest nests were at 2.4 and 2.6 m (7.9 and 8.6 ft) and the highest nest was at 24 m (79 ft). Nests in gallery forests were generally higher than those in more open regions. They were rarely built in the very tops of tall trees. Bournonville (1967) reported up to 8 nests per tree but he could not tell whether they had been occupied on the same night. Most lodge trees bore only one or 2 nests.

Albrecht and Dunnett (1971) remarked that Guinean chimpanzees usually rested near the last feeding spot of the day. They generally lodged in deciduous trees, but one nest was in a palm (*Elaeis*).

Baldwin et al. (1981, 1982) have provided the most comprehensive data on the locations of chimpanzee nests based on a 40-month study of Senegalese *Pan troglodytes verus* at Mt. Assirik. It is particularly valuable for documenting seasonal changes in nesting habits in a habitat that is characterized by a 7-month dry period, in which 4 months are usually rainless, and a 5-month period of modest (950 mm) rainfall. Overall, 32% of the 4,478 nests occurred in gallery forests, 56% in woodlands and 12% in grasslands (Baldwin et al., 1982). Most (93.6%) nests were located between 5 m and 22 m (16.5-72.6 ft) above ground; only 3 nests were between 39 m and 44 m (129-145 ft) above ground. Up to 18 nests were grouped in a nesting area (Baldwin et al., 1981).

During the wet season and at the beginning of the dry season, 67% of nests were found in woodlands and the remaining nests were about equally divided between grasslands and gallery forests. In mid dry season, 53% of nests were in the woodlands and the number (41%) increased in gallery forests. At the end of the dry season, 69% of nests were in gallery forests, 30% were in woodlands, and only one percent were in grasslands (Baldwin et al., 1982).

Nests were higher (median = 12 m [40 ft]) in the wet season than in the dry season (median = 10 m [33 ft]). In the dry season they were less sheltered by overhanging branches and were more clumped (7 per group versus 2 per group in the wet season). The median minimum distance between nests (4 m [13 ft]) was the same year-round. But nests were closer in woodlands than in grasslands and furthest apart in the forests. Forest nests were higher than grassland nests and woodland nests were the lowest. In all 3 habitats, the chimpanzees built their nests in the crowns of the tallest trees with suitable leafy branches. The lions and especially the leopards at Mt. Assirik probably encourage the chimpanzees to nest as high as possible (Baldwin et al., 1981).

During a 2-month reconnaissance in the Sapo Forest region of southeastern Liberia, Anderson et al. (1983) located 67 chimpanzee nests. Most (n = 22) of them were isolated. The largest cluster contained 10 nests. Groups of 2-8 nests were found once each. Anderson et al. (1983) estimated that the Sapo Forest is composed of 63% primary and mature secondary forest, 13% swamp, 13% seasonally inundated forest, and one percent secondary forest.

The chimpanzees lodged mostly (83.3% of nests) in the primary and mature secondary forests, and much less often in the seasonally flooded forest (13.3%) and swamps (3.3%). The median altitude of Sapo chimpanzee nests is 12 m (40 ft). Most of them (81%) were between 6 and 20 m (20-66 ft) above ground. Seventy-nine percent of the nests were overhung by foliage.

The median height of Sapo lodge trees is 18 m (59 ft). Most (92%) of them were between 6 and 25 m (20-82 ft) tall. Although many trees taller than 30 m (99 ft) were available, none were used by the Sapo chimpanzees for nesting. The median girth of the Sapo lodge trees is 52 cm (20 in).

During 77 days in the Tai Forest of southwestern Ivory Coast, Boesch (1978) found 146 arboreal chimpanzee nests. Eighty-two percent of them were at altitudes between 3.5 and 15 m (12-50 ft). Thirteen percent (n = 19) of them were higher than 15 m and 4% were between 2 and 3.5 m (7-12 ft) above ground.

In Equatorial Guinea *Pan troglodytes troglodytes* nested mostly in mature forests and their daily ranging patterns seemed to be influenced by the need for suitable lodge trees. Occasionally, they slept at the same spot on 2 or 3 consecutive nights (Jones and Sabater Pi, 1971).

Jones and Sabater Pi (1971) noted that their subjects selected relatively small trees instead of the largest ones as nesting sites. Only 14.6% of nests were in areas that sloped westward and northward, the remainder faced eastward and southward or were in flat regions.

In Equatorial Guinea, 96.4% of 195 nests were at heights between 0 and

20 m (66 ft); the median height is 10 m (33 ft). None of the trees exceeded
1.6 m (5.3 ft) in girth. The chimpanzees tended to lodge in the first forks
instead of the crowns of trees. Therefore, they commonly had branches
overhead. On average the nests were higher and more open during the wet
season though the climate was much less variable in Equatorial Guinea.
Single and pairs of nests were most common in Equatorial Guinea and
usually there was only one nest per tree (Baldwin et al., 1981).

Hladik (1974, 1977) reported that the chimpanzees (*Pan troglodytes
troglodytes*) at Ipassa, Gabon, preferred certain lodge trees that were 40-50
m (132-165 ft) tall. If they were undisturbed during the day, they built nests
near the afternoon feeding site. They never occupied the same site on
consecutive nights.

Near the western shore of Lake Kivu, Zairean chimpanzees (*Pan
troglodytes schweinfurthi*) generally nested at heights of 10-25 m (33-82.5
ft). They selected umbrella trees (*Musanga*) even when sturdier and higher
trees were nearby. Members of a group built nests in one or more trees that
were not widely scattered (Rahm, 1967).

Reynolds (1965) reported that in the Budongo Forest, Uganda, where
maximum canopy height is typically 42 m (140 ft), chimpanzees (*Pan
troglodytes schweinfurthi*) most commonly built night nests only 9-12 m (30-
40 ft) above ground. He explained that the placement of nests low in the
forest canopy would provide more young, pliable (2.5 cm in diameter)
branches and protection from strong night winds. Further, the "saplings"
that served as lodge trees would move if a leopard attempted to climb the
trunk, thereby alarming the chimpanzee, or would collapse with both beasts
if the cat leapt into a nest from above (Reynolds, 1964, 1965). From the
Reynoldses' accounts it is difficult to discern the extent to which Budongo
chimpanzees nested in the canopy. Reynolds (1965a, p. 130) stated that
15% of 259 nests were over 90 ft high, and some were "a good 150 feet
high." But Reynolds and Reynolds (1965, p. 387, fig. 11-11) showed the
highest nests at only 100 ft. The lowest tree nests were not less than 3 m (10
ft) above ground (Reynolds, 1965).

Bolwig (1959) found that 25 chimpanzee nests in the forests near Fort
Portal and Ruwenzori, Uganda, varied between 4.5 m and 42 m (15-150 ft; $\bar{x}$
= 16.5 m [54 ft]) from the ground; 76% of them were lower than 21 m (70 ft).
Many nests were widely dispersed, but 5 were together in one tree and at
another site 3 nests were close together, though not in the same tree.

Schaller (1963, pp. 181 and 374) reported that 39 chimpanzee nests in
the Impenetrable Central Forest Reserve, Uganda, ranged between 3.3 m
and 24 m (11-80 ft; $\bar{x}$ = 11 m [36 ft]) above the ground. The heights of about
half of them were between 6.3 m and 12 m (21-40 ft).

Ghiglieri (1984a) estimated the heights of chimpanzees nests at 2
localities in the Kibale Forest, Uganda. The mean altitude of 372 nests at
Ngogo is 12.2 m (40 ft; r = 2-35 m [7-116 ft]); and the mean height of 63

nests at Kanyawara is 10.8 m (36 ft; r = 5-23 m [16-76 ft]). There was a greater variety of trees in the Ngogo forest, where the chimpanzees nested in 21 of the available species. Ninety-four percent of Ngogo nests were built in only 5 of the 21 tree species. Ghiglieri (1984a) recorded only one instance of an Ngogo chimpanzee reusing a night nest. Nests lasted between 15 and 202 days ($\bar{x} = 110.8$ days; n = 28 nests).

At Kanyawara, 83% of 148 nests were located in 5 of the 14 tree species in which the chimpanzees lodged. Nests were often clustered, e.g. around large fruiting *Ficus mucuso*. Additional nests were built nightly in the vicinity until the fruit was depleted (Ghiglieri, 1984a).

Goodall (1962; Lawick-Goodall, 1968) provided an excellent account on the nesting habits of Gombe chimpanzees (*Pan troglodytes schweinfurthi*) before the provisioning facility was installed. They lodged in gallery forests, lakeside trees, and *Brachystegia* woodlands. Thirty-four percent of 384 nests were 9-12 m (30-40 ft) above ground; 56.5% were at heights between 12 m and 24 m (80 ft); and the heights of 9.4% of them were between 4.5 m (15 ft) and 7.5 m (25 ft). The lowest tree nests were in branches that overhung gullies and streams so that they would be difficult to reach from below. The high nests were near streams in tall valley forest trees that branched near their tops. Intermediate nests were in the tops of lower trees or in lower branches of tall trees. A male nested one night on the ground near a palm tree in which his female consort was lodged. There appeared to be no suitable arboreal nesting site near her (Lawick-Goodall, 1968, p. 195).

Gombe chimpanzees usually make new nests each evening though they occasionally build them in the same tree on consecutive evenings (Goodall, 1962). Over a span greater than 40 months, Lawick-Goodall (1968) documented only 20 instances of nest reuse, 2 of which were by the nest-makers and 6 by freeloaders. None of the nests was more than 2 weeks old and all were refurbished by the new lodger. Goodall (1962) concluded that the daily food quest chiefly determined where the chimpanzees would sleep because they built nests near the final food trees of the day. They used any species of non-fruiting tree that provided adequate height, foliage and supple branches. However, *Brachystegia bussei* was commonly selected. In 1961, the Gombe chimpanzees began to nest in palm trees (*Elaeis guineensis*).

Members of foraging parties nest together, split into smaller lodging groups, or aggregate with neighboring foraging parties for the night. One slumber party produced 17 nests at a site and another resulted in 10 nests in one tree. Independent males tend to lodge 90 m (100 yd) or more from females and juveniles, which nest within a few meters of one another (Lawick-Goodall, 1968).

In the extensive region south of Gombe National Park, Tanzanian chimpanzees nest predominantly in woodlands and forests. Kano (1972) found no evidence that they slept in areas of savanna (though they contained trees tall enough for nesting), grasslands, swamps, or bamboo bush. Fifty-one percent (n = 512) of 1,003 nests were in woodlands and 63% of these (n = 322) were in the areas dominated by *Brachystegia bussei*. Forty-nine percent (n = 488) of nests occurred in forests, of which *Cynometra, Carapa*, and mixed types were preferred by the chimpanzees. Kano (1972) remarked that because nests are more easily sighted in woodlands, they were overrepresented in his survey. He concluded that the chimpanzees "undoubtedly" utilized forests more than woodlands as sleeping locales. He presented no data on seasonal variations in nesting habits.

During a 7-day dry season survey in the arid Ugalla region, Tanzania, Itani (1979) located 125 chimpanzee nests. Eighty percent of them were in woodlands and 20% were in riverine forests. Several of the woodlands proffered hard fruits of *Brachystegia bussei* as a major dry season food source (Itani, 1979).

During 4 months of the wet season in 1963 and 1964, Izawa and Itani (1966) found 491 chimpanzee nests in the Kasakati Basin, Tanzania. All of the night nests were in trees. The majority of Kasakati nests were in (58.3%) or near (24.4%) the tops of trees. Nest heights above the ground ranged from 5 m (16.5 ft) to 40 m (132 ft); the average height is 19 m (62.7 ft).

Although they observed only 5 sessions of nesting by 8 subjects and saw one chimpanzee lodge in an old nest, Izawa and Itani (1966) concluded that they usually make new nests each night. Most (90.4%) nests were in riverine forests and the remainder (9.6%) were in woodlands. Izawa and Itani (1966) suggested that the Kasakati chimpanzees preferred trees of *Cynometra* and *Brachystegia* because they provide supple branches and dense clusters of small leaves. They probably varied the location of their nests according to seasonal availability of food in the forests and woodlands.

Suzuki (1969), who succeeded Izawa and Itani, located 1,323 chimpanzee nests during a 15-month ecological study. He did not present data on their seasonal distribution in different vegetational zones. Overall, 53.5% of them were in woodlands and 46.7% were in riverine forests. Suzuki (1969) confirmed that Kasakati chimpanzees lodge more often in *Cynometra* than in other types of riverine forest (40%, *Cynometra*; 29.1%, *Cordia*; 22.6%, *Albizzia*; and 8.2%, *Vitex*).

BONOBOS

During a 2-year period in forests near the western shore of Lake Tumba, Zaire, Horn (1980) found 106 bonobo (*Pan paniscus*) nests. They seemed to prefer young saplings as lodge trees but showed no clear preference among available species. Typically, lodge trees had bare sloping trunks with crowns in the top quarter or eighth. Only 5 tree nests were lower than 7 m .

and 7 nests were between 22 m and 27 m (73-89 ft); 59% were at 7-12 m (23-40 ft) and 30% were at 13-21 m (43-69 ft). Ninety-one percent of lodge trees had trunk diameters that were less than 38 cm (15 in) one meter above the ground (Horn, 1980). Nishida (1972) found 4 bonobo nests at heights of 15-20 m (49.5-66 ft) in a swamp forest west of Lake Tumba.

Kano (1979) observed 1,292 bonobo nests in the Bolomba forests (Wamba), Zaire; 94.5% of them were arboreal. They were rarely in tall trees. Instead most of them were less than 15 m above ground in trees that were less than 20 cm (8 in) in diameter.

In primary forests between the Lomako and Bolombo Rivers bonobos nested at great heights. MacKinnon (1976) reported that 46 nests had a mean height of 24.6 ± 6.7 m (82 ± 22.2 ft). Badrian and Badrian (1977) found that in Lomako forests 174 bonobo nests ranged between 5 m and greater than 35 m (115.5 ft) above ground: 20% were at 5-14 m (16.5-46 ft); 64% were at 15-29 m (49.4-96 ft); and 15.5 % were above 30 m (99 ft). Twenty-one percent of them were in trees with edible fruits or leaves. Badrian and Badrian (1977) saw them lodged "quite often" in trees in which they had taken afternoon meals.

Kano (1983) located 2,380 bonobo nests during 3.5 months in the primary, secondary and swamp forests and swamp grasslands of Yalosidi, Zaire, near the southeastern extremity of the species distribution. Only 3.7% of them were terrestrial. Kano (1983) doubted that all of them were used at night. Of 2,030 arboreal nests, approximately 78% were at 2-14 m (6.6-46.2 ft); 12% were at 15-19 m (49.5-62.7 ft); and 10% were at 20-<50 m (66-<165 ft). Most commonly (63.3%) arboreal nests were built in forks of medium and tall trees. But 32.1% were in the crowns of low trees (sometimes incorporating branches of neighboring taller trees) and 4.6% were built by interweaving the ends of branches from 2 or more sizeable trees. Only about 2% of 986 nesting trees were 40 cm (16 in) or more in diameter and the majority (85%) had trunk diameters less than 25 cm (10 in).

Bonobo nests occurred in 103 species of trees, 19 of which accommodated 81% of them. *Leonardoxa romii*, a dominant species in the lower stratum of the primary and seconday forests, bore 25% of the arboreal nests. Kano (1983) concluded therefrom that the Yalosidi bonobos have specific preferences for lodge trees. Of 2,116 nests, 60% were isolates in a single tree; 14.6% were singles that incorporated branches from 2 or more trees; and 25.4% were clusters of between 2 and 7 in one tree (Kano, 1983).

GORILLAS

Among apes, only the gorillas commonly lodge at night on the ground (Figure 13b). There are no reports of hylobatid apes sleeping on the ground. Terrestrial nesting by orangutans is very rare (Galdikas, 1978, 1982).

Rehabilitants are more disposed than forever free orangutans to build nests on the ground (Harrisson, 1969). Terrestrial nests have been attributed to common chimpanzees and bonobos at several African localities. It is not always clear whether they were used diurnally or nocturnally.

Kortlandt (1962) reported that some Zairean chimpanzees slept on the ground in a plantation that was free of leopards. Reynolds (1964, 1965) found 2 ground nests during 8 months in the Budongo Forest. Bolwig (1959) ascribed 5 terrestrial nests on a narrow mountainside terrace in the Impenetrable Central Forest Reserve, Uganda, to chimpanzees. In Tanzania, Lawick-Goodall (1968) noted only one instance of nocturnal ground nesting by a Gombe chimpanzee and Izawa and Itani (1966) concluded that Kasakati chimpanzees occasionally built terrestrial nests, but only for diurnal rests. Kano (1979) found 71 bonobo nests on the ground in the Bolomba forests and inferred that, unlike the few terrestrial nests that he had observed in the habitat of *Pan troglodytes*, they were used nocturnally.

In Equatorial Guinea approximately 84% of 410 gorilla (*Pan gorilla gorilla*) nests were terrestrial; 11% were within 2 m (6.6 ft) of the ground; and the remainder were at heights between 2 m and 8 m (26 ft). Jones and Sabater Pi (1971) inferred that nests were used on only one night. But lodge sites were sometimes occupied on 2 consecutive nights and during subsequent visits to an area. The gorillas used only 9 species of plants in their nests. The herbs *Aframomum* and *Sarcophrynum* were, respectively, in 95.6% and 54.1% of the nests. Most of them were located in thickets of *Aframomum*. Arboreal nests were most commonly on sturdy branches of *Musanga* in secondary forests. They seemed to prefer flat terrain and areas that sloped eastward and southward. Most (90.5%) nests were completely exposed to the sky. Usually they were at least 2 m from surrounding stands of vegetation and 10 m (33 ft) or less from one another (Jones and Sabater Pi, 1971).

While proudly procuring 115 Cameroonian gorilla specimens for European museums, Merfield (with Miller, 1956) concluded that they always nested on the ground, usually in open glades among secondary growths of saplings and canes. Bützler (1980) located 6 gorilla nesting sites in southern Cameroon. Half of them were near Moloundou in the southeast. The nests were composed entirely of *Aframomum* (2 species), which also served as food, and were presumably terrestrial.

Schaller (1963, p. 373) found that of 3,012 eastern gorilla nests in 8 areas of eastern Zaire, southwestern Uganda and northwestern Rwanda, nearly 88% were on the ground; 7% were less than 3 m (10 ft) above the ground; 3% were at 3.3-6.0 m (11-20 ft); and 2% were higher than 6.3 m (21 ft). However, there was notable variation in the frequencies of terrestrial nests among the areas inhabited by *Pan gorilla graueri* (Utu, 22% of 110 nests; Mt. Tshiaberimu, 33% of 85 nests; Mwenga-Fizi, 65% of 23 nests; Mt. Kahuzi, 100% of 7 nests) and by *Pan gorilla beringei* (Kabara, 97% of 2,488 nests;

Kisoro, 45% of 106 nests; Visoke-Sabinio saddle, 57% of 14 nests; Impenetrable Central Forest Reserve, 54% of 179 nests).

The highest frequency of non-terrestrial nests (78% of 110) occurred in the Utu region, Zaire, where ground cover was sparse (Schaller, 1963, p. 171). In dense secondary forests the trees (e.g. *Musanga*) had sufficient numbers of small branches near crotches and sturdy horizontal branches to support the heavy sleepers. In the primary forests of Utu, gorillas nested mostly with saplings. Schaller (1963, p. 181) speculated that the Utu gorillas might be more habituated to arboreal lodging than other gorillas are. Their nests were up to 18 m (60 ft) above ground (Schaller, 1965).

The high incidence (67%) of nests above ground on Mt. Tschiamberu is because the gorillas built springy, sturdy structures in tall (3-6 m) stands of bamboo, where undergrowth was sparse. Indeed at Mt. Tschiamberu 72% of nests were bamboo (Schaller, 1963).

Casimir (1979) greatly augmented Schaller's information on the nesting habits of Grauer's gorillas in the Kahuzi region of Zaire. During a 15-month period Casimir and Butenandt (1973, Casimir, 1975) located 128 lodge sites. Most (68.7%) of them had been used once, but 20.3% had been used twice and 8.6% were occupied thrice. One site had been used 5 times and 2 other sites had been used 4 times each.

Of 964 nests at 63 lodge sites, 663 (69%) were on the ground (Casimir, 1979). The Kahuzi gorillas evidenced no clear-cut preference for specific plants as nesting materials, though in secondary forests 37% of the terrestrial and bush nests included *Brillantaisia nyanzarum*. *Dombeya goetzenii* (10%) and *Neobourtonia macroclayx* (15%) were most often used as lodge trees and they were the most common species in secondary forests.

Most (87.7%) non-terrestrial nests were at 1-10 m (3.3-33 ft). The remaining (12.3%) raised nests were at >10-15 m (49.5 ft). Casimir (1979) found that most nests were poorly sheltered from rainfall and rarely lay strategically to receive maximum morning sunshine. The Kahuzi gorillas showed no clear preference for nesting on slopes that were oriented in a particular direction. The mean horizontal distances between terrestrial and bush nests at 59 lodge sites ranged between 8.2 and 9.7 m (27-32 ft). Casimir (1979) concluded that the spatial distribution of nests at a site was random with regard to age and sex of the lodgers.

Schaller (1963, p. 172) concluded that where ground cover was abundant at Kabara, the mountain gorillas nested wherever they happened to be at nightfall. However, they sometimes seemed to show preference for particular lodge sites over several consecutive nights. And in the giant scenacio zone, where ground cover is sparse, they preferentially nested in clumps of *Senacio erici-rosenii*. About 65% of 2,439 nests were made chiefly of herbs and lobelias, 15% were of branches, 10% were of *Galium*, and 10% were on nearly bare ground.

Schaller attributed the rarity of arboreal nests to the sparsity of suitable lodge trees at Kabara. Indeed one nest, 6 m in a *Hypericum*, fell in the night when the trunk gave way. He found no evidence that Kabara silverbacks nest above ground. Juveniles nested above ground about twice as often as females or blackbacked males did.

On the basis of mapping 400 lodge sites, Schaller (1963, p. 182) concluded that there is no consistent pattern of nest placement which would optimize the defensive position of the silverbacks. They slept furthest apart ($\bar{x} = 10.2$ m [34 ft]); juveniles nested close ($\bar{x} = 0.9$ m [3 ft]) to adult females or blackbacks, which nested near one another ($\bar{x} = 1.6$ m [5.4 ft]). Silverbacks were somewhat further separated from juveniles ($\bar{x} = 6.2$ m [20.7 ft]) and medium-sized ($\bar{x} = 4.0$ m [13.4 ft]) group members (Schaller, 1963, p. 375).

Bingham (1932) reported that approximately one-third of mountain gorilla nests at 2 localities in the vicinity of Mt. Mikeno (Parc des Virungas), Zaire, were arboreal. Some were as high as 18 m (60 ft) above ground. He suggested that in the intermountain saddle forests cold winds and rain induced them to nest on the ground. Schaller (1963, p. 180) disputed this interpretation. Dixson (1981, pp. 124-125) found that approximately 98% of 282 gorilla nests at Mt. Mikeno were on the ground or no more than 3 m (10 ft) above ground. None of the arboreal nests was higher than 9 m (30 ft). Usually they were made by small gorillas.

Fossey and Harcourt (1977) noted that gorillas on Mt. Visoke (Parc National des Volcans), Rwanda, usually lodged in species of plants that they did not feed upon because the food species provided unsuitable nesting materials.

Elliott (1976) observed no consistent pattern of nesting by a small, predominantly male group of Visoke gorillas. They nested on westward slopes; but most of their range was on the western face of Mt. Visoke.

Observations of nesting by gorillas of the eastern Virungas have been focussed chiefly in the Uganda Gorilla Sanctuary on the northern slopes of Mts. Muhavura, Gahinga and Sabinio. Near Kisoro, Osborn (1963) found that 44% of the 100 night nests were terrestrial and another 23% were in squashed bushes and bamboo near the ground. Twenty-two percent of night nests were in bamboo and bushes about 2.4 m (8 ft) above ground and 11% were at arboreal heights less than 7.5 m (25 ft). Most nests were terrestrial or on bushes in the saddle region but ground nests were rare in the bamboo zone. Adults lodged about a meter apart. Osborn (1963) speculated that the tree nests of young gorillas may protect them from pneumonia (due to damp ground), leopard attacks, and disturbances by nocturnal pigs.

Donisthorpe (1958) reported that 54% of 225 Kisoro gorilla nests were terrestrial. Most (77%) nests in the saddle area were terrestrial, probably because sturdy bushes were spare there. In the bamboo forest, 70% of nests were above ground because the dense grass stalks left little uncluttered surface. In lower forests, nests were about equally distributed between terrestrial and elevated lodge sites. Six nests were 6-9 m (20-30 ft) in small trees. However, here the gorillas seemed to prefer clumps of bamboo, which were either bent to the ground or used for platform nests. Donisthorpe (1958) found no evidence of nest reuse.

Bolwig (1959) seconded the suggestion that eastern gorillas do not reuse their nests and was skeptical that silverbacks placed their nests strategically in order to protect other group members from predation. Indeed he found that they also lodge in elevated nests where there is vegetation that can support their bulk.

Kawai and Mizuhara (1959) noted that 42.5% of nests were terrestrial in the Uganda Gorilla Sanctuary. Of the 210 nests above ground only 16 (7.6%) were arboreal. The vast majority (89.5%) of non-terrestrial nests were in bamboo. The remainder (2.9%) were in tangles of vines and creepers. At most lodge sites individual nests were between 2 m and 6 m (6.6-20 ft) apart. Kawai and Mizuhara (1959) concurred with earlier observers that eastern gorillas nest where they happen to have arrived at the end of the daily food quest and that they do not reuse nests on consecutive nights.

Schaller (1963, p. 373) found that of the 58 nests built above ground by Kisoro gorillas, only 5 (8.6%) were at 3.3-6.0 m (11-20 ft). The remainder (91.4%) were at 0.6-3.0 m (2-10 ft). His sample value for terrestrial nests (45%) is close to the value of Osborn (1963) and Mizuhara and Kawai (1959). Bamboo was the prime nest material. It was used in 68% of Kisoro nests (Schaller, 1963, p. 186).

Pitman (1935) reported that mountain gorillas in the Impenetrable Central Forest Reserve, Uganda, lodge terrestrially and arboreally at heights of 1.8-15.0 m (6-50 ft). According to Schaller's (1963, p. 373) data, of 83 non-terrestrial nests, 76% were at 0.6-3.0 m (2-10 ft), 19% were at 3.3-6 m (11-20 ft), and 5% were 6.3-15 m (21-50 ft). He commented (p. 180) that they rearely nested higher than 9 m (30 ft) even though there seemed to be many suitable tall lodge trees in the reserve.

WHAT DETERMINES THE CHOICE OF LODGE SITES BY APES?

No generalizations about the determinants of lodge site selection emerge from an extensive survey of hylobatid and pongid field studies. Because of the paucity of nocturnal and crepuscular observations we have almost no information on factors that might perturb an ape's sleep or give it permanence in an Asian or African forest. Most field workers and theorists emphasize the food quest (sometimes combined with social factors) as the major feature of hylobatid and pongid ranging behavior. The night is spent at or near the place where the subjects happen to have eaten lately. This inference would be more trustworthy if we knew a good deal more about (1) the frequency of predatory attempts by felids, humans, and perhaps large snakes; (2) how busy various trees and terrestrial spots are at night because of nocturnal foragers; (3) which plants accommodate ectoparasites and the intermediate hosts of endoparasites that infest apes; and (4) the fluctuations in temperature and moisture horizontally, vertically, and temporally in the forest.

Whitten (1982c,d) and Tenaza and Tilson (1985) are outstanding among researchers for systematically testing factors that potentially affect the choice of lodge trees by one species of ape. But here more direct observations are needed in order to establish the empirical reality of statistical correlations between the lodge trees of Kloss gibbons and the deterrent effects of biting ants. For example, Whitten did not mention whether in fact the ants are active at night. Perhaps they prevent the gibbons from entering their host trees before nightfall.

Although it is reasonable to assume that pongid apes nearly always build new nests each night because their foraging rounds end at different locations, it is also possible that their transience confounds competitors and would-be killers. There are no reliable reports of chimpanzees having been killed by Carnivora (Teleki, 1973; Tutin et al., 1981). Frequent changes of bedding also might reduce the parasite loads of apes and the possibility of attacks by nocturnal pestiferous arthropods.

IS PONGID NEST-MAKING INSTINCTIVE OR LEARNED?

Köhler (1959, p. 83) concluded that chimpanzee nest-building is "the manifestation of a special and elaborated 'instinct'." Yerkes and Yerkes (1929, p. 122) left unanswered whether pongid nesting behavior is primarily instinctive though Yerkes (1943, p. 16) suggested that it was. More than 3 decades later Bernstein (1962, 1967, 1969) conducted a series of experiments on wild-born captives and lifers at the Yerkes Laboratories (and RPRC). In his initial study of 25 11-40 year old chimpanzees, the 7 wild-born subjects always built nests during the first hour of exposure to test materials. Most of them were scored "good" by 4 fellow psychologists. Only 3 of the 18 never-free subjects constructed good nests: 5 built crude ones; and the rest remained nestless. On average, it took 16 min (r = 6-43 min) to produce good nests. The nest builders preferred to sleep in their handiwork and some of them cleverly kept it from human nest robbers. They tended to build better night nests than day nests with the same materials. Because all of the subjects engaged in similar manipulations of the test materials Bernstein (1962) concluded that nesting is dependent on the experience of chimpanzees and must be a learned behavior.

Next Bernstein (1967) provided 10 of the naive chimpanzees with nest-making cagemates or neighbors and plenty of pine straw, burlap strips, crumpled newspapers, lengths of hose, rope, etc. None of the never-free adults built nests either while with or away from the nesters. Three out of five 3 year old chimpanzees that had been captured during the first year of life built crude nests. Bernstein (1967) concluded therefrom that early opportunities and models facilitate nesting behavior in chimpanzees. If these factors are missed, as adults they will be retarded nest-makers despite their interest in nesting materials, normal manipulatory capacities, and exposure to proficient conspecific nesters.

Goodall (1962; Lawick-Goodall, 1968) observed that although 1.5-2.5 year old Gombe chimpanzees slept nocturnally with their mothers, 4 of them made day nests as a form of play activity. Youngsters continue to freeload in their mothers' nests until they are 3 or 4 years old. They only rarely attempt to help by bending down a few twigs. The close bond and relatively long contact between mothers and infants and the attention paid to the mother's activities greatly facilitate a youngster's learning the skills that are essential for survival and comfort. In this context, there seems to be little need for a nesting instinct unless we assume that what they learned could be quickly forgotten were the mother to be absent.

Bernstein (1969) observed that 8 young (2.5-3.5 yr old) gorillas did not construct nests in their large outdoor enclosure, which contained "many" trees less than 6 m (20 ft) tall. Six older juvenile and adolescent gorillas, caged in pairs at Yerkes RPRC, initially failed to build good nests from proffered materials. Instead they slept huddled together on the bare or littered floor. In later trials only 3 of the oldest subjects managed to construct good nests. And one 7 year old female performed the task quite admirably. However, they often slept draped over the rim.

Infant Virunga gorillas sleep with their mothers until at least 3 years of age and some individuals continue to do so until they are 5 years old (Fossey, 1979, p. 151). Schaller (1963, pp. 194-195) described the bending down of stalks by an 8-month old subject as "traces of nest building." Further, a 1.5 year old built a ground nest. And a 2 year old spent 10-12 min constructing a crude arboreal nest before joining its mother for the night.

Schaller (1963, p. 193) observed juveniles of unspecified ages building (1) a terrestrial day nest in 45 sec, (2) an arboreal day nest in 4.5-5.0 min, and (3) a terrestrial night nest in 1 min. He (1963, p. 66) designated 3-6 year olds juveniles. Schaller (1963, p. 195) concluded that "the general pattern [of African pongid nesting] is innate but that the specific adaptations are learned."

At Yerkes RPRC 2 young orangutans, in a large enclosure like that of the 8 young gorillas, spontaneously intertwined the crowns of small trees. However, it took 3-5 trials before they constructed nests from test materials. Thirteen adult orangutans all constructed good or excellent nests quickly upon receiving materials. They slept fully within them at night. Eight of them also covered themselves. Infants up to 11 months old showed no nesting behavior but slept instead with their mothers in well-knit nests (Bernstein, 1969).

Davenport (1967) noted that Bornean orangutan youngsters that at some time still clung to their mothers shared their nests at night. MacKinnon (1974, p. 12) reported that wild orangutan infants (<2.5 yrs) sleep with their mothers and young juveniles (<7 yrs) either build their own nests near the mother or, if she does not have an infant, may sleep with her. Horr (1977) observed the weaning period between an infant-laden mother and a 3.5 year old, which after noisy disputes, sometimes succeeded in joining the pair. But on other nights, she had to stay put in her own nest. Rijksen (1978, p. 247) noted that although even one year old Sumatran orangutans can build crude nests, they continue to sleep with their mothers as juveniles.

Thus, like wild chimpanzees and gorillas, by about 3 years of age, orangutans seem to have developed adequate skills for lodging independently. Although Galdikas (1978) emphasized the relatively long and close relationship of mothers and their youngsters, which would permit them to learn a great deal about their social and physical environments, she suggested that orangutan nest-building "may reflect inherent manipulative capabilities." (1982, p. 27). Galdikas also provided some support for Bernstein's (1969, p. 402) speculation that critical periods may be involved in the acquisition of nest-building. One of her subadult rehabilitants, which had become a proficient tool-user while living with humans, did not master nesting until 1.5 years after his reintroduction into the forest.

The 2 juvenile rehabilitant orangutans that Harrisson (1969) studied intensively in Sarawak exhibited quite different preferences for nesting. A male that had been captured at 1.5 yrs and released at 2.5 yrs regularly built arboreal night nests until he was transferred to another locality where he nested on the floors of rock shelters. A female that had been captured at one year of age and also released a year later slept on the ground, with and without nests, in both areas.

Rijksen (1978, p. 153) reported that a 6-week old rehabilitant constructed a crude nest from artifacts but that some adolescents and sub-adult males could not build good tree nests upon release into the forest. Eventually, most of them learned to build their own nests instead of occupying abandoned ones or trying to supplant industrious orangutans from their handiwork.

In summary, there appears to be little evidence for full-fledged nesting instincts in pongid apes. Further controlled experiments and naturalistic observations are needed to elucidate how nesting is learned, whether it is best learned during a particular stage in the youngster's ontogeny, and whether it is linked to special or general cognitive and motor capabilities. It might also be wise to study each species more thoroughly before declaring that the acquisition of nest-building is the same in chimpanzees, gorillas and orangutans.

NESTS AS TOOLS AND EXEMPLARS OF PONGID (PROTO-) CULTURE

With noteworthy exceptions (Lancaster, 1968; Galdikas, 1978) the general consensus is that pongid nests should not be classified as tools. In an authoritative compilation and synthesis of information on nonhuman tool behavior, Beck (1980, pp. 11, 124-125) excluded nests from the tool kit because they are not held or carried. He acknowledged that there appears to be no biological dimension in which nests differ fundamentally from tools as per his definition: unattached environmental objects that are used to alter more efficiently the form, position, or condition of other objects, other organisms, or the users themselves when they hold or carry them during or just before use and are responsible for their proper and effective orientation (Beck, 1980, p. 10). There is no reason to expect that the motor, sensory and cognitive processes that underpin nesting behavior are different from those that permit tool behavior by the Pongidae. Further, because observation learning appears to be an important component in the acquisition of nest manufacture the behavior can be viewed as intelligent to the extent that pongid tool behavior may be considered to be an expression of their intelligence.

Except for gross features like the construction of overhead shelters by some orangutans and the greater tendency for gorillas to nest nocturnally on the ground and to poop in their beds, little is known about variations in nest construction and use. Therefore, a truly informed discussion about social traditions that might be independent of the physical limitations of nest materials and lodge sites is forestalled.

Lawick-Goodall (1973b, p. 152) listed nest-making first in a list of "cultural elements" that characterize Gombe chimpanzees. To her, traits that the chimpanzees acquired by observational learning and imitation could be considered cultural. As we have seen, there is persuasive evidence that these modes of learning are important in the ontogeny of pongid nesting behavior. However, because of the absence of evidence for symbolic processes few anthropologists are likely to agree that nesting is a manifestation of chimpanzee culture.

5

Tool Behavior

"If you throw a spear at the *njina* [gorilla], . . . he will spring out of its way; but if you throw one at the *nchingo* [chimpanzee], he will catch it in his hand and throw it back at you." (Reade, 1864, p. 184)

". . . chimpanzees in the wild often spy on human activities. We wonder, therefore, why they have not adopted some of our cultural achievements." (Kortlandt and Kooij, 1963, p. 62)

"To document the discovery scientifically, the history of the animal must be known sufficiently to conclude that it has never before performed the tool behavior in question or observed the behavior performed by others. Further, the very first performance must be observed and documented to illuminate the dynamics of acquisition. No case of the discovery of tool behavior that meets all of these requirements is known." (Beck, 1980, p. 135)

Because of human dependence on technology, the tool behavior of nonhuman primates has been of special interest to comparative psychologists, ethologists and evolutionary anthropologists as it might reveal the sorts of mental and manipulatory capabilities upon which the development of human intelligence was premised (e.g. Parker and Gibson, 1977; Vauclair, 1982, 1984). Yerkes and Yerkes (1929) reported on the use and construction of implements by captives and throwing and nesting by wild and incarcerated apes. In the synoptic comparative table (pp. 557, 561) they classified the use and manufacture of implements under "intelligence" and throwing and nesting as "modes of life."

INSTRUMENTATION BY CAPTIVES

Yerkes and Yerkes (1929) gave chimpanzees the highest marks among apes for the use, including throwing, of implements and noted their ability to construct them. The orangutan scored a distant second in these abilities and the gorilla was ranked the least promising artisan and pitcher. In the synoptic table, Yerkes and Yerkes (1929, p. 561) indicated that there were no reports on tool behavior by hylobatid apes. However, earlier in the text, they had excerpted a report on the use of a food rake and a food box by a captive gibbon. Its performance supported the conclusion that behaviorally the gibbon more closely resembled the monkeys than the great apes (Yerkes and Yerkes, 1929, p. 99). More recently, Parker (1973) and Deriagina (1982) came to somewhat similar conclusions.

Yerkes and Yerkes (1929) compiled conflicting accounts of offensive and defensive throwing by wild orangutans but concluded, largely on the basis of Wallace's (1856) report, that they occasionally did so (p. 119). Their inferences about the capacities of orangutans to use implements were based on very few subjects because most of the studies that they reviewed lacked proper experimental controls. Though quick to ape some human tool use, one of their own subjects, named Julius, showed little potential to become a Caesar (Yerkes, 1916). Yerkes and Yerkes (1929, p. 190) were impressed by his versatility and resourcefulness. Julius sometimes obtained food incentives by using proffered implements in ways that were unanticipated by the observers. But he had to be shown solutions to many experimental tests before he could command them.

The pre-Depression Era literature on tool behavior by gorillas is even more impoverished than that on orangutans. Indeed, Yerkes and Yerkes (1929) relied solely on Congo, a 4 or 5 year old mountain gorilla, to assess the mentality and implemental capabilities of gorillas. She slowly learned to use a stick and a few other simple objects as implements after considerable human demonstration and tuition (Yerkes 1927a,b; 1928).

In addition to citing their own confirmatory experiments, Yerkes and Yerkes drew heavily upon the elegant classic studies of Köhler (1925) to support their high opinion of the instrumental talents of captive chimpanzees. Köhler had included several implemental tasks among the experiments whereby he hoped to discern the relative intelligence of chimpanzees. His perceptive reports on the spontaneous use and manufacture of tools and the results and inferences from his implemental experiments provided a rich backdrop for subsequent discussions of field observations on tool behavior and the possibility that at least some tool behavior of wild chimpanzees might have been learned initially by the observation of humans, who were not intent upon their tuition.

Köhler (1925, pp. 67-68) noted that the stick is a multi-purpose tool for captive chimpanzees. Because of the chimpanzee's versatility the functions of sticks expand if subjects are allowed ready access to them and to contexts in which they might be employed profitably. In addition to swatting down, probing and drawing in food incentives with sticks, Köhler's young (3-8 year-old) subjects used them to dig up buried fruits, as brandished, stabbing and thrown weapons, as prizing levers, and as upstanding substrates from which to jump to overhead food incentives. They also used straws, wires, ropes, stones and pieces of cloth as tools.

They placed sticks together and stacked boxes in order to reach distant incentives. They tore splinters from boards, gnawed and peeled the ends of sticks, and trimmed other objects orally and manually so that they could use them as tools. Köhler's chimpanzees used straws to dip fluids, which were then ingested, and to fish for ants that crawled outside the wire-netting of their enclosure. They sought twigs, rags and bits of paper with which to wipe feces off their feet (Köhler, 1925, p. 74).

Köhler candidly discussed the possibility that various modes of implemental behavior could have been learned by observing humans. He eliminated this possibility in a number of cases because no one who had contact with the subjects on Tenerife could have served as a model. However, Yerkes and Yerkes (1929, p. 355) wisely warned that in no case was the previous history of a subject so thoroughly known that human influence could be denied unequivocally. A variety of situations could have affected Köhler's and Yerkes's subjects during their capture, initial detention, and lengthy transport by ship and land vehicles to the experimental stations.

Modern laboratory experiments and other observations on the tool behavior of captive great apes have not led to discoveries that are as dramatic and provocative for evolutionary theorists as those of field researchers. The major advances have been largely empirical.

Seven of 8 wild-born chimpanzees at the Delta Regional Primate Research Center unobtrusively invented or otherwise acquired a variety of implemental methods for escaping from their large outdoor compound, which initially contained a number of pine trees. Over a period of 5 years, they had been provided with artifacts to reduce boredom and vegetation from nearby woods to supplement their supermarket fare and to serve as nesting materials. When they were between 6 and 7 years old, they began to use wooden poles as ladders with which to enter and trash the humans' observation booth and to escape over a high (5.5 m) fence. This followed a long period during which they had climbed up free standing poles and sometimes vaulted on them. But they had not been seen to prop poles against surfaces and then ascend them (Menzel, 1970, 1972, 1973a). Neither had they used human ladders as humans would, even when they had access to them and had seen people climbing on them (McGrew et al., 1975).

Once the Delta chimpanzees had discovered pole ladders, they began to disassemble the trees and manmade structures in their compound in order to extend their range outside it. They even circumvented the hot wires that were supposed to keep them out of the treetops (Menzel, 1972, 1973a). After they had depleted materials suitable for pole ladders, they began to employ short sticks as pitons to facilitate their escapes. The chimpanzee would stick a piton into a crack in the upper section of a wall, stand on it, insert one or more fingers into a higher chink, and hoist itself to freedom. McGrew et al. (1975) did not observe the chimpanzees deliberately making pitons. Instead they used sticks that had been modified during other activities.

While the status of the common chimpanzee as a pongid tool whiz has been underscored by contemporary workers (e.g. Nash, 1982; Kitahara-Frisch and Norikoshi, 1982), the gorilla, and particularly the orangutan, have steadily earned more esteem as implemental apes (Döhl and Poloczak, 1973; Ellis, 1975; Lethmate, 1977a, 1979; Parker, 1968, 1969; Rensch and Dücker, 1966; Galdikas, 1982). This work has been summarized and discussed thoroughly by Beck (1980) and Maple (1980). Bonobos are also proving to be very versatile tool-makers and users (Jordan, 1982; Kano, 1982b; McGrew, 1982).

Parker (1968, 1969) found that 3 orangutans at the San Diego Zoo were more responsive and quicker to solve a dip problem (i.e. rope into a hole, ultimately for food incentives) and a hoe problem (eventually to rake in food incentives) than 3 gorillas and 3 chimpanzees were. Based chiefly on his own observations of captive orangutans, Lethmate (1976a et seq.) demonstrated that they are quite capable of solving instrumental problems of the sort with which Köhler and Yerkes had challenged chimpanzees and that they are remarkably clever and innovative when manipulating objects and making tools outside the context of structured experiments. They stacked boxes and assembled composite poles in order to reach food incentives. They also twisted cloth into ropes and masticated leaves to sponge up fluids and made "fishing" probes out of a wide range of materials (Lethmate, 1976a, et seq.).

With considerable human tuition and assistance one orangutan even learned to use a tool to make a tool (Wright, 1972, 1978). Abang, a 5.5 year old male at the Bristol Zoo, imitatively learned to strike flakes from nodules of flint with a 3 pound (1.4 kg) hammerstone and to use them to cut nylon cords, thereby gaining access to food in a metal box. His mentor strapped the flint core to a board, padded with Plasticine, in order to stabilize it. Abang might not have been wholly satisfied with this arrangement as he once urinated over the core after securing some food incentives.

Wright (1978) did not state how many bangs Abang used to make a flake. Interestingly, Wright admitted that he was wrong to have been biased against orangutans versus African apes as subjects for this experiment. If the latter had been available at the Bristol Zoo he would not have used

Abang. However, Abang's sterling success induced him to conclude that "We have perhaps been misled by associating the arboreal state with low manipulative capability and the terrestrial state with a high one." (Wright, 1978, p. 232). Wright (1978, p. 224) chose Abang over his female cagemate "on the hunch that he had more potential . . .". Another genre of prejudice might be detected therein.

INSTRUMENTATION IN THE WILD

The highly informative period of experimentation by Köhler and Yerkes was followed by a long hiatus in reports on pongid tool behavior. It was punctuated sparely by tantalizing anecdotes from persons who had found circumstantial evidence and had brief encounters with apes in the forest. Pitman (1931) saw a Kayonza gorilla use a stick to obtain distant fruit. Beatty (1951) watched an adult male Liberian chimpanzee cracking oil palm nuts between a manually held rock and a flat rock in the ground. He also noted piles of broken and intact nuts scattered about other rocks in the forest. For more than a half hour, Merfield watched 8 Cameroonian chimpanzees using long twigs to extract honey from a subterranean apian nest (Merfield and Miller, 1956, p. 43).

Nineteen sixty-three was a landmark year in pongid tool behavioral studies. Hall and Kortlandt and Kooij published major reviews and theoretical discussions of the topic and Goodall reported on tool use and manufacture by Gombe chimpanzees.

Hall (1963) and Kortlandt and Kooij (1963) sent questionnaires to persons who had opportunities to observe the implemental behavior of wild and captive primates. Hall gleaned that wild orangutans (Wallace, 1869; Schaller, 1961) and gibbons (Carpenter, 1940) broke off and threw or dropped branches toward human observers and that disturbed gorillas (Merfield and Miller, 1956; Schaller, 1963) variously threw branches as part of their agonistic displays. He concluded that nonhuman instrumental behavior probably developed initially as part of agonistic behavioral complexes that served primarily to inhibit direct attacks among opponents. Later it extended into certain supplementary foraging activities of a few species. He noted that dietary tool-use was quite rare in wild nonhuman primates, having been reliably reported among apes, only in the chimpanzee.

On the basis of data from captive chimpanzees and gorillas, Kortlandt and Kooij (1963) inferred that terrestrial lifeways in semi-open habitats were prerequisite for the development and refinement of their agonistic throwing (both species) and clubbing (chimpanzee only). They proposed that in wild chimpanzees, these activities were directed predominantly at large carnivores and perhaps other natural enemies, especially in savanna habitats where refuges were rare. To explain the fact that incarcerated chimpanzees from rain forest habitats (e.g. Köhler's subjects) also learn to throw objects and to wield sticks, Kortlandt and Kooij (1963) speculated that the African apes are dehumanized descendants of savanna hominoids that were forced to retreat into the forest in the wake of hominid hunters, who had invented the spear.

Whereas they ascribed the agonistic use of objects by certain monkeys to genetic factors, Kortlandt and Kooij (1963) acknowledged that analogous behaviors in the great apes were only partly instinctive and otherwise were learned from conspecifics. Because their correspondents provided relatively few descriptions of non-agonistic object use in the wild, like Hall, Kortlandt and Kooij (1963, p. 80) stressed that agonistic employments of objects were by far the most common form of tool behavior in nonhuman primates. But citing the prior development of non-agonistic tool use in Köhler's Cameroonian captives, Kortlandt (1966) doubted that agonistic tool use had had much to do with the development of non-agonistic tool use in chimpanzees.

Bountifully endowed with patience, perceptiveness and perseverance, the intrepid Jane Goodall (1963a,b) provided the first major field evidence for regular tool manufacture and use by wild apes. Hall would have had to amend his thesis considerably if Goodall's study had been published before 1963.

Levers, Probes, Sops and Wipes

During the wet season, when alate termites (*Macrotermes bellicosus*) prepare for their nuptial flights, subadult and adult Gombe chimpanzees spend considerable time foraging and fishing for them with sticks, stems, slender twigs, strips of bark fiber, and sections of vine (Lawick-Goodall, 1968; Teleki, 1974; Figure 14a). Whereas Goodall (1964) stated that the tools were usually between 6 and 12 inches (15.2-30.5 cm) long, McGrew et al. (1979) found that the average length of 145 terminting tools from Gombe was 30.7 cm (r = 7-100 cm).

Chimpanzees regularly inspect termite mounds during their daily rounds. They indexically scratch into passages near the surface of a mound, insert their tools, withdraw them, and labially nibble off the clinging prey. Tools are selected variously before, during, and after inspection of a mound. Goodall (1964) saw no individuals move more than 10 yards from a feeding site for new tools. But some chimpanzees initially picked tools which were used at termitaria that were out of sight and up to 100 yards from the tool source. Certain individuals were more selective than others in their choice of tools and they sometimes took several tools to the mound for sequential use. Fishing tools were modified as needed; that is to say, stems were denuded of leaves, overwide grasses and bark fiber were divided longitudinally, and frayed or bent ends were nipped off (Goodall, 1963a,b; 1964, 1965; Lawick-Goodall, 1965, 1968, 1970). Lawick-Goodall (1967, p. 33) astutely noted that tool use allowed chimpanzees to prey upon termites

Figure 14: (a) Adult male Gombe chimpanzee fishing for termites. Note prey clinging to tool. (b) Adult female Mahale chimpanzee probing for arboreal ants *(Camponotus)*. The tool is to the right of her left fingertips. (Photos courtesy of C.E.G. Tutin; © C.E.G. Tutin.)

before the birds, baboons (Teleki, 1974, p. 589), and other monkeys, which wait until they fly out of the mounds.

By holding sturdy sticks bimanually and moving them forwards and backwards, the Gombe chimpanzees enlarged the entrances of subterranean bees' nests and then dipped out honey with their hands. One enterprising individual broke off a stick and swatted a proffered banana from a human's hand (Lawick-Goodall, 1970). Gombe chimpanzees also use slender sticks to extract 2 species of ants from subterranean (*Dorylus* (*Anomma*) *nigricans*) and arboreal (*Crematogaster* (*Atopogyne*) sp.) nests. The sticks, which Goodall (1964) estimated to be between 1.5 and 3.5 ft long, were broken off nearby vegetation or picked up from the ground.

During the first 45 months in Gombe National Park, Goodall observed only one episode of chimpanzees using sticks while preying upon driver (=safari) ants (*Dorylus nigricans*). Subsequently, students and other observers followed habituated subjects away from the banana boxes and recorded naturalistic feeding behavior at close range. McGrew (1974, 1977, 1979) focussed on their insect predation and manipulation of objects. During this 4-year period, Gombe researchers observed 29 different chimpanzees engaged in a total of 104 ant dipping episodes at 45 locations (McGrew, 1974).

The chimpanzees attacked nests that had been raided before, followed ants to pristine nests, and spotted surface perturbations that indicated the presence of unraided subterranean nests. They dug rapidly with their hands into previously undisturbed nests, thereby enlarging the entrances and causing the inhabitants to swarm forth in its defense. Similarly, they cleared the entrances of previously raided nests and removed debris from them several times during a dipping session (McGrew, 1974).

Chimpanzees used old tools that had been left in the vicinity of the nests or they made new ones. One individual made a tool and then walked apace over 75 meters to an active nest. To make a tool, the chimpanzee breaks off a long ($\bar{x}=66$ cm; r=15-113 cm; n=13), straight, and sturdy, but pliable, stick from a nearby sapling or shrub, manually denudes it of foliage, and sometimes orally and manually strips off its bark, especially if it is rough. They repair damaged tools orally or replace them during a feeding bout (McGrew, 1974).

As many humans can attest, safari ants are formidable foes. To avoid their bites, ant dippers often stay back from the hole and stand bipedally if they retain a grip on the tool. They prefer to perch on elevated substrates such as saplings, which they have pushed into position, rocks, tree trunks, and fallen logs. They commonly dash forward, insert a tool, move away, and return to retrieve it when it is coated with ants. The dipper pokes the smaller end of a tool into a nest and sometimes swirls it slightly. When the ants have swarmed up about three-quarters of its length, she withdraws it, holds it upright with the distal end near her mouth, and rapidly slides her other hand along the tool, thereby sweeping a mass of ants into her mouth. She quickly munches the wiggly bolus, which is about the size of a hen's egg and may contain as many as 300 ants (McGrew, 1974).

Some Gombe chimpanzees placed grass stems or twigs momentarily into columns of safari ants and then picked off and ate their catch. Foraging chimpanzees also used tools, like those employed for termite-fishing, to explore insect nests and holes in branches. They sniffed the free ends of probes after they were withdrawn. An infant picked its nose with a twig and an adult appeared to use a twig to pick her teeth (Lawick-Goodall, 1970). But the latter individual cannot hold a candle to Belle, an 8 year-old wild-born female at the Delta Regional Primate Research Center, which became a proficient dental hygienist for 4 gaping cage mates. Though not working by the book, Belle employed her fingers and a variety of handy wooden objects, one of which she manufactured by denuding a leafy red cedar twig, to dislodge foreign particles from the groomees' teeth and gums. She even extracted a deciduous molar from the mouth of a patient supine male (McGrew and Tutin, 1972, 1973).

The Gombe chimpanzees used leaves as drinking sops and toilet wipes. Instead of finger or hand dipping water directly, they inserted wads of masticated leaves into tree holes and then sucked out the steeped liquid. Individuals plucked and used leaves to wipe mud, blood, urine, feces, and sticky food residues, such as honey and fruit juice, from various parts of their bodies (Lawick-Goodall, 1965, 1970).

Although long-term monitoring of chimpanzees at Gombe National Park proffered an opportunity for longitudinal observations on the ontogeny of tool behavior, the reports (Lawick-Goodall, 1968, 1970, 1973a; McGrew, 1977) mostly contain cross-sectional data on the ages at which various components of implemental behavior appeared and were successfully integrated into mature sequences. Probably because termite fishing, leaf sponging and toilet wiping are virtually hazardless, chimpanzees practice them and exhibit proficiency at younger ages than is the case with ant dipping (McGrew, 1977). Lawick-Goodall concluded (1968, 1970, 1973a,b) and Beck (1974, 1975, 1980, p. 176) and McGrew (1977) concurred that observational learning is probably an important mechanisms whereby young chimpanzees acquire tool behavior from adults.

Lawick-Goodall (1968, p. 209) observed that by 6 weeks of age chimpanzees reach for leaves and branches and from 8 months of age they contact and manipulate potential tool materials. She saw no infants under 2 years old using tools to obtain termites, even though they sometimes watched their fishing mothers intently (Lawick-Goodall, 1970, p. 229), played on the termitaria, and tinkered and played with tool materials and discarded implements. Infants between 2 and 3 years old attempted to fish for termites but they selected inappropriate tools and used them clumsily. Between their third and fourth years, novice termite fishers improved technically but they were not as persistent as older individuals (Lawick-Goodall, 1970, pp. 230-231). However, infants between 3 and 4 years old exhibited full-fledged leaf sponging (Lawick-Goodall, 1973a, p. 158).

Instead of romping on a driver ants' nest, young infant chimpanzees cling to their mothers and older youngsters usually discreetly stay away from the hurtful hymenopterons (McGrew, 1977). Infants sometimes exhibit certain components of ant dipping behavior away from the nests. The youngest successful dipper was 46 months old. But generally infants and juveniles are not persistent dippers and their clumsy efforts yield few prey. Ant dippers achieve proficiency during adolescence, which begins at about 7 years of age (McGrew, 1977; Lawick-Goodall, 1975, p. 78).

Schiller (1952) found that captive chimpanzees exhibited a maturational gradient in the solution of instrumentation problems that ranged from relatively simple to complicated. Older animals were increasingly successful as they progressed through the series, probably due to their experience with sticks in a nonproblem situation. Similarly, Davis et al. (1957) reported that adult chimpanzees (n = 2) were more successful than youngsters (n = 5) were in solving a series of 40 bent-wire problems wherein Life Savers had to be manipulated over and off variably kinked wires within a time limit (Zimmermann and Torrey, 1965).

Female Gombe chimpanzees exhibited more dietary tool use than the Gombe males did. McGrew's (1979) analysis of records on 16 female and 14 male subjects revealed that during each of 19 months, females fished for termites more frequently than males did (4.3% versus 1.4% of total observation time). Whereas 75% (45/60) of females dipped for ants, only 45% (30/67) of males did. The females engaged in slightly longer bouts of ant dipping than the males did and they engaged in more bouts of termite fishing that lasted an hour or more (McGrew, 1979).

Studies outside of Gombe National Park have revealed notable regional variations in the tool behavior of chimpanzees. For example, no dietary tool behavior has been recorded for the well studied Budongo chimpanzees and some populations south of Gombe have exhibited relatively little termite fishing. During a 15-month study in the Kasakati Basin, Suzuki (1966) and Itani recorded only one instance of implemental termite foraging. They recovered 2 tools, both of which were approximately 45 cm long.

The chimpanzee watchers at Kasoje recovered 133 bark, twig, grass, bamboo and sedge probes from termite (*Macrotermes herus*) mounds in the range of the unhabituated B-group, sometimes immediately after the subjects had fled from them. Per contra, although Nishida and Uehara (1980) commonly observed chimpanzees of the neighboring, well habituated K-group consuming alate termites after manually destroying the emergent towers of their mounds or licking workers from slabs of rotten bark, they recorded only one instance of K-group members fishing for termites (*Pseudacanthotermes spiniger*). They reasoned that K-group chimpanzees have little need for termiting probes because they live in a relatively moist region where the alate prey species can be obtained simply by wrecking their mounds.

Uehara (1982) described 16 termiting tools that were collected after he watched 8 members of K-group fishing at the base of a mound (*P. spiniger*) with techniques like those of Gombe chimpanzees. The mean length of 12 bouts was 12.5 min (r = 2-47 min). Thirteen tools were made of bark and 3 were denuded liana shoots (*Paullina pinnata*). The mean length of the tools is 51.5 cm (20.6 in; r = 29.5-97.2 cm). Uehara (1982, p.55) also collected 97 tools from a mound, "judged to be that of *Macrotermes*," in the range of B-group. Their mean length (54.6 cm [21.8 in]; r = 21.7-125.6 cm) is similar to that of the termiting tools of K-group and is greater than that of the Gombe chimpanzees, which fish for *Macrotermes*. Most of the termiting tools attributed to B-group were made of bark and a few consisted of stripped liana shoots and unmodified or peeled stems (Uehara, 1982).

Over a 4-month period, spanning the end of the dry season and half a wet season McGrew and Collins (1985) studied the tools (n = 290) and termiting behavior (n = 4 bouts) of the unhabituated B-group, which ranged in the driest region of the Mahale Mountains. Three-quarters of the termite probes were made of bark from shrubs and lianas, 15% were of sedge (*Cyperus pseudoleplocladus*), and 10% were segments of vine, denuded twigs, and an unrolled young leaf. The Mahale chimpanzees used them to extract *Macrotermes herus* from their mounds much as Gombe chimpanzees employed similar probes to fish for termites. About half the Mahale tools had been used at both ends and some of them were honed dentally when the ends became frayed (McGrew and Collins, 1985).

The mean length of the Mahale termite probes is 37.7 cm ± 14.7 cm (15.1 ± 5.9 in; r = 12-84 cm [4.8-33.6]); the median is 35.9 cm (14.4 in). Tools were made about the same length regardless whether bark, vines, sedges or twigs were the raw material. Mahale chimpanzees seemed to prefer tools that were 3-5 mm wide (McGrew and Collins, 1985).

Termite fishing was a seasonal activity; 98% of the tools and associated litter were found in November and December. This indicated that the subjects of B-group started termite fishing soon after the rainy season was fully underway. They stopped in January (McGrew and Collins, 1985). Almost all of the sedge tools were found in December, when *Cyperus* matures to suitable length. Tools of the other 3 main classes of raw material peaked in November (McGrew and Collins, 1985).

Unlike termites in the range of K-group, the ants (*Camponotus maculatus, C. vividus, C. (Myromotrema)* sp., *C. brutus*) are often difficult to obtain directly by hand or mouth. They nest variously inside the trunks and large boughs of standing and fallen trees (Nishida and Hiraiwa, 1982). In order to extract ants from sizeable entrance holes, which are nonetheless too small for hand-dipping, the chimpanzee breaks off a neighboring branch, trims it dentally, and inserts the probe into the nest (Figure 14b). If the entrance hole is very small, the chimpanzee more carefully selects and trims strips of grass, bark from vines, the midribs of ovate leaves, or vine stems to use as extractive probes. Sometimes ant fishers induce prey to bite and to move towards the entrance hole by vigorously shaking the tool in the passage and slapping or bipedally pummeling the tree trunk at a lower level (Nishida, 1973; Nishida and Hiraiwa, 1982). The average length of the 37 probes is 33.2 cm (13.3 in; r = 10-80 cm) (Nishida, 1973). The mean length of 28 additional probes is 21.4 cm (8.6 in; r = 10.5-58.8 cm). Uehara (1982) concluded that fishing for *Pseudacanthotermes* requires tools that are much longer than those used for *Camponotus*.

Mahale chimpanzees made anting tools predominantly from vines and secondarily from arboreal bits. At least 22 species of plants were exploited for tools. Among 206 probes, 57.3% were strips of bark, 15.5% were modified branches, 13.6% were vines or grasses, 5.8% were unbarked sticks, and 3.9% each were midribs of leaves and unmodified branches (Nishida and Hiraiwa, 1982).

The chimpanzees most commonly made tools from materials in the ant-infested trees. Most other tool materials were gathered in close proximity to the anting trees. However, they characteristically made tools and took them to the anting holes instead of checking the holes first to see whether they would be productive. The lag time between tool selection and anting usually spanned only a few minutes though it could be as long as 17 minutes. Mahale chimpanzees transported tools over more than 10 m on only 9 occasions. The longest distance is 100 m (Nishida and Hiraiwa, 1982).

In order to negotiate the meandering tunnels of ant nests, tools must be flexible, long and durable. The average tool is used for 4.8 min (r = 0.5-19 min; n = 75). Chimpanzees rarely use the discarded tool of another or snatch one from its owner probably because tools are easily made (15 sec or less) and materials are plentiful. Competition can be keener for anting sites. Individuals have been recorded waiting between 0.5 and 51 min ($\bar{x}$ = 16 min; n = 10) for a spot to be vacated so that they could try their luck. Occasionally a fisher is displaced from a site by a gesturing, posturing or vocalizing group member. The average span of 98 fishing bouts is 33.2 min (r = 1-136 min). Single probing actions yield between 0 and 15 ants (Nishida and Hiraiwa, 1982).

Analyses of stools from known individuals in M-group revealed that adult females consumed more *Camponotus* than adult males did. Whereas 39% (56/143) of female stools contained remains of ants, only 5% (8/151) of male stools evinced that they had probed a formicary. The neighboring K-group showed no sex difference in ant-eating (Uehara, 1984).

The youngest Mahale chimpanzee that was seen to use a tool was 32 months old. He was slow and clumsy. He prepared simple tools when he was 37 months old. Tool assisted fishing for *Camponotus* is probably not fully refined until Mahale chimpanzees are between 6 and 8 years old. The overall timing of its development is slower than that described for Gombe chimpanzees that fish for *Macrotermes*. *Camponotus* are more hurtful than *Macrotermes* (Nishida and Hiraiwa, 1982).

McGrew and Rogers (1983) reported the first evidence for termite fishing by *Pan troglodytes troglodytes*. While censusing apes in the primary forests of Belinga, Gabon, they found 28 probes (27 denuded twigs and one of liana) and 2 denuded sticks atop subterranean nests of *Macrotermes* ?*nobilis*. Recall that remains of *Macrotermes* were found in the feces of Belinga chimpanzees but not in the stools of sympatric gorillas (Tutin and Fernandez, 1985; Chapter 3).

The mean length of Belinga probes is 37.8 cm (15.1 in; r = 17-59 cm [6.8-23.6 in]; n = 23); the median length is 38 cm (15.2 in). The mean and median proximal and distal diameters of the probes are 4 and 3 mm, respectively. The sticks, which presumably were for gouging the nests, were 68 and 78 cm (27 and 30 in) long and much stouter than the probes (McGrew and Rogers, 1983).

Jones and Sabater Pi (1969; Sabater Pi, 1974) collected 220 sticks from the vicinity of termite mounds (*Macrotermes muelleri* and *M. lilljegorgi*) at 3 localities in Equatorial Guinea. The sticks were rigid and varied between 19.5 and 86.7 cm (7.8-34.7 in) long and between 1 and 15 mm in diameter. They were variously stripped of twigs, leaves and bark and had worn ends. They had been broken from small saplings, the stubs of which were located between 1 and 24 m (3.3-79.2 ft) from the mounds. None of the sticks were found during the period of drought. Most of them (202) were found in Okorobiko and none were found near Mt. Alen, where chimpanzees and the 2 species of *Macrotermes* also occurred together (Sabater Pi, 1974). On the basis of sparse observations on Okorobiko chimpanzees, Sabater Pi (1974) concluded that instead of termite fishing, they use stick tools to open termite nests and then dig up and retrieve the riled insects with their hands.

Struhsaker and Hunkeler (1971) found stick tools, similar to those from Equatorial Guinea, lying atop and stuck into termite mounds in the forests of West Cameroon. Chimpanzees were common there.

Sugiyama and Koman (1979b) observed a variety of tool behaviors by members of a non-provisioned group of 21 chimpanzees (*Pan troglodytes verus*) that they studied for 6 months at Bossou, in southeastern Guinea. The Bossou chimpanzees spent considerable time and effort to modify long, thorny sticks with which to beat, hook, and lever the branches of a fruit laden fig tree within reach so that they could climb into it. The trunk of the fig tree was too thick to climb up directly so they had to enter it from neighboring trees. The tools were made de novo from branches of the base tree; they were never retrieved from the ground (Sugiyama and Koman, 1979b).

The Bossou chimpanzees stripped twigs between 5 and 15 cm (2-6 in) long and placed them in the nest entrance of a termite infested tree. They pounded the trunk below the hole, removed the tool, and ate the few termites that were on it. Youngsters dentally stripped twigs between 10 and 20 cm (4-8 in) long and used the tools to probe resin from a tree hollow (*Carapa procera*). An adolescent female dipped water from a tree hole with an unmodified leaf (*Aningueria robusta*) (Sugiyama and Koman, 1979b).

During the rainy seasons of 2 years (1976-1978), McGrew et al. (1979b) collected evidence for termiting by far western chimpanzees (*Pan troglodytes verus*) at Mt. Assirik in the Parc National du Niokolo-Koba of southeastern Senegal. They found remains of one species of termites (*Macrotermes subhyalinus*) in the feces of their subjects and collected tools at 13 mounds on 22 occasions. They did not observe tool behavior directly.

The chimpanzees of Mt. Assirik made termiting probes from twigs (47%), leaf stalks (31%), vines (19%) and grass (3%). Ninety-four percent of them were made from materials collected within arm's reach (i.e. 2 m) of the mounds. However, tools made from vines (*Cissus*) were probably brought from distances greater than 5 m (McGrew et al., 1979b).

During the termite fishing season (May-August) of 1979, McBeath and McGrew (1982) conducted a more intensive study on 54 termitaria and tool distributions by Assirik chimpanzees. They collected 323 tools from 25 assemblages at 15 mounds, which were unevenly distributed among the 5 vegetational zones of the chimpanzees' range. The majority (67%) of tools were found in ecotones between plateau and woodland areas even though termitaria were equally or more abundant in the woodland, forest and grassland. Eighty percent of the tools were made from *Grewia lasiodiscus*. It commonly grows near mounds in the ecotonal zone. Therefore, McBeath and McGrew (1982) concluded that the Assirik chimpanzees probably concentrate their termite fishing in ecotonal localities because reliable tool material are handiest there.

Nut-cracking

Anderson et al. (1983) found evidence that chimpanzees (*Pan troglodytes verus*) in the Sapo Forest, Liberia, use stone hammers to break open 4 species of nuts (*Coula edulis, Panda oleosa, Parinari excelsa*, and *Sacoglottis gabonensis*) on laterite boulders and tree roots (anvils) (Figure 15). The observers were attracted by distinctive pounding noises but the subjects fled upon their approach.

The mean weight of 8 laterite hammers from sites where nuts of *Coula* and *Parinari* were processed is 3.3 lb (1.5 kg; r = 0.8-5.6 lb [0.35-2.55 kg]). Stone anvils were more common than root anvils at sites with *Coula* (83% vs. 17%; n = 18), *Parinari* (71% vs. 29%; n = 14), *Sacoglottis* (100%; n = 10) and *Panda* (80% vs. 20%; n = 5). Anderson et al. (1983) found no evidence that Sapo chimpanzees employ wooden clubs as nutting hammers.

Stones were plentiful in some nutting areas but probably had to be

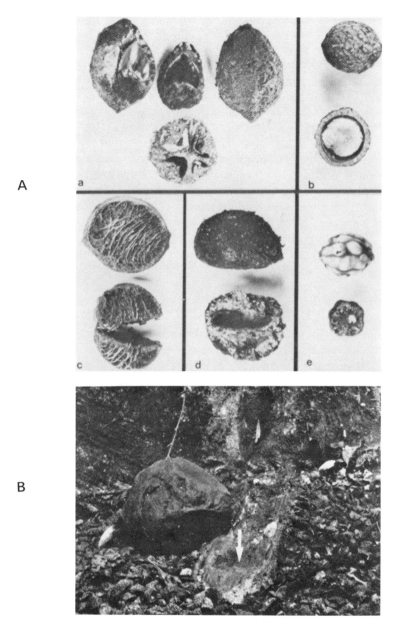

Figure 15: A. Five species of nuts, which are eaten by chimpanzees in the Tai National Park, Ivory Coast: (a) *Panda oleosa;* (b) *Coula edulis;* (c) *Detarium senegalense;* (d) *Parinari excelsa;* (e) *Sacoglottis gabonensis.* B. Nut-cracking site of chimpanzees in the Tai National Park. Arrow indicates depressed area of a surface root that is used as an anvil. A 6.6 kg granite hammer lies to the left of the anvil. The ground is littered with nut shells. (From Boesch and Boesch, 1982.)

transported to others. Further some Sapo nutting sites were hammerless when the observers located them, suggesting that the subjects had taken them away (Anderson et al., 1983).

Chimpanzees on Tiwai Island, in neighboring Sierra Leone, hammer hard pods of *Detarium senegalense* to retrieve the single seeds therein. Via hammering noises, Whitesides (1985) located a pod littered site beneath a large tree of *Detarium*, the roots of which were lacerated from recent blows. At another nutting site there was a stone anvil. Whitesides (1985) found no stones that could serve as hammers though several serviceable sticks were present. He saw no hammers in the hands of fugitives from a site.

Sugiyama and Koman (1979b) initially located 29 nut smashing sites under palm oil trees (*Elaeis guinensis*) in the Bossou forest, Guinea. In a later study, Sugiyama (1981) found evidence for nut smashing under 31% (39/127) of mature oil palm trees in the range of the group.

The Bossou nutting apparatus consists of platform rocks, generally weighing more than 1 kg (2.2 lbs) (Sugiyama and Koman, 1979b), and hand held striking stones that weigh between 500 g and 1.5 kg ($\bar{x}=790$g; n=50) (Sugiyama, 1981). Most of them were gneiss. The working faces of the stones are relatively flat but have depressions approximately 0.5 cm deep and 3-5 cm in diameter at their centers.

Bossou nut smashers took turns crouching before a platform rock, placing dry palm kernels singly upon it, raising a stone between 5 and 20 cm above it, and then striking the nut. On average it took 2 or 3 blows to open a prize, after which its contents were extracted digitally and eaten. Nutting sympatric humans either carried palm kernels back to their village for processing or opened and ate them on the path. Apparently, they did not use the chimpanzee nutting sites in the forest (Sugiyama and Koman, 1979b).

Struhsaker and Hunkeler (1971), Rahm (1971) and Boesch (1978) reported that chimpanzees in the Tai Forest of western Ivory Coast employ hefty sticks and stones to smash fruit kernels for their edible contents. Struhsaker and Hunkeler (1971) and Rahm (1971) were attracted to nut smashing sites by the sounds of chimpanzees engaged in the activity. Struhsaker and Hunkeler (1971) twice saw chimpanzees fleeing from nut smashing sites. One adult fled with a large rock in his hand. Apparently, the chimpanzees placed kernels of *Coula edulis* and *Panda oleosa* in cuplike depressions in the exposed roots of the fruiting trees and smashed them with clubs and rocks. Struhsaker and Hunkeler (1971) noted that humans use machetes to open the nuts, thereby leaving incisions on the substrates. This distinguishes human from chimpanzee nutting places. Sticks were used at 15 places, stones were employed at 7 places and the kinds of implements are unknown for 4 places. Stones were present in all 5 places where kernels of *Panda oleosa*, the hardest of the 2 species, were smashed (Figure 15). Remnants of *Coula edulis* occurred at 21 places with sticks or stones. One stone tool was estimated to weight 16 kg (35.2 lb); the average weight of 5 others is 5 kg (11 lb; r = 3.5-5.75 kg).

Boesch (1978) initially located 107 nut smashing sites and found that Tai chimpanzees used tools also to open kernels of *Parinari excelsa*. They utilized rocky outcrops, as well as exposed tree roots, as hammering substances. He confirmed that clubs and small stones were used to smash kernels of *Coula edulis*, whereas the harder nuts of *Panda* and *Parinari* were opened with rocks that were commonly heavier than 5 kg. Although rare in the Tai forest, granite appeared to be chosen more often than laterite or quartzite to smash the hardest nuts, viz. those of *Panda oleosa*. Granite hammerstones seem to have been transported more than 100 m (330 ft) from their sources to the fruit trees.

From an intensive 3-year follow-up, Boesch and Boesch (1981, 1982, 1984a-c) provided much valuable quantitative data on the nut-cracking activities and tools of Tai chimpanzees. Further, they made the remarkable discovery that in 2 of 3 methods for harvesting kernels of *Coula* and *Panda*, adult females are generally more adept and efficient than adult males are. Only after 2.5 years did some of the unprovisioned subjects, mainly males, tolerate human observers at the nutting sites (Boesch and Boesch, 1984b). Visibility is poor in the Tai forest. The study community included about 70 individuals in a range of 27 km² (Boesch and Boesch, 1984b).

The nuts are only available seasonally. All 5 species are high in calories, mostly from fats; and, kernels of *Panda* are especially proteinaceous (Boesch and Boesch, 1982, p. 267; 1984a). Coula trees are common and fruitful in the Tai forest. *Parinari* and *Sacoglottis* are also relatively abundant. Trees of *Detarium* and *Panda* are rarer. Boesch and Boesch (1984a) concluded that Tai chimpanzees depend or at least rely heavily on coula nuts during 4 months per year.

The subjects carried handfuls and mouthfuls of nuts that they had collected arboreally or off the ground to anvils, upon which they carefully positioned and hammered them one at a time. Terrestrially they commonly employed hammers that were lying near the anvil. For arboreal coula nut-cracking, the chimpanzee transported a hammer up the tree and held on to it while it gathered nuts. Then it set to work on a horizontal coula bough. Whether the nuts were hard or pliable, rocks were preferred over exposed roots as terrestrial anvils. However, because they were much more common in the forest, roots were used predominantly by the Tai chimpanzees (97% of 1,434 work sites) (Boesch and Boesch, 1982).

Panda nuts are very hard. Therefore, the chimpanzees employed stone hammers to open them (90% of 78 work sites). Indeed Boesch and Boesch (1982) suspected that 8 clubs that were near panda nut-littered anvils had been left there by naive subadults.

Sticks were more commonly employed as hammers on the other 4 species (*Coula*, 92%, n = 557 sites; *Detarium*, 86%, n = 51 sites; *Parinari*, 61%, n =

31 sites; *Sacoglottis*, 100%, n = 2 sites). Three-quarters of the wooden hammers for coula nuts were 8-32 in (20-80 cm) long. Eighty-nine percent of coula clubs had diameters between 1.6 and 4 in (4-10 cm). Most (77%) weighed less than 4.4 lb (2 kg); 15.5% were between 4.4 and 8.8 lb (2-4 kg). Wear patterns indicated that a club was held so that it contacted target nuts near its center of gravity. Experiments showed that this prevents banged nuts from bouncing off the anvil (Boesch and Boesch, 1982).

Tai chimpanzees preferred granite among the 3 kinds of stone that they utilized for hammers. It was the hardest of the 3 and quite rare in the forest. Stone hammers generally weighed between 2.2 and 53 lb (1-24 kg). But an outstanding one weighed 92 lb (42 kg). Stone hammers with panda litter were significantly heavier than those at coula sites (Boesch and Boesch, 1982).

By regularly monitoring nutting sites and marking stone hammers, Boesch and Boesch (1982) were able to document their transport between sites. The Tai chimpanzees carried stones, including choice heavy ones, from greater distances to the panda (versus coula) nutting sites.

Among Tai chimpanzees, adult females (>13 yr old; n = 23) were significantly more proficient and persistent than adult males (> 15 yr old; n = 13) at opening coula nuts arboreally and panda nuts terrestrially (Boesch and Boesch, 1981, 1984b). Arboreal coula nutting requires considerable coordination and some forethought. A hammer is located on the ground before the subject climbs the tree. It is held manually or pedally during the 50 ft (15 m) or longer vertical ascent, while picking nuts, and between extractions as the subject eats kernels. The chimpanzee eats kernels one at a time while balancing herself on a bough and juggling her collection of unopened nuts between foot and mouth (Boesch and Boesch, 1981).

Although panda nut-cracking may be somewhat less challenging to the cerebellum, it too requires notable skill and cognition (Chapter 6). Stone hammers may have to be transported over several hundred meters and the cracker must precisely control the forces of a series of blows so that the delicate kernels, encased by the hard shell, are not pulverized to inedibleness. Panda nut-cracking is more energetic and time-consuming than cracking coula nuts. Females are more adept than males with the heavier hammers (Boesch and Boesch, 1981, 1984b).

Unlike ant-dipping and especially termite-fishing, in panda nut-cracking and arboreal coula nutting, the learning process extends well into adolescence in both sexes and even into adulthood in males. Most males never master either technique (Boesch and Boesch, 1981, 1984b). Further, even on the ground, males crack coula nuts more when they are dry and easier to open. Females surpass them with the tougher fresh nuts (Boesch and Boesch, 1984b).

Boesch and Boesch (1981, 1984b) thoroughly explored a number of possible explanations for, and, within the limits of their data, tested hypotheses on the causes of sex differences in the nutting behavior of Tai chimpanzees. They concluded that males do not master panda nut-cracking and arboreal coula nutting because they are too busy monitoring the activities of other group members and are inclined to move away from nutting sites in response to others instead of staying to feed as individual females do. They can see farther on the ground than in trees. Moreover, in order to crack coulas in a tree they would have to concentrate in order to keep their balance and not lose their nuts and the hammer. This makes good sense in view of the facts that in chimpanzees intermale relationships are very complex and they must also be attuned to predators and other chimpanzees groups that might attack them or trespass their range (Chapter 8).

That brawn is not linked closely to effective nut-cracking in chimpanzees and that females are commonly more frequent and effective tool-users (McGrew, 1979; Uehara, 1984; Boesch and Boesch, 1981, 1984b) should smash the silly sexist notion that males *must* have been the first hominoid technologists.

Missiles and Clubs

Sometimes during thunderstorms and on other occasions when they were excited, adult male Gombe chimpanzees broke off or snatched up and brandished branches as they dashed about precipitately (Goodall, 1963b; Bygott, 1979, p. 411). Occasionally they hit at animal targets with a hand held stick. Often it was released just prior to contact (Lawick-Goodall, 1970). During periods of social excitement in the banana feeding area, adult and subadult males threw sticks, stones, handfuls of vegetation and artifacts as part of displays directed at other chimpanzees, baboons and humans. Initially, most of this throwing was unaimed. In a few instances where aim could be inferred, the objects usually missed the mark. The apes stood bipedally or tripedally and could throw overarm as well as underarm (Goodall, 1964; Lawick-Goodall, 1968, 1970). Because of the initial rarity of aimed throwing and the fact that less than half of the 17 male subjects exhibited it, Lawick-Goodall (1968, p. 203) concluded that, prior to setting up the feeding area, exhibitionistic offensive and defensive throwing were not highly developed in the Gombe population. However, during the banana bonanza, the frequency of aggressive episodes, including the throwing of objects, increased; more males engaged in it; and some youngsters and an adult female joined the rookie pitchers. The percentage of hits versus misses increased too (Lawick-Goodall, 1970).

Plooij (1978) described a rare instance of tool use by wild chimpanzees while hunting mammalian prey. A Gombe adult named Mike threw a sizeable (25 cm × 10 cm × 10 cm) rock, which hit one of a group of bushpigs that he and several companions had surrounded. Later, when the pigs had panicked and fled into the brush, the chimpanzees captured and ate a piglet. Previously, with the aid of 4-gallon kerosene cans, which he innovatively kicked and slung about the camp, Mike had bluffed his way past Goliath and other high ranking males to the pinnacle of the Gombe community (Lawick-Goodall, 1970, 1973a; Hamburg, 1971, p. 389). However, he was no longer the upmost ape when he beaned the bushpig.

Sugiyama and Koman (1979b) recorded 69 episodes in which Bossou chimpanzees threw branches or an epiphyte, mostly at them. Initially during the study, adult males were the principal attackers. They broke off hefty (100-900 cm [40-76 in] long and 1-3.5 kg [2.2-7.7 lb]) nearby branches, ran along supports above the humans, brandished and then pitched them underhandedly towards the inquisitive targets. Sugiyama and Koman (1979b) judged that the throwing of adult males was well controlled. The researchers' bodies and equipment bore brutal testimony to their effectiveness. After the group had become accustomed to human presence, throwing was performed mainly by youngsters. They selected less fearsome missiles (40-100 cm [16-40 in] long and weighing 1 kg) and stayed farther away from their targets. Twice an adult male threw sidearm and several times youngsters threw overhanded.

In Guinea and eastern Zaire, Kortlandt (1962, 1965, 1967, 1968) and his associates (Kortlandt and van Zon, 1969; Albrecht and Dunnett, 1971) conducted the most extensive naturalistic and experimental studies of weapon and other tool use during intimidation displays by free-ranging chimpanzees. From a towering tree platform (and presumably undetected by the subjects) Kortlandt (1962) observed adult male Zairean chimpanzees brandishing and occasionally throwing branches after cautiously entering a plantation to feed. The displays seldom seemed to be directed towards particular members of the group. Kortlandt also confronted captive and free-ranging Guinean and Zairean chimpanzees with caged large felines and a dummy leopard, respectively, to test their reactions to potential predators and to elucidate how they might have been influenced by previous natural experiences versus instinctive fear of the beasts. When sticks or other potential clubs and missiles were available the chimpanzees often snatched them up and brandished or threw them toward the cats.

Largely on the basis of 16 tests with a dummy leopard at 3 sites in Guinea (Bossou and Kanka Sili) and Zaire (Beni Chimpanzee Reserve, northern Kivu Province) and accounts in the literature, Kortlandt (1972) concluded that only savanna-dwelling chimpanzees engage in true approximations of human weapon use and that the brandishing of sticks and poorly controlled hurling of missiles by forest chimpanzees are really only extensions of their intimidation displays. Apparently his own team's records on savanna chimpanzees were limited to one test in Guinea (Kortlandt and van Zon, 1969). He erroneously included Gombe National Park as a savanna locality. A good deal more research, especially including observations on naturalistic encounters between wild leopards and chimpanzees (Gandini and Baldwin,

1978) and precise ecological profiles, will be needed before we can generalize the extent to which chimpanzees in different kinds of habitats are more of less predisposed to take up arms against errant cats.

THE QUESTION OF CULTURE

Some authors have designated regional differences in the tool behavior and feeding habits of chimpanzee populations cultural variations (Lawick-Goodall, 1973b; Warren, 1976; McGrew et al., 1979b; McGrew, 1983; Nishida et al., 1983). As we have noted, whereas the Budongo chimpanzees appear to be toolless, the Gombe chimpanzees are practiced insect probers and leaf spongers and certain western chimpanzees are nut smashers. Although the fruit pulp of oil palms (*Elaeis guineensis*) is a staple food of Gombe chimpanzees (Lawick-Goodall, 1973b) implemental nutting is unreported for eastern chimpanzees. Insect probing, albeit by a variety of methods, has been observed widely in western, as well as in eastern Africa. It is not possible to pinpoint the particular localities where each tool behavior might have been discovered or to discern how many independent discoveries of the same technique have occurred. Because of insufficient longitudinal observations on free-ranging individuals we do not know the extent to which innate manipulatory proclivities, trial-and-error learning, observational learning, social facilitation, insight, or other mechanisms are involved in the development and transmission of particular tool techniques (Birch, 1945; Schiller, 1952, 1957; Menzel et al., 1970; Lawick-Goodall, 1973a, 1973b; Warren, 1976; Beck, 1980). It is generally acknowledged that the playful manipulation of natural objects and extensive use of the hands during feeding, nesting and positional behavior facilitate the ontogenetic development of chimpanzee tool behavior. Indeed Kitahara-Frisch (1977) argued that generally even the adult chimpanzees are using objects as toys instead of as tools or weapons in manners that are comparable to human utilitarian usages. While one is tempted intuitively to dismiss wide application of this suggestion, we cannot fairly assess it until we have more precise information on the nutritional and other adaptive advantages that tool behavior might confer on the individuals that engage in it.

Lawick-Goodall (1973b; p. 144) concluded that Tylor's (1871) definition of culture—"that complex whole which includes knowledge, belief, art, morals, law, custom, and any other capabilities and habits acquired by man as a member of society"—can accommodate chimpanzee tool behavior, particularly because observational learning of group members plays a significant role in its acquisition by youngsters.

McGrew and Tutin (1978) and McGrew et al. (1979b) were more conservative about designating aspects of chimpanzee tool behavior cultural. McGrew and Tutin (1978) operationally defined culture with 8 empirically verifiable criteria: innovation, dissemination, standardization, durability, diffusion, tradition, non-subsistence, and natural adaptiveness. They expressed skepticism that chimpanzee tool-use is a manifestation of nonhuman culture because it failed to meet the last criterion in their definition. Natural adaptiveness of traits can occur only in nonhuman populations that live where direct human interference is absent and indirect human influences do not exceed levels exerted by human gatherer-hunters (McGrew and Tutin, 1978, p. 247).

Based on comparisons of termiting tools and observed and inferred tool-making and use by the chimpanzees of Mt. Assirik, Gombe, and Okorobiko, McGrew et al. (1979b) concluded that there are cross-cultural differences, particularly between the Assirik and Gombe chimpanzees. The differences between Okorobiko tool behavior and those at Gombe and Assirik are chiefly shaped by the physical environment instead of social custom and therefore do not qualify as cultural traits. Indeed only certain differences in peeling bark from sticks during tool-making, the variable employment of vines, and the use of one or both ends of termiting tools could be attributed to social custom versus environmental determinism. Like Lawick-Goodall (1973b), McGrew and his associates did not broach the sticky topic of symbolism as a sine qua non of culture.

INTERSPECIFIC LEARNING

Equally knotty is the question of wild apes having initially acquired tool behavior by unobtrusively watching people who were engaged in maintenance tasks. Captive Sumatran and Bornean orangutans in rehabilitation camps learn a remarkable variety of tool behaviors, probably to some extent by watching their custodians or other orangutans that have learned by aping them (Rijksen, 1978; Galdikas, 1978, 1980, 1982a; Beck, 1980). A number of these behaviors are similar to those of wild chimpanzees. For instance Rijksen (1978) observed Ketambe rehabilitants using sticks to prise objects; to probe termite (*Copotermes curvignatus; Natusitermes matagensis*) and ant (*Camponotus* sp.) nests, a wounded monitor lizard, and a caged clouded leopard; and to stab into a durian fruit. Ketambe rehabilitants also swatted away swarming bees with leafy twigs and brandished sticks and threw stones and vegetation at camp pets, people, and living, dead and dying reptiles. One orangutan became a proficient catcher, though no one dared to test whether he can catch and return a spear as Reade's (1864) "nchingo" was supposed to do.

Galdikas's (1982a) charges were even more versatile than Rijksen's were. One adult female folded weeds so that they extended over one side of a river, stood on them, caught weeds on the other bank, united them with the first section, and then traversed her bridge. Rehabilitants also positioned vines and logs as bridges and occasionally commandeered a raft or dugouts to return to camp or to go for a ride. In camp and at the feeding station, they commonly used sticks (the most common genre of tool) as digging, pounding, prying, probing, raking, stirring and climbing implements and to strike and poke other creatures. But in the forest, they did not regularly employ tools to exploit natural food sources (Galdikas, 1980, p. 306).

Forever free Tanjung Puting orangutans exhibited tool behavior in agonistic contexts and otherwise when they were disturbed. Arboreal individuals detached, waved and released branches when pestered by nuisances on the ground. Adult males toppled dead tree trunks as part of displays. A juvenile detached and brandished a leafy branch at wasps that flew about him and other individuals scrubbed their faces with crumpled leaves, presumably out of frustration induced by noisy people. A male broke off a bit of dead branch and scratched his perineum with it (Galdikas, 1978, 1980, 1982a,d; Galdikas-Brindamour, 1978).

Thus, it is clear that the rehabilitant orangutans acquired a number of unique implemental skills while in human society. Like Yerkes and Yerkes (1929, p. 355), Beck (1980, pp. 172-173) considered it possible that some tool behavior of apes might have been learned from humans who did not intend to instruct them. This would apply particularly to the rehabilitant orangutans in the Tanjung Puting Reserve and at Ketambe.

Galdikas (1982a) agreed that they probably learn some tool behavior by watching humans and she cited several amusing examples. However, she pointed out that other behaviors, e.g. draping objects on the head, probably reflect inherent manipulative capabilities. Further, the extent to which particular tool behaviors were learned by trial-and-error is often unclear. The rehabilitants had considerable leisure and an abundance of appropriate tool materials and artifacts to manipulate. Galdikas (1982a) also mentioned that, like chimpanzees, rehabilitant orangutans are more terrestrial and therefore have freer hands for tool behavior. Although her further statement that "most chimpanzee tool-use takes place on the ground" (Galdikas, 1982a, p. 30) is correct for the apes of Gombe, Assirik and Okorobiko, it is much less applicable to the Mahale chimpanzees.

Galdikas (1982a) noted that orangutans that had been raised by humans were more inclined to imitate human tool behavior than those with less extensive human contact were. If indeed apes imitate humans only when they are raised by them then it is unlikely that wild chimpanzees would have acquired tool behavior by spying on human gatherers, hunters and horticulturists. It is unlikely that youngsters in several areas would have acquired tool behavior while being held captive in villages and then escaped and adapted their skills to naturalistic settings. Perhaps the potential for wild chimpanzees to learn tool behavior by observing humans could be tested by a person who would learn insect probing and nut-cracking from chimpanzees in Tanzania and western Africa and then regularly ply these skills in a locality like the Budongo Forest where the chimpanzees are toolless. Adventurous students should be warned, however, that even after spending months with chimpanzee termite fishers, Teleki (1974) remained as inept as a young juvenile ape.

6

Brains and Mentality

"Does configuration of the brain as a whole, or of the cerebral hemisphere, indicate mental status, or is size of organism an important factor?" (Yerkes and Yerkes, 1929, p. 563)

" . . . the great challenge is to analyze the rapid evolution of brain size in the later hominids, using the same principles that apply to other . . . mammals." (Jerison, 1973, p. 411)

"It is performance, not size, that is important." (Anonymous lady)

"Indeed, in my present thinking there is no question about the reality of chimpanzee mind, individuality, personality." (Yerkes, 1939, p. 97)

ANATOMY

Yerkes and Yerkes (1929) relied on *The Brain from Ape to Man* (Tilney, 1928) for their brief discussion of hominoid brains. However, they disagreed with Tilney's conclusion that gorillas should be preeminent among brainy apes simply because they have the biggest organs. Instead they favored chimpanzees because of their impressive laboratory performances and the fact that their brains are heavier in proportion to body weight than gorilla brains are (Yerkes and Yerkes, 1929, pp. 476-480, 563).

Since 1929 there have been notable advances in the description of hominoid brains and anthropoid neurophysiology. Unfortunately for our knowledge of neuropsychology and pace the early experiments of Leyton and Sherrington (1917) on the motor cortex, apes have not been subjects in refined ablation or other neurophysiological studies (Passingham and Ettlinger, 1974). As compassion increasingly counterbalances curiosity, it is unlikely that the empirical chasm will be closed soon. A summary of the massive neuroscientific literature on other primates is outside the scope of this book. Instead we will focus selectively on relative brain size in the Hominoidea and several other features that have figured in paleoanthropological discussions.

Brain Size and Sexual Dimorphism

The mean absolute size of brains increase from common gibbons through siamang, bonobos, common chimpanzees and orangutans to gorillas (Table 2). Among adult Hylobatidae, cranial capacities range from 70 cc to 152 cc (Schultz, 1933). Kloss gibbons may be the smallest brained apes (Tobias, 1971) though the small sample size and wide range of values make this inference suspect. The most reliable population samples are provided by Schultz (1944, 1965), who found that mean cranial capacities of male and female *Hylobates lar carpenteri* are 104 cc and 101 cc, respectively. Siamang have mean cranial capacities that are about a fourth larger than those of other hylobatid apes (Tobias, 1971). Gibbons and spider monkeys (*Ateles geoffroyi*) of the same body weight have very similar cranial capacities (Schultz, 1941).

As in body weight, there is low sexual dimorphism of brain size in the Hylobatidae. The average cranial capacity of female *Hylobates lar carpenteri* is 97.1% of the male mean. In *Hylobates syndactylus* the mean cranial capacity is 97.6% of the male value (Table 2).

Pongid cranial capacities are between 3.5 and 5 times larger than those of hylobatid apes. With the notable exception of *Pan paniscus*, there is greater sexual dimorphism of brain size in the Pongidae than in the Hylobatidae. Cramer (1977) showed that *Pan paniscus* is probably the least dimorphic ape in regard to cranial capacity. The female mean is a whopping 98.9% of the male mean (Table 2) and the *t*-test revealed no significant difference between them (Cramer, 1977).

On average, the cranial capacities of female *Pan troglodytes* are only about 25 cc greater than those of *Pan paniscus* and the mean value for male *Pan troglodytes* is approximately 55 cc larger than mean values for *Pan paniscus* (Cramer, 1977; Table 2). *Pan troglodytes* is moderately sexually dimorphic in cranical capacity; female means are 92.4% and 92.8% of the male means in the samples of Schultz (1969) and Cramer (1977), respectively.

Gorillas and orangutans exhibit considerably greater sexual dimorphism in brain size than chimpanzees do. Even infant and juvenile males have larger brains than their female peers (Schultz, 1965). Still the extent of endocranial dimorphism does not match the degree of dimorphism in the body weights of orangutans and gorillas. In Schultz's (1965) hefty samples the amount of endocranial sexual dimorphism is closely similar in the biggest Asian and African apes (Table 2). The female mean is 81.2% of the male value in *Pongo pygmaeus* and 82.8% in *Pan gorilla*.

Table 2: A Sampler of Adult Hominoid Cranial Capacities
(and Brain Weights*)

Species		No.	Mean (cc or gm*)	Range (cc or gm*)	Source	Dimorphism (F x 100/M)
Hylobates klossii		10	87	78-103	Schultz, 1933	–
Hylobates lar carpenteri	M	95	104	89-125	Schultz, 1965	97.1%
	F	86	101	82-116		
Hylobates concolor		69	101	82-136	Schultz, 1933 Tobias, 1971	–
Hylobates syndactylus	M	23	126	100-150	Tobias, 1971	97.6
	F	17	123	105-152		
Pan paniscus	M	6	356	334-381	Schultz, 1969	92.4
	F	5	329	275-358		
	M	29	352	140	Cramer, 1977	98.9
	F	30	348	160		
Pan troglodytes	M	57	381	292-454	Schultz, 1969	92.4
	F	59	352	282-418		
	M	33	404	140	Cramer, 1977	92.8
	F	34	375	260		
Pongo pygmaeus	M	57	416	334-502	Schultz, 1965	81.2
	F	52	338	276-425		
Pan gorilla	M	72	535	412-752	Schultz, 1965	82.8
	F	43	443	350--523		
Homo sapiens Swiss	M	70	1,463	1,250-1,685	Schultz, 1965	91.7
	F	40	1,314	1,215-1,510		
Danes* 19-25 yrs.	M	724	1,440	–	Pakkenberg &	89.0
	F	302	1,282	–	Voigt, 1964	
Hungarians* 20-40 yrs.	M	80	1,386	1,050-1,670	Tóth, 1965	90.4
	F	120	1,253	900-1,750		
40-60 yrs.	M	405	1,375	1,000-1,900		90.2
	F	383	1,241	800-1,775		
60-80 yrs.	M	571	1,335	1,000-1,700		90.5
	F	462	1,208	900-1,700		
80-100 yrs.	M	54	1,254	1,000-1,500		94.0
	F	57	1,179	1,000-1,400		

Whereas on average the cranial capacities of female orangutans are smaller than those of chimpanzees, those of male orangutans are larger than the chimpanzee male values. Average female gorillas have cranial capacities that are larger than those of average male chimpanzees and orangutans. Few adult male gorillas have brains that are as small as those of chimpanzees and orangutans (Table 2). The largest pongid cranial capacity (752 cc) was recorded from an Equatorial Guinean male gorilla (Schultz, 1962b). This is most unusual because no other pongid specimen has provided a value close to 700 cc (Tobias, 1971). Hydrocephalus occurred in a captive infant gorilla at Yerkes Regional Primate Research Center. The condition was corrected surgically so we do not know how capacious his cranium could have become.

No extant ape can rival *Homo sapiens* in brain size. Average human cranial capacities are roughly 10-17 times larger than the average values of the Hylobatidae and 2.4-4.2 times larger than average values of the Pongidae (based on data in Table 2). In humans, brain weight correlates significantly with stature, but not on body weight (Pakkenberg and Voigt, 1964).

We do not know whether the brain occupies similar percentages of space in human, pongid and hylobatid skulls. Because slucal impressions are more common on the endocranial vault in the hylobatid apes than in larger hominoids (Hirschler, 1942; Radinsky, 1972), it is possible that hylobatid crania are more closely packed by the brain than pongid and human crania are.

The brain sizes of *Homo sapiens* are sexually dimorphic to about the same degree as those of *Pan troglodytes* (Table 2). Interestingly, Tóth (1965) discovered that during the eighth and ninth decades human brains are less dimorphic than at previous times, i.e. although the brains of both sexes shrink during the golden and leaden years, those of men seem to contract more than those of women. Still, ancient humans do not approach the unisex condition of prime bonobos and hylobatid apes.

As Yerkes and Yerkes (1929) suspected, brain size is to a large extent determined by overall body size in primates (Stephan, 1972; Passingham, 1982; Lynch et al., 1983). Apes follow the general primate pattern. But humans have brains that are three times larger than one would expect for primates of their size. This abundance is especially pronounced in the cerebellum and cerebral neocortex. These areas are associated with the motor coordination of fine movements and the integration of information from the several senses (Passingham, 1973, 1975b, 1982; Jolicoeur et al., 1984).

Like Tilney (1928), when Stephan (1972) and Jerison (1973) compared brain size to body size, gorillas and orangutans fell below many monkeys. However, Passingham (1975a) reinstated them alongside chimpanzees on the nonhominid summit by relating brain size to size of the medulla. The medulla is a reasonable indirect indicator of lower level activities in the brain vis-à-vis routine bodily functions (Passingham, 1978). In primates, big brains in relation to the medulla indicate well developed cognitive capacities over and above mere receipt of sensory inputs and the production of motor outputs (Passingham, 1982).

Cells and Functions

But is there anything special about the composition of large hominid brains? Do bigger cortices have proportionately more nerve cells and connections among them? These questions have been lucidly addressed in several thoroughgoing reviews (Holloway, 1968; Passingham, 1973, 1975b, 1981a, 1982; Passingham and Ettlinger, 1974) albeit without definitive results.

Shariff's (1953) pioneer empirical study of single cerebral hemispheres from a human, a chimpanzee, a guenon, a marmoset, and a tarsier demonstrated that increased cerebral volume is accompanied by decreased density of nerve cells in the neocortical grey matter. Further, the cells increase in size as cortical volume increases. From Shariff's half-brained and other studies Holloway (1968; 1979) hypothesized that during the evolutionary emergence from apehood, hominid brains had undergone reorganization. He emphasized the proliferation of dendritic processes which would facilitate interconnections among the cerebral nerve cells. Current histological techniques do not permit the detection of dendritic structure with the resolution that can be achieved for the nerve cells and their axons. Thus, we do not know whether humans have more finely wired cerebra than apes do or how the apes compare structurally with one another and with monkeys in this feature. Needless to say, an expanded and modernized study that would build on Shariff's (1953) singular work is imperative.

Observations that apes can versatilely employ a variety of objects as tools (Chapter 5) and communicate via artifactual symbols according to simple human rules (Chapter 7) have sparked questions about possible high-level neurological links between them and us. That is to say, are there specific areas of ape central nervous systems (CNS) that underpin these capabilities and are they homologous with areas in the human CNS which are thought to make speech and other symbolically mediated behaviors possible? We are greatly hampered in the exploration of these questions by the fact that, to some extent, the "ape condition" must be interpolated from experiments on monkeys and clinical observations on humans. As mentioned earlier, ape brains are only crudely charted by direct studies. Further, the functional significance of human cerebral asymmetries vis-à-vis object manipulation and language is still a question for refinement and debate.

Cerebral Asymmetry

Human brains are notable for hemispheric dominance as reflected in the localization of linguistic functions, usually on the left side (Figure 16), and

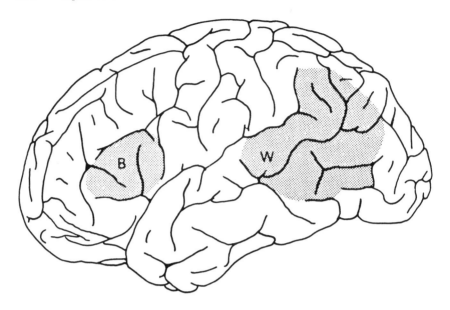

Figure 16: Speech areas of the human brain. Stippled area B depicts Broca's area. Stippled area W depicts Wernicke's area. (Courtesy of R. Passingham.)

handedness (Penfield and Roberts, 1959; Darley, 1967; Geschwind, 1971; Levy, 1982, 1983; Levy et al., 1983; Levy and Reid, 1978; Heller and Levy, 1981; Bradshaw and Nettleton, 1983). But the relationship between laterality of hand preference and hemispheric dominance for speech is not clear-cut. Exceptions occur: for instance, left-handers who are left brain dominant instead of right brain dominant for speech (Rossi and Rosadini, 1967; Levy, 1980; McGlone, 1980). Further, LeMay and Culebras (1972) reported that a right-hander may have had right hemispheric dominance for speech.

Von Bonin (1962) concluded that the gross structural asymmetries of human cerebra are not substantial enough to account for their dramatic functional differences. If neuroscientists were to accept this pronouncement, and many did (Ettlinger, 1984), it would be trivial to search for anatomical asymmetries in anthropoid brains and fossil endocasts that might represent precursors for human structures and abilities.

Geschwind and Levitsky (1968) gave new impetus to comparative anthropoid anatomy by documenting the existence of marked asymmetries of the planum temporale in adult human brains. Subsequently, Wada et al. (1975) confirmed their results and discovered comparable asymmetry of the planum temporale in infant human brains as well as an asymmetry of the frontal operculum. The former, containing Wernicke's area, is related to understanding speech and the latter, including Broca's area, to its articulation (Geschwind, 1972; Passingham, 1981b). Gur et al. (1980) showed experimentally that most of 36 right-handed young men had more grey matter in the left versus right cerebrum, especially in frontal areas involved with the production of speech and abstract reasoning. Their noninvasive method of xenon-133 inhalation might be used profitably on apes.

Despite negative reports from several of Yerkes's contemporaries (Economo and Horn, 1930; Pfeifer, 1936), LeMay (1976) and her associates (LeMay and Geschwind, 1975; LeMay et al., 1982; Galaburda et al., 1978) sought structural asymmetries in the cerebra of nonhuman anthropoid primates. They found them, particularly in great apes.

Among a dozen orangutans, 9 chimpanzees and 7 gorillas, 17 brains (61%) showed asymmetries in the positions of the Sylvian fissures and points, which bound the temporal lobe superiorly. The Sylvian asymmetry was most marked in the brains of *Pongo pygmaeus* and least conspicuous in specimens of *Pan gorilla* (LeMay and Geschwind, 1975). As in *Homo sapiens*, the right Sylvian point was higher than the left in most Pongidae that exhibited the asymmetry. Only 9 out of 41 monkeys and lesser apes (22%) evinced notable Sylvian asymmetry after their brains had been enlarged to the size of an orangutan's brain. And in one-third of them the left Sylvian point was higher than the right one (LeMay and Geschwind, 1975).

Yeni-Komshian and Benson (1976) measured the length of Sylvian fissures in brains from 25 each of *Homo sapiens, Pan troglodytes* and *Macaca mulatta*. The left fissure was significantly longer than the right one in humans and chimpanzees but not in the rhesus monkeys. The mean intercerebral difference was greater in humans (10.2 mm) than in chimpanzees (2.0 mm). Falk (1978) found statistically significant Sylvian asymmetry in 4 colobine genera but not in 4 cercopithecine genera, including *Macaca*. The Sylvian fissure was longer on the right side in cercopithecine and colobine specimens. Falk (1980) concluded that this was caused by expansion of the left parietal cortex over the caudal extremity of the left superior sulcus and Sylvian fissure, thereby foreshortening them. This must be confirmed on actual brains.

LeMay et al. (1982) documented that, like humans, great apes exhibit a tendency for the left hemisphere to extend farther posteriorly and for the left occipital lobe to be wider than the right one. Further, the right frontal lobe tends to be wider and the right hemisphere to project farther anteriorly than the left one in Pongidae and *Homo sapiens*. The asymmetries in length of the frontal and occipital lobes are reflected externally by the frontal and occipital bones of the skull (LeMay et al., 1982). This is contrary to the view (Coolidge, 1929; Groves and Humphrey, 1973) that cranial asymmetry in *Pan gorilla beringei* might be ascribed to left-sided chewing.

Whereas the occipital and frontal asymmetries can be related to handedness in humans, they cannot be so linked in great apes because there is insufficient evidence that they possess comparable degrees of laterality for hand preference (Finch, 1941a; Warren, 1980; Brésard and Bresson, 1983). The observations of Fischer et al. (1982) on 4 adult female *Pan gorilla gorilla* suggest a strong dextral preference in a species of the Pongidae. They used the right hand to retrieve food rewards 96% of the time and, while tripedal, 3 mothers supported their infants with the left forelimb 64% of the time. Lockard (1984) also noted hand preferences in 6 out of 8 captive *Pan gorilla gorilla*. But before we could conclude that western gorillas exhibit a division of labor between manipulatory and helping hands more subjects must be tested.

As most people have heard, apes lack speech. Nevertheless, the asymmetries of their temporal regions might have something to say about precursors of this high-flown human attribute. Research on cercopithecine monkeys indicate that pongid studies could be telling (Steklis, 1985).

In *Macaca fascicularis*, Dewson (1978) produced a specific auditory cognitive deficit with lesions of the left temporal lobe. Because this is in an area comparable to Wernicke's area of the human brain and because in both species it serves as an auditory association area it seemed reasonable to infer that it represents an evolutionary precursor and homologue of the human structure (Dewson, 1978; Denenberg, 1981). After surgery, the conditioned monkeys retained the ability to discriminate between a pure tone and a burst of noise and visually between numerals 4 and 6. But they had lost the ability to distinguish between the human utterances /a/ and /u/.

Peterson et al. (1978, 1984; Marler, 1983) used naturalisitic vocalizations of Japanese macaques (*Macaca fuscata*) to test for laterality in the receipt of salient communicative features of their vocal signals. All 5 of the Japanese macaque subjects exhibited right-ear advantage in discriminating between 2 classes of coo sounds. One vervet (*Cercopithecus aethiops*) also did so; but the other macaques (*M. nemestrina* and *M. radiata*) failed to show right-ear advantage (Peterson et al., 1978). If, as it seems reasonable to infer, the discernment of specific communicative features is located in the left temporal region of the Japanese macaque brain, then we should not be surprised to find that laterality for comprehension of communicative utterances is a basic catarrhine feature. Accordingly, it should not be considered unique to humans. Now apes and more species of monkeys (Pohl, 1983) must be tested if we are to move to higher levels of speculation on the anthropoid substrates of human vocal capabilities.

Nonhuman counterparts of Broca's area have been more elusive than avatars of Wernicke's area. Indeed Passingham (1981b; 1982) concluded that none have been established in monkeys and apes. Ablation of the area of frontal cortex that might correspond with Broca's area did not affect a chimpanzee's capacity to vocalize spontaneously (Leyton and Sherrington, 1917).

Pyramidal Tract and Dexterity

According to Passingham (1981a), the number of fibers in the pyramidal tract, which conducts motor impulses from the neocortex to nerve cells in the spinal cord that control muscles, reflects the capacity for complex movements of the limbs, and especially of their distal extremities. The number of fibers is greater in mammals that are manually dextrous than in clawed and hoofed species. It and the motor cortex (Goodwin and Darian-Smith, 1985) are particularly important for independent control of the five fingers. Pinnipeds are a puzzling exception because they have as many corticospinal fibers as monkeys do (Passingham, 1981a).

Verhaart (1970) presented a concise comparative anatomy of the pyramidal tract in humans, chimpanzee, orangutan, common gibbon, siamang and several monkeys and prosimians. He concluded that given that greater fiber size and caudal extent of the pyramidal tract in the spinal cord indicate neurological advancement, the hominoids rank above monkeys (and pinnipeds) which, in turn, are more developed than the prosimians.

Heffner and Masterson (1975) demonstrated that over a broad sample of mammals corticospinal fiber size did not correlate significantly with estimates of dexterity. They showed that large body size can muddle this and other pyramidal characteristics that could underpin manual skill. Like Verhaart (1970), they concluded that refined dexterity occurs only in mammals in which the pyramidal tract extends beyond the lumbar spine and synapses more or less directly with the efferent (motor) neurons. None of these studies presented distinctions between apes and humans in the pyramidal tract. Studies on the relative dexterity of apes, monkeys and humans are crude and incomplete (Napier, 1960, 1961; Bishop, 1964; Parker, 1969; Tuttle, 1969b, 1970, 1972). We need more systematic behavioral tests, particularly those that would quantify relative independent control of the five fingers. Only then can we challenge anthropoid primates with psychological test apparatus that will limit confusion between dexterity and intelligence.

BEHAVIOR

Prime objectives of Yerkes's research were to establish the relative mentality of apes and to find affinities with human wit. Yerkes and Yerkes (1929, p. 80) expressed aversion to overuse of the word intelligence in *The Great Apes* because "it is more profitable to deal specifically with the facts of behavior and experience than to hide our ignorance in general terms." Accordingly, in the synoptic comparative table they placed the heading intelligence between quotation marks and then listed 17 characters for which gibbon, siamang, orangutan, chimpanzee and gorilla were ranked hierarchically (Yerkes and Yerkes, 1929, pp. 560-561). Their data base was spotty. For certain apes, particularly the hylobatids, they listed no information beside some of the characters. Quantification and tests for statistical significance were nil. Tentatively, they concluded that gorillas and chimpanzees were highest overall in psychological resemblance to humans, with a slight edge for the former. Although chimpanzees were near to us affectively, gorillas might be more akin cognitively. The red apes were embarrassed with third place and the hylobatids hung notable distances below them.

The Yerkeses were most mindful of the inadequate information that was used to compare the overall intelligence or, as they preferred, behavioral adaptivity of the apes. The Hylobatidae were represented by a single female *Hylobates* (*Nomascus*) *concolor leucogenys* that had been challenged with several box opening tasks. Interestingly, the experimenter (Boutan, 1914) had reported "learning with voluntary attention" by his subject. Yerkes and Yerkes (1929, p. 96) noted that this process is like the "insightful solution of problems" by Köhler's chimpanzees and what they had independently described as "ideational behavior" by an orangutan.

In addition to insight, they were keen to discover the relative abilities of apes in mechanical skill, memory, and foresight. Only common chimpanzees, which Yerkes, Köhler and Kots had challenged with a variety of problems, were sampled in numbers. Julius and Congo were the sole experimental subjects from which they had to assess the mentality of orangutans and mountain gorillas, respectively. Their fumbling attempts to acquire food incentives with tools led to low marks for their species (Yerkes, 1916, 1927a,b; 1928).

Among the notable conclusions from the early experiments on chimpanzees were that they (1) had better memory for the physical locations of objects than for colors, (2) generally lacked cooperative planning, and (3) rarely exhibited and possessed only rudimentary abilities to abstract, generalize and infer from test experiences (Yerkes and Yerkes, 1929, pp. 368, 374-375). Nevertheless, they speculated that "Indeed were it capable of speech and amenable to domestication, this remarkable primate might quickly come into competition with low-grade manual labor in human industry." (Yerkes and Yerkes, 1929, p. 376). This reflects more an unsupported snobbish opinion on the mentality of blue-collar workers and booboisie than firmly established knowledge of chimpanzees.

During the next three decades, hominoid psychologists focussed on chimpanzees largely because the pioneering studies of Kots, Köhler and Yerkes had made them appear to be the most promising exemplar of an ape mind that could be compared with our own and those of other vertebrates. However, in founding his chimpanzee colony, Yerkes (1943, pp. 289-301) was not as much interested in the natural chimpanzees as to determine how its behavior might be altered, thereby serving as a prototype for human biological engineering. Toward this Orwellian and purer scientific ends, considerable effort has been devoted to designing apparatus and research protocols to test their perceptual (especially visual), discrimination learning, mathematical, problem-solving, and mnemonic skills (Birch, 1945a,b; Braggio et al., 1982; Brown et al., 1978; Buchanan et al., 1981; Carpenter and Nissen, 1934; Clark, 1961; Cowles, 1937; Cowles and Nissen, 1937; Crawford, 1937, 1941; Crawford and Spence, 1939; Davenport and Menzel, 1960; Davenport and Rogers, 1968; Davenport et al., 1969; Döhl, 1970, 1973; Dooley and Gill, 1977a,b; Farrer, 1967; Ferster, 1958a,b, 1964; Ferster and Hammer, 1966; Finch, 1941b, 1942; Forster, 1935; Garcha and Ettlinger, 1979; Gardner and Nissen, 1948; Gellerman, 1933a,b; Gillan, 1981; Gillan et al., 1981; Gonzales et al., 1954; Grether, 1940a-d, 1941, 1942; Grilly, 1975; Hall et al., 1982; Harrison and Nissen, 1941a,b; Haselrud, 1938; Hayes and Hayes, 1953; Hayes and Nissen, 1971; Hayes et al., 1953a,b; Hayes and Thompson, 1953; Jarvik, 1953, 1956; Jenkins, 1943; Kelleher, 1957a,b, 1958, 1965; McClure and Helland, 1979; McColloch and Nissen, 1937; McDowell and Nissen, 1959; Menzel, 1964, 1969, 1971a, 1973a; Menzel and Davenport, 1961; Menzel et al., 1961, 1972; Meyer et al., 1965; Nissen, 1942, 1951a,b; Nissen et al., 1936, 1938, 1948, 1949; Nissen and Elder, 1935; Nissen and Harrison, 1941; Nissen and Jenkins, 1943; Nissen and McColloch, 1947a,b; Nissen and Taylor, 1939; Premack, 1975, 1976a, 1983; Premack and Premack, 1983; Premack and Woodruff, 1978a,b; Premack et al., 1978; Prestrude, 1970; Razran, 1961; Rensch and Döhl, 1968; Reynolds and Farrer, 1969; Riesen, 1940; Riesen and Nissen, 1942; Robinson, 1955, 1960; Rogers and Davenport, 1971; Rohles, 1961; Rohles and Devine, 1966, 1967; Schiller, 1952; Schusterman, 1963, 1964; Smith et al., 1975; Spence, 1937, 1938, 1939, 1941, 1942; Spragg, 1936; Strong, 1967; Strong and Hedges, 1966; Thompson, 1954; Tinklepaugh, 1932; Vauclair et al., 1983; Voronin, 1962; Warren, 1965; Welker, 1956a-c; Wilson and Grunzke, 1969; Wood et al., 1980; Woodruff and Premack, 1981; Woodruff et al., 1978; Wolfe, 1936; Yerkes, 1934, 1939; Yerkes and Yerkes, 1928).

Chimpanzees have proven to be particularly capable performers on learning sets, albeit after rather extensive discrimination-reversal training. Often on the first trial, they grasp all the information that is necessary for problem solution and succeed with a "win-stay; lose-shift" strategy (Schusterman, 1962; Miles, 1965, p. 77; Schrier, 1984). Older subjects generally do not respond more rigidly in discrimination problems than youngsters do (Bernstein, 1961a,b; Riopelle, 1963; Riopelle and Rogers, 1963, 1965). However, learning set skills and performance on oddity discrimination tasks may diminish with age (Rumbaugh, 1971, 1974; Riopelle and Rogers, 1965).

A major practical triumph of operant conditioning methods (Skinner, 1938) was achieved when Enos, a right stuff chimpanzee, masterfully performed his duties in a Mercury space capsule as it orbited Earth (Belleville et al., 1963; Rohles, 1966, 1970; Rohles et al., 1961, 1963). Withal, unlike his high-flying human successor, this was not sufficient to launch his political career.

Concurrent with intensified laboratory research on chimpanzees, several comparative psychologists tested the effects of human environments on chimpanzees by raising them as bourgeois children in their homes. The most famous cases are Gua, a female that spent 9 months with the Kellogg family (1933, 1968, 1969), and Viki, a female that lived for 6.5 years with the Hayeses (1951, 1952, 1953, 1954; Hayes, 1951; Hayes and Nissen, 1971; Hayes et al., 1953a,b). These efforts generated considerable lay interest but they were not very productive scientifically.

The Kelloggs (1933) detailed the physical, motor, sensory and behavioral development of Gua (7.5-16.5 mo) in comparison with that of their son Donald (9-19 mo). Because Donald aped the ape more than the ape aped him Gua had to go. She not only failed to develop speech but also might have delayed Donald's linguistic development. Viki's oral achievements were little better than Gua's (Chapter 7). And she too remained every inch an ape despite frequent exposure to human playmates, travel, drive-in movies, parties and humdrum household routines (Hayes, 1951). After reading these accounts one more fully appreciates human children on their bad days.

Viki seemed to perceive some photographs and line drawings accurately (Hayes, 1951; Hayes and Hayes, 1953). Two decades later Davenport and Rogers (1971) concluded from systematic experiments that photographically naive chimpanzees and orangutans can proficiently match objects with color and black-and-white photographs of them and also perform fairly well with silhouettes and line drawings. This cast doubt on the conclusion of Segall et al. (1966) that persons in traditional societies cannot initially perceive photographs as representations of actual objects. Further studies on humans and apes are needed; Winner and Ettlinger's (1979) chimpanzees responded to photographs as though they were meaningless 2-dimensional objects. However, Malone et al. (1980) found that rhesus monkeys could cross-modally match objects with photographs thereof, though not at first sight.

The chief function of the comparative method is to establish the generality of phenomena (Scott, 1973). Especially prior to the late 1960s comparative psychologists had not tested enough anthropoid species (or indeed vertebrates) and devised a sufficient battery of methods to allow definitive scientific statements about how the chimpanzee ranked mentally. This is manifest in the incisive reviews of primate learning by Harlow (1951), Warren (1965, 1973, 1974), Rumbaugh (1970, 1971a), Riopelle and Hill (1973), Fobes and King (1982) and King and Fobes (1982).

Harlow (1951, p. 234) lamented that the barest beginning had been made in the psychological study of the nonhuman primates. Only the chimpanzee and the rhesus monkey had been subjects of detailed investigations and these were not conducted under comparable conditions. Nevertheless, he concluded that "In both variety and complexity, the problems that fall within the repertory of the primate contrast sharply with those testable on rodent or carnivore."

Warren (1965, p. 274) tried to temper this primate supremacist view. He concluded that "No one can yet provide a definitive answer to the question 'In what ways does learning in primates differ from learning in other vertebrates?'." At best, he could list a few ways in which primate learning does not differ qualitatively from that of other vertebrates. He conceded that chimpanzees are quantitatively superior to selected nonprimate mammals in the speed with which they form learning sets, generalize and transfer learning sets, and solve exceptionally difficult discrimination problems. But he immediately followed with a caveat that additional experiments might level the apparent distinctions between catarrhine primates and other mammals.

In 1974, Warren was more certain that in comparison with other mammals catarrhine primates possess quantitatively unique learning capabilities. This conclusion was supported by studies on intertask transfers of knowledge, which indicate that the subjects had learned to generalize strategies on the basis of prior test experiences.

The Transfer Index

Rumbaugh (1969, 1970, 1971a, 1974, 1975) and associates (Rumbaugh and Gill, 1973; Rumbaugh et al., 1973; Gill and Rumbaugh, 1974; Essock and Rumbaugh, 1978; Rumbaugh and Pate, 1984a,b; Davenport et al., 1973; Wilkerson and Rumbaugh, 1979) provided a major advance in the study of comparative primate learning and intelligence by developing the transfer index and applying it to statistically significant samples of apes, monkeys and lemurs. Rumbaugh's culminant application of the transfer index to apes was facilitated by the addition of a colony of gibbons and 15 orangutans and 15 gorillas to the Yerkes Regional Primate Research Center in the early 1960s (Bourne, 1971, p. 124).

Apart from Rumbaugh's ambitious project, only a smattering of psychological tests have been conducted on apes other than common chimpanzees. The siamang remains sadly neglected by comparative psychologists. In some experiments, there were too few subjects for statistical tests on the significance of intertaxonal comparisons. Many of the subjects were housed in zoological parks (Abordo, 1976; Beck, 1967; Berkson, 1962; Bernstein, 1961a,b; Davis and Markowitz, 1978; Döhl, 1972, 1975; Döhl and Podolczak, 1973; Essock-Vitale, 1978; Fischer, 1962; Fischer and Kitchener, 1965; Gossette, 1973; Jordan and Jordan, 1977; King, 1973; Kintz et al., 1969; Parker, 1974, 1978; Patterson and Tzeng, 1979; Redshaw, 1978; Riesen et al., 1953; Robbins and Bush, 1973; Robbins et al., 1978; Rumbaugh, 1968, 1971b; Rumbaugh and McCormick, 1967, 1969; Rumbaugh and Rice, 1962; Rumbaugh et al., 1972; Rumbaugh and Steinmetz, 1971; Schusterman and Bernstein, 1962; Snyder et al., 1978; Wilson and Danco, 1976).

Rumbaugh (1969) devised the transfer index in order to compare the complex learning skills of primates while minimizing perturbations from intertaxonal (and ontogenetic) differences in motivational, perceptual and motor attributes of the subjects. Transfer indices measure the capacity for reversal after holding constant the amount learned during the pre-reversal acquisition period in a series of two-choice discrimination problems. It is computed by dividing percentage correct in the second through tenth post-reversal trials (R) by the percentage correct during acquisition training (A). Generally, values of 67% and 84% are used for A.

Pan gorilla, Pan troglodytes, and *Pongo pygmaeus* performed significantly better than *Hylobates lar, Cercopithecus aethiops, Miopithecus talapoin* and *Lemur catta* on both the 67% and 84% schedules (Figure 17). *Pan gorilla* and *Pongo pygmaeus* were significantly superior to *Pan troglodytes* when tested with the 67% pre-reversal schedule. There was no statistically significant difference among the three great apes when tested with the 84% schedule (Rumbaugh, 1974, 1975; Rumbaugh and Gill, 1973). Thus Rumbaugh (1971a) was probably quite right to challenge the myth of mental supremacy by chimpanzees among apes. From a broader ethological perspective on vertebrate cognitive skills, Beck (1982) seconded the call for an end to chimpocentrism in comparative studies on intelligence.

Rumbaugh et al. (1973) used the transfer index in tests on the extent to which great apes attend to cues in the foreground of their visual fields. They found that the performances of orangutans were more disrupted by irrelevant foreground cues than those of African apes were. Further, chimpanzees were more distracted than the gorillas were. Thus, the tendency to attend to cues in the immediate foreground is most pronounced in arboreal apes and least evidenced in the most terrestrial apes. Rumbaugh (1974) speculated that attendance to objects in the foreground is adaptive to protect the eyes from injury by profuse projections in the canopy. This, of course, needs to be tested systematically as nettles, brambles, sharp grass blades and other hazards dot the eastern gorilla's landscape. We should recall too that adult African apes generally have more prominent brow ridges than orangutans do.

The transfer index has been employed strategically to measure persistent cognitive deficits caused by extreme early environmental privation in chimpanzees (Davenport et al., 1973) and to assess whether relatively bright and dull gorillas learn the same task by different processes (Gill and Rumbaugh, 1973). The latter study revealed only differences in competence. The dullards and the wizzes employed the same processes to earn their marks.

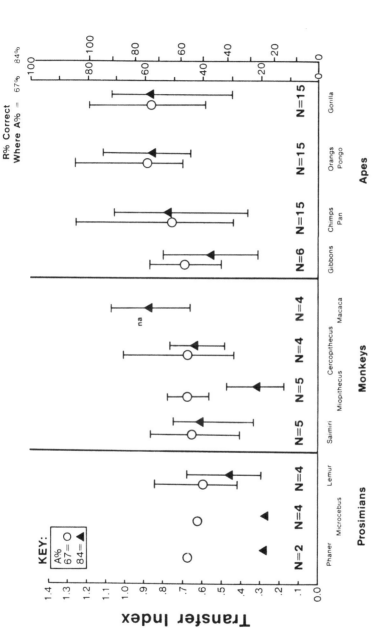

Figure 17: Transfer index values, means and ranges, for the 67% and 84% criterion levels of learning. Responses correct on the reversal trials for each of the two levels can be accessed through use of the axes at the right of the figure. (From Rumbaugh and Pate, 1984b; Courtesy of D.M. Rumbaugh.)

Abstraction

Rumbaugh (1971b) ingeniously used a variety of discrimination-reversal tasks to test whether there are qualitative differences in the learning processes among three anthropoid species. He concluded that whereas talapoin monkeys learn primarily through stimulus-response processes, the learning of gorillas can be abstractive. The gibbons performed better than the talapoins but less well than the gorillas. On reversal trials, the gibbons were so attracted to novel objects that the test failed to reveal the nature of their learning process. However, their performance provided little evidence that abstractive learning was involved.

Patterson and Tzeng (1979) retested subjects from Rumbaugh's (1971b) study and concurred that gorillas learn by abstractive strategies. Further, they demonstrated that gorillas can retain the abstract concept for at least 2.5 years. This long-term memory is not contingent upon the specific stimuli with which the subjects were challenged.

Like gorillas, orangutans promise to adapt well in a wide variety of learning situations as shown by their performance on a complex discrimination-reversal task in which they strategically employed a win-stay; lose-shift hypothesis. Macaques (*Macaca mulatta* and *M. arctoides*) performed well on similar tests though probably by somewhat different processes (Essock-Vitale, 1978).

Chimpanzees have been subjects of the most extensive studies that reveal abstract reasoning and novel problem-solving skills by apes. David Premack's group at the University of Pennsylvania Primate Facility have been especially creative and thorough in documenting the cleverness of *Pan troglodytes*.

Sarah, a 16-year-old star manipulator of plastic symbols, completed analogies by choosing the correct alternative (B') in forced-choice problems with the design, A:A' same B:? And she could correctly note analogies or absences thereof by choosing symbols for same or different in the problems A:A' ? B:B' and A:A' ? B:C, respectively. The tester left the room during each trial and returned to score Sarah when she rang a bell to signal that she had made a choice (Gillan et al., 1981). Sarah's stunning success was facilitated somehow by extensive language training (Chapter 7). Chimpanzees that had not been trained to communicate with plastic symbols could not pass even the simplest tests for analogical reasoning (Premack and Premack, 1983, p. 128).

Language-ignorant juvenile chimpanzees showed that they can draw transitive inferences about relative amounts of food. For example, they reckoned that $D > B$ in a transitive series where $E > D > C > B > A$. A double-blind procedure rules out the possibility that accurate performance on the tests was due to inadvertent cues from the experimenter (Gillan, 1981). The mechanisms whereby chimpanzees make transitive inferences about amounts of food and draw analogies are obscure. However, Gillan (1981) concluded that the two types of processes are so different that it is probably not advisable for psychologists to lump them together under the general term reasoning.

Super Sarah successfully assembled a veridical face from the picture of a blank-faced chimpanzee and cutouts of its eyes, nose and mouth (Premack and Premack, 1975). Seven other chimpanzees botched the visage (Premack and Premack, 1983). After delightedly trying on hats in front of a mirror, Sarah transformed the face by placing the mouthpiece upside down and bottomside up on its head. Sometimes she kept the face veridical and placed fruit peels or candy wrappers on the head. Sarah's face making was probably predicated on her memories of chimpanzee faces and the hat fitting sessions; that is to say, she used imaginal instead of abstract representation (Premack, 1983). Because chimpanzees have failed consistently to draw, paint or sculpt representational figures, Premack and Premack (1983, p. 108) concluded that they are characteristically unable to analyze complex objects into their parts and to understand the relations among them.

Premack and Woodruff (1978a) tested Sarah's problem-solving comprehension by showing her videotapes of people in predicaments, stopping the tape short of solution, and proffering her photographs of correct and incorrect means to resolve the situations (Figure 18). Although Sarah was quite familiar with television as a pastime, this was the first time that it was used in experimental sessions with her. She chose photographic solutions in solitude so there was no chance for human cues.

Sarah quickly and consistently solved 7 of the 8 problems. It is unlikely that she had encountered some of them previously, though she was acquainted with the objects (and functions thereof) that appeared on the tapes. For her, the most problematic problem was to predict removing blocks from an otherwise portable box in order to gain access to a reward. But she was probably strong enough herself to drag the loaded box out of the way. These studies show that a chimpanzee can recognize the nature of certain problems and infer solutions for them, i.e. it can deal with them abstractly.

Intentionality

Premack and Woodruff (1978b; Premack and Premack, 1983) expanded their video games with Sarah in order to discern whether she could attribute mental states to actors on tapes. That is to say, does Sarah have a theory of mind? They decided that she did. In addition to choosing constructive solutions to problems for a trainer to whom she had shown affection, Sarah elected predominantly hurtful or otherwise negative outcomes for a person whom the experimenters believed that she had disliked. And Sarah was no wimp. In youth she defaced a trainer and thenceforth had to show her mettle in a cage (Premack, 1976a).

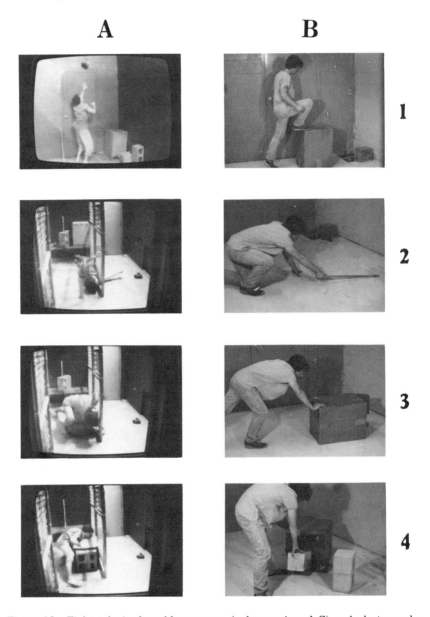

Figure 18: Eight televised problem scenes (columns A and C) and photographs of solutions to them (columns B and D), which were used by Premack and Woodruff (1978a) to demonstrate that a chimpanzee can infer the nature of problems and recognize potential solutions to them. A person (1) tries to reach suspended bananas; (2) falls short of reaching bananas behind a vertical barrier; (3) tries to reach around a box for bananas; (4) tries to push aside a loaded box in order to gain access to bananas;   (continued)

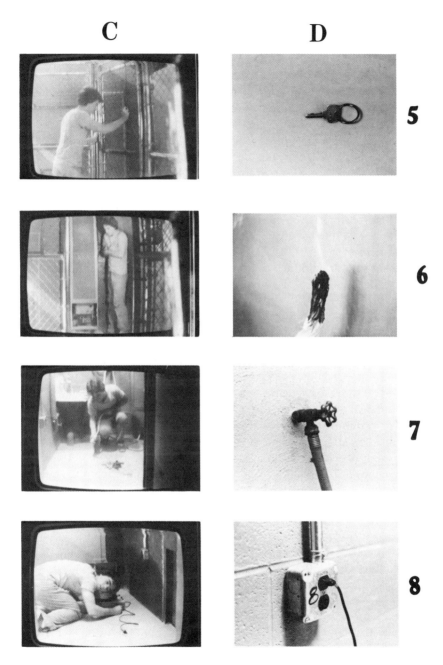

Figure 18 (continued): A person (5) is locked in a cage; (6) is shivering from cold; (7) attempts to wash a soiled floor; (8) tries to operate a phonograph. The solution to 6 is a flaming torch. (Photos courtesy of David Premack.)

Sarah failed tests in which she was to indicate which photograph another chimpanzee would choose in order to solve a familiar problem that confronted the person in a videotape. Hence it appears that the attribution of attribution is beyond the capability of an intentional ape (Premack and Premack, 1983, p. 67).

In a separate set of experiments, young language-ignorant chimpanzees also demonstrated intentional behavior by directing a well regarded trainer to food rewards; misdirecting an untrustworthy one; and even suppressing the urge to look toward the baited container after duping the bogus villain (Premack and Premack, 1983, pp. 51-57). Lawick-Goodall (1971) and Menzel (1973b, 1974, 1979; Menzel and Halperin, 1975) reported similar instances of field chimpanzees withholding responses to preferred objects, while overpowering chimpanzees were watching them, and misdirecting others before sneaking back to an incentive. The 4-year-old chimpanzee in Lefebvre's (1982) food exchange experiment also illustrated intentionality.

As Putney (1985, p. 60) put it, the experimental apes often appear to know what they are aiming at. If taken, Putney's call for additional rigor in research designs and theoretical language should further advance this compelling field.

Conservation and Calculation

Sarah proved to be a remarkable conservationist by passing Piaget's (1973) classic tests in which matched quantities of fluid are placed in differently shaped containers and solids are deformed without loss of mass. She flunked pretests in which she was to discern whether the numbers of buttons on two trays were the same or different. With thorough experimental controls, Woodruff et al. (1978) showed that Sarah's successful conservations were based on inference instead of mere perceptual acuity. But the nature of her inferences are obscure. Muncer (1983a) found that a 4-year old female chimpanzee could solve Piagetian problems, including number conservation.

Later Sarah redeemed herself as a calculating ape. Whereas four juvenile chimpanzees failed match-to-sample tests for number (1, 2, 3, and 4) and proportionality (1/4, 1/2, 3/4, and 1), Sarah shined even though the samples and alternatives were dissimilar in shape, color, mass, area, length and other features (Woodruff and Premack, 1981; Premack and Premack, 1983). It is unclear how she managed these feats. Analogical reasoning is a reasonable possibility.

Several other studies, using different methods, demonstrate that chimpanzees can discriminate more from fewer objects, especially when the maximum number per sample is not much greater than 7 (Brown et al., 1978; Dooley and Gill, 1977a,b; Ferster, 1966; Ferster and Hammer, 1966; Hayes and Nissen, 1971; Matsuzawa, 1985a; Muncer, 1983a). Confusion most commonly occurs when the alternative cardinal numbers are adjacent ones above 3. Brown et al. (1978) concluded that chimpanzees have more difficulty with the concept of some than with none, one and all. But King and Fobes (1982, p. 350) suggest that this might have been determined by the schedule for rewarding subjects instead of their conceptual abilities. In brief, if chimpanzees do have latent mathematical abilities, they are either well masked or very rudimentary by comparison with the ecumenical human measure. Little more may be expected from other apes, though it would be prudent to test them.

Cognitive Mapping

Menzel (1973c, 1974, 1978, 1979), Premack, and associates (Menzel and Halperin, 1975; Menzel et al., 1978; Premack and Premack, 1983) have made notable progress testing the abilities of chimpanzees to exploit their captive environments efficiently through cognitive mapping (Tolman, 1948). They quickly learn the locations, distances, directions, relative quantities and other features of desired and fearful objects; can approach series of them by economical routes of their own design; and can remember them over respectable spans. Further, small groups of chimpanzees will subdivide practically in order to harvest different sized caches of preferred food (Menzel, 1974). While direct views of the objects lead to greatest success, young chimpanzees performed significantly above chance after viewing the placement of incentives via a small black-and-white television monitor (Menzel et al., 1978). Needless to say, mapping is an invaluable cognitive skill for free-ranging apes that must make a living and raise their young in vicissitudinous natural habitats (Boesch and Boesch, 1984c; Chapter 5). Still, however adept chimpanzees might be at mapping familiar tracts from direct views and motile television screens, they apparently cannot orient themselves toward incentives in a distant room on the basis of simple small-scale models (i.e. symbolic maps) like ones comprehensive to 5-year-old humans (Premack and Premack, 1983, pp. 102-105).

Cross-Modal Skills

The ability to generalize representations of phenomena that are received via one sensory modality so that they are recognizable from stimuli in another modality would surely enhance learning and adaptability. Before 1970 some scientists held that cross-modal representational abilities were probably mediated by language. Consequently, they would not be found among the skills of nonhumans (Ettlinger, 1967; Geschwind, 1965; Lancaster, 1968; Myers, 1967). Early experiments on monkeys seemed to support this view (Blakemore and Ettlinger, 1968; Ettlinger, 1967; Ettlinger and Blakemore, 1966, 1967; Milner and Ettlinger, 1970; Rothblat and Wilson, 1968; Wegener, 1965; Wilson and Shaffer, 1963). Then Davenport and Rogers (1970, 1971; Davenport, 1976, 1977; Davenport et al., 1973, 1975; Rogers and Davenport, 1975) provided a major breakthrough in research

on nonhuman cross-modal skills. With innovative match-to-sample procedures they evidenced that chimpanzees and orangutans can abstract and
exchange information from visual and haptic characteristics of familiar and
novel objects.

Four of Premack's (1976a) language-trained subjects (Sarah, Peony,
Elizabeth and Walnut; Chapter 7) exhibited cross-modal transfer between
gustatory and visual senses. After concealed samples of fruits were placed
in their mouths they not only matched examples of the fruits visually but
also selected the correct plastic words for them (Premack, 1976a, pp. 304-
305). Further, Sarah and Peony showed cross-modal transfer between
auditory and visual modalities (Premack, 1976a,b).

There followed a number of experiments which showed that several
species of monkeys and very young humans also can match or recognize
objects across sensory modalities (Bolster, 1978; Bryant, et al., 1972;
Cowey and Weiskrantz, 1975; Elliott, 1977; Ettlinger and Garcha, 1980;
Ettlinger and Jarvis, 1976; Gunderson, 1983; Hewett and Ettlinger, 1978,
1979; Jarvis and Ettlinger, 1977). Indeed, after demonstrating that
monkeys can recognize objects bidirectionally between visual and haptic
perceptions, Jarvis and Ettlinger (1978; Ettlinger, 1983) concluded that
there is no indication that the cross-modal skills of chimpanzees are
markedly superior to those of monkeys. These studies confirmed the
inference, based on preschool children (Blank et al., 1968) that verbalization
is not prerequisite for cross-modal recognition and problem-solving. Of
course, the opposite may well be true; cross-modal capabilities are
probably important enhancers if not actual prerequisites for human
language (Wright, 1970).

Neuropsychologists have only scratched the surface of problems on the
neural substrates of cross-modal skills. Ettlinger (1973) speculated that
apes and humans may possess a higher-order cortical system which
mediates their cross-modal performances whereas prosimians (Ward et al.,
1970) and non-primate mammals (Yehle and Ward, 1969) have only a
lower-order subcortical system which mediates cross-modal transfers of
specific learning. However, in pursuit of cortical mediators for cross-modal
skills, surgical ablations and probes have been applied to the brains of
expendable monkeys instead of precious apes (Ettlinger and Garcha, 1980;
Bolster, 1981).

THE PROBLEM OF SELF-AWARENESS

Of all the arrogant conceits in which humans have indulged, among the
greatest must be counted the claim that only we are aware of ourselves, have
individual identities, and think about what we are doing, have done, and will
do. It has been argued that consciousness and its concomitant, mind, are

premised on human language. According to this dogma, thoughts are impossible in the absence of human language.

During the past decade, these entrenched beliefs have been challenged vigorously by scientists from diverse disciplines. One of the main spokesmen for reopening the question of awareness in animals is Griffin (1976, 1978, 1981, 1984a,b), an experimental zoologist reknowned for his studies on echolocation in bats and bird migration. He argues that we must search for evidence of awareness and mind not only in our nearest relatives, the great apes, but also in much more distant species, including almost brainless creatures like the social insects.

The first edition of *The Question of Animal Awareness* (Griffin, 1976) is little more than a polemic leading to a plea for "an agnostic reservation of judgement" and greater open-mindedness (p. 68) by the scientific community. Griffin's (1976, p. 96; 1981, p. 157) naive suggestion that experimenters don chimpanzee disguises and dab themselves with pheromonal perfumes so that they can conduct participant observations among them is unlikely to be tried by informed students of great apes. Even if they tolerated the presence of such oddballs, I doubt that the subjects would be fooled and that humans could do what chimpanzees do in the forest (Chapters 2, 3, 4, 7 and 8) without rapid deterioration of their health and the costumes.

Griffin (1976, pp. 103-105) resolved that the complex, versatile and limited symbolic behavior of dancing honeybees and signing chimpanzees (Chapter 7) were especially compelling evidence that nonhuman animals have mental experiences and conscious intentions.

By 1981, Griffin had become aware of many more studies that seemed to support his argument for animal consciousness, thinking and intentionality. Notable among them were the elegant experiments by Gallup (1968, 1970, 1971, 1975, 1977a,b, 1979, 1982) and associates (Gallup and McClure, 1971; Gallup et al, 1971, 1977; Hill et al., 1970; Suarez and Gallup, 1981) on self-recognition as revealed by mirror-image stimulation (MIS).

Most animals react to their reflections in mirrors as if they are viewing other individuals of their species. Social reactions to mirror-image stimulation, commonly in the form of aggressive or sexual displays, have been elicited in a variety of fish, birds, monkeys and other mammals (Gallup, 1966 et seq.; Lethmate and Dücker, 1973; Anderson, 1983a,b, 1984b,c; Behnaret al., 1975; Bertrand, 1969; Menzel et al., 1985). Mirrors also serve as social facilitators in diverse creatures, including humans. For instance, chickens lay more eggs and students can do better on examinations in bemirrored rooms (Gallup, 1975).

Like monkeys, apes initially react socially to their mirror images. But after several days, young and adult chimpanzees begin to use the mirror to explore parts of their bodies that are normally out of view or difficult to position for direct viewing. They also use the mirror to inspect food wadges in their mouths, clean between their teeth, pick their noses, blow bubbles and make faces. This led Gallup (1970 et seq.) to conclude that they must recognize themselves in the mirror and that they may have a self-concept. To test these hypotheses he anesthetized subjects that had some and had no prior experience with mirror-image testing and placed spots of red dye on the superior aspects of their brow ridges and ears. The dye could not be felt locally after it had dried. Once alert and introduced to a mirror, the experienced subjects directed many responses toward the spots, including touches that were followed by visual and olfactory inspection of their exploratory fingers. The mirror-naive subjects reacted socially to the mirror.

Menzel et al. (1978) reported that a male chimpanzee, which had spontaneously learned to recognize himself in the steel cover over an electrical outlet in his cage, commenced to groom his heel while watching it projected in a monitor within one minute of his first exposure to a television system.

In an ingenious follow-up, Menzel et al. (1985) showed that two adult male language-trained chimpanzees, Sherman and Austin (Chapter 7), could track the images of their hands in a mirror and on closed-circuit television pictures in order to locate a hidden object, which was located on the exterior wall of the enclosure. They quickly adjusted to special effects generated rotations and reversals of images on the monitor and readily distinguished live performances from playbacks of earlier palpations on a second screen.

Adult orangutans have also passed the dye test after exposure to mirrors, but gibbons and gorillas flunked (Gallup, 1975 et seq., Suarez and Gallup, 1981; Lethmate and Dücker, 1973; Ledbetter and Basen, 1982). Whereas the hylobatid failure is not too surprising, considering their other resemblances to monkeys, the results for gorillas are puzzling.

Further, we do not know whether self-recognition in mirrors truly reflects self-awareness of a human sort (Ristau, 1983b; Epstein et al., 1981) or that lack of evidence for mirror-image recognition demonstrates absence of self-consciousness (Fox, 1982). Gallup (1975) found that in chimpanzees, self-recognition must develop through exposure to reflecting surfaces and that it is influenced by early experience. It appears that awareness of others, particularly ones with which the individual has social relations, is an important feature in the development of self-identity.

Isolation-reared chimpanzees did not show other-directed responses to mirrors even though they spent considerable time passively viewing their images. And they failed the dye test. After a period of social rehabilitation some isolation-reared chimpanzees showed signs of self-recognition; control subjects that had remained isolated did not (Gallup, 1975).

Again, gorillas are puzzlers because they naturally live in more coherent social groups than orangutans do (Chapter 8). Could it be that having been paired in cages is not enough social stimulus for them to develop the capacity for self-recognition in mirrors? On the other hand, it is possible that the capacities that were revealed with MIS are needed more by the highly social great apes that engage in frequent fission and fusion of subgroups but still must remember their places in a bounded community and section of the habitat.

Further studies, especially with young gorillas from naturalistic social groups, are urgently needed. Before attempting generalizations about self-awareness in great apes, we also need mirror-image studies with bonobos. Bleached or other bright spots should be used on both the basically black species of the Pongidae.

SUMMARY

Robert M. Yerkes probably would be more pleased than surprised by what his successors have discovered about ape mentality. The record has been set straight regarding their abilities to discriminate colors. Believing that chimpanzees might be color-blind, Yerkes and Yerkes (1929) had underestimated their chromatic sense (Matsuzawa, 1985b). Insofar as we know, the visual (Riesen, 1970) and other senses of apes are at least as keen as our own, though much research remains to be done on comparative hominoid olfaction, gustation, touch and audition (Jackson et al., 1969; Prestrude, 1970; Elder, 1934).

Regrettably, after the classic studies of Yerkes (1916) and Köhler (1925) and attempted refutations by Spence (1938), Birch (1945b) and Schiller (1952), the question of insight in apes was largely relegated to a back burner (Rumbaugh, 1970, p. 15; King and Fobes, 1982, p. 333). We can safely agree with Mason (1979, p. 287) that chimpanzees have insight as per Köhler's definition: ". . . the appearance of a complete solution with reference to the whole layout of the field." But it would be most helpful to have more studies on a greater variety of species to establish the role of insight in the total repertoire of catarrhine problem-solving and adaptation.

Menzel has upgraded our information on the capacity of chimpanzees to coordinate their efforts to achieve goals. But we still do not know the extent to which this represents cooperative planning. Further examples of cooperation between laboratory subjects will be presented in Chapter 7.

Clearly the greatest advance in pongid comparative psychology has come with the ingenious demonstrations, via Rumbaugh's fair-minded transfer index and the Premack group's two-way communication with Sarah, that great apes can indeed generalize and infer from test experiences to the point that abstract reasoning is manifest.

On the down side, apes chronically fail to wow us with mathematical puzzle-solving, foresightful behavior and futuristic sense though we have learned that their memories are better than many people had imagined (Mason, 1979, p. 286).

Although universally applauded for their ingenuity, Gallup's experiments have not persuaded all skeptics that chimpanzees and orangutans have a self-concept or more particularly that the level of their self-consciousness is established vis-à-vis the human condition. For instance, Ristau and Robbins (1982b, p. 218) succinctly commented that MIS data "do not require an interpretation that the chimpanzee has an awareness of self as a mind; an awareness of self as a body will suffice."

Despite the herculean efforts that must be exerted in order to explore the minds of apes we should develop even more refined psychological probes and press on to establish the actual limits of their cognitive abilities. We need to learn whether their particular adaptive zones have endowed them with special mental skills that are less developed in humans. Here knowledge from the field must penetrate into the laboratory. Perhaps now that the drama of pongid field studies is winding down and animal welfare groups are limiting terminal studies on captives there will be a resurgence of noninvasive comparative psychological studies that will culminate in a deeper understanding of ape mentalities.

7

Communication

"All social life depends on the ability of individuals to coordinate and regulate their actions with respect to each other and to the environment. A process by which this coordination is achieved is communication." (Menzel, 1973b, p. 192)

"I am inclined to conclude from the various evidences that the great apes have plenty to talk about, but no gift for the use of sounds to represent individual, as contrasted with racial, feelings or ideas. Perhaps they can be taught to use their fingers, somewhat as does the deaf and dumb person, and thus helped to acquire a simple, nonvocal, 'sign language'." (Yerkes, 1925, p. 180)

"give orange me give eat orange me eat orange give me eat orange give me you." (Nim Chimpsky in Terrace et al., 1979, p. 895)

Yerkes and Yerkes (1929) summarized available descriptions of the natural vocalizations, percussions, facial expressions, and gestures of selected apes and discussed several futile attempts to teach pongids to speak (Furness, 1916; Garner, 1896; Yerkes and Learned, 1925). Most of the early accounts were based on captives or naturalistic subjects that were heard or encountered briefly but not studied systematically. They concluded that while no ape speaks in a human sense, they are disposed to use "the voice as means of expressing feelings, desires, and ideas" increasingly from the hylobatids through orangutans and gorillas to the chimpanzees. The same order was maintained for the extent of "intercommunicational complexity and biological value" of their vocalizations (Yerkes and Yerkes, 1929, p. 546). But, they concluded, in comparison with humans, the vocalizations of all apes must be considered secondary to facial expressions, gestures and postures as means for intraspecific communication (Yerkes and Yerkes, 1929, p. 569).

In the area of communication the dichotomy between field and laboratory studies of apes is beginning to break down. Still whereas field behavioralists are interested mainly in the natural repertoires of signals and their functions, comparative psychologists have mostly endeavored to establish two-way humanoid communication with great apes in order to discern their cognitive skills (Chapter 6). Anthropologists and linguists have drawn unevenly upon both realms in order to speculate on the evolution of human language. Somewhat better balance between field and laboratory efforts has been achieved regarding the dramatic vocalizations of hylobatid apes. Hence at the outset of this review we must acknowledge that although great empirical progress has been made, well-set comprehension of the apes' communicative abilities and the extent to which they represent the condition of our prelinguistic ancestors is still lacking.

ARTIFACTUAL SYMBOLS AND RULES

Although some early naturalists believed that apes could talk (but they discreetly kept silent in the presence of humans), it became apparent that this was not so when humans spent considerable time with them. Further, on the basis of "the results of varied and long-continued training experiments," Yerkes and Nissen (1939, p. 587) had concluded that while symbolic processes occasionally occur in the chimpanzee, they are relatively rudimentary and ineffective and probably do not increase in frequency and functional value with increase in experience and age. Undeterred by these observations, a number of psychologists conducted projects to establish humanoid communication with great apes.

Projects Gua and Viki

The Kelloggs and the Hayeses fostered infant female chimpanzees in order to document the extent to which their homes could humanize them (Chapter 6). At the time no greater psychological triumph might be imagined than for them to say mama and papa in a prelude to more complex linguistic accomplishments.

The results are pitiful. Gua did not even babble, let alone recite nursery rhymes. Viki produced only 4 hoarse, breathy utterances (mama, papa, cup, up) after much human tuition, including the molding of her lips. These vocalizations would decay if her trainers did not practice with her regularly.

Gua's vocalizations were predominantly emotional responses to provocations that were readily apparent (Kellogg and Kellogg, 1933). Viki was also mum between emotional calls (Hayes, 1951; Hayes and Hayes, 1951, 1954). It was concluded that chimpanzees do not easily imitate human sounds even when they are immersed in human society and are tutored endlessly. Other approaches had to be sought.

Project Washoe et al.

In 1966, Gardner and Gardner (1969, 1971, 1972, 1974, 1975, 1978, 1980, 1984; Drumm et al., 1986) took up Yerkes's (1925) suggestion that chimpanzees might handle gestural language more adeptly than speech. They obtained a nearly yearling female chimpanzee and ensconsed her in a house trailer in suburban Washoe County, Nevada. Because chimpanzees are naturally sociable and human language develops in a social environment the Gardners wanted a humanoid foster home for Washoe, complete with furniture (primary among which was a potty chair to which she was trained by 31 mos.), tools, magazines, toys, a well stocked, locked pantry of supermarket goodies, and numerous signing human companions. She stayed alone in the trailer at night and exercised in a spacious treed yard that was equipped with a swing and other apparatus.

The Gardners elected American Sign Language (ASL) to establish two-way communication with Washoe. ASL is a mixture of iconic and arbitrary signs, notable for the absence of finger spelling. It is supposed to contain units (cheremes) that are counterparts of phonemes in human speech (Stokoe, 1960, 1970), viz. hand-sign positions, configurations and motions. The ASL signs are analogous to spoken English words. Combinations of signs translate very roughly and with difficulty into English sentences. Translations are commonly quite telegraphic.

Washoe's first teachers were learning ASL themselves and there was considerable turnover of her playmate-instructors. She was encouraged to ask for and to name objects, and to ask questions about them with ASL. Trainers minimized verbal utterances in Washoe's presence, though other sounds were allowed if Washoe might be able to mimic them.

Some of Washoe's signs were acquired by imitating her tutors while others were shaped and molded by her trainers. Molding was the most effective method of training (Fouts, 1972). Signs were adjusted to her anatomy and positional behavioral preferences when necessary. This has led some commentators to label it "pidgin" sign language (Terrace, 1981; Ristau and Robbins, 1982b).

Washoe's first sign was "come-gimme." During the first 7 months of training she added "more," "up," and "sweet." By the end of the twenty-second month of training (i.e. when she was about 3 years old) she had mastered 34 signs. One year later she reliably used 85 signs. As a five-year-old, Washoe had mastered 132 signs and understood several hundred more that were directed toward her by humans (Gardner and Gardner, 1969, 1971, 1972, 1974).

Washoe used signs for classes of referents, e.g. "dog" for the animals of many breeds, sizes etc., pictures of dogs, and barking sounds. She could innovatively apply signs learned in one context to new situations. For example, she signed "open" not only for doors, rooms and cupboards but also for containers and a water faucet (Gardner and Gardner, 1971).

When she was about 21 months old, Washoe began to combine signs, mostly in pairs, in order to make simple statements which have been compared with the early sentences of verbal children. Her positioning of pivot words were similar to those of human children. For instance, "this" and "that" were in initial position. Washoe could answer Wh- questions. Her signs were comprehensible to persons who could not see what her referents were (Gardner and Gardner, 1971; Van Cantfort and Rimpau, 1982).

Baby Washoe manually babbled and signed to herself while leafing through magazines and settling in for the night. She even corrected herself. Further, she signed "quiet" to herself while creeping to a forbidden part of the yard and "hurry" while dashing to the potty chair (Gardner and Gardner, 1974).

In short, Washoe, reestablished the expectance that the communication gap between apes and us would be narrowed notably if not actually bridged (Linden, 1974, 1981).

Because Washoe was aproximately a year old at the beginning of the study and because her tutors initially were not fluent in ASL, Gardner and Gardner (1975, 1978, 1980, 1984) ran a new project with 4 newly born chimpanzees to determine the development of two-way communication in chimpanzees compared with that of human infants and to learn whether they would communicate among themselves with ASL. (Figure 19).

Two of the subjects (Moja and Pili) began to sign when 3 months old (Gardner and Gardner, 1975). They commanded 50 signs each by the time they were 21 and 23 months old (Gardner and Gardner, 1978). On average, human children have 50 words when 20 months old. The Gardners' new chimpanzee subjects progressed more rapidly than Washoe had. Like human infant vocabularies, nominals were predominant in those of the chimpanzees (approximately 50%). Negative signs were mastered by 15 months of age (Gardner and Gardner, 1978).

Moja was schooled during her third year. At 3.5 years she sat at a desk, drew with chalk, crayon and pens, and named her masterworks. For instance, she chose an orange pen and included circular squiggles when asked to "draw berry there." She then signed "berry" when asked "what that." She signed her first chalk sketch "bird" (Gardner and Gardner, 1978, p. 71). Viewing these works I suspect that the idea of Moja the artist would fly before most critics.

In 1970, one of Washoe's teachers (Fouts, 1972) took her to Norman, Oklahoma, where studies on her were continued and new studies were inaugurated with other young chimpanzees (Fouts, 1973, 1974, 1975a,b, 1977, 1978; Fouts and Couch, 1976; Fouts et al., 1976; Fouts and Rigby,

Figure 19: Common chimpanzees communicating with American Sign Language. (a) 6-year-old Moja signing COME to 3-year-old Tatu. (b) Tatu signing DRINK to Moja. (Photos courtesy of R.A. and B.T. Gardner.)

1977; Fouts and Budd, 1979; Fouts et al., 1982). Fouts aimed to determine the range of individual variation for the acquisition of ASL by chimpanzees and to establish intraspecific ASL communication in the colony. He hoped that the chimpanzees eventually would acquire signs from one another. Initially, several of the subjects were kept in human homes where, without the distraction of natural chimpanzee communication, they learned ASL more quickly (Linden, 1974, p. 89).

Two young males (Booee and Bruno) learned to sign to one another about food, mutual comfort and play, particularly tickling. A female, Lucy, and Washoe have combined signs presumably to describe novel objects for which they had not been given the ASL signs. For example, Lucy designated oranges, grapefruit, lemons and limes "smell fruits." After tasting a radish she signed "cry hurt food." Watermelon was "drink fruit" and "candy drink." Washoe signed "water birds" for swans, "dirty monkey" for a feisty macaque, and "dirty Roger" for her trainer after he refused her requests. Dirty was her sign for feces and soiled objects (Fouts, 1974, 1975a). Before Fouts (1975a) learned the ASL sign for leash Lucy invented an iconic sign to indicate that she wished to go outside on her leash. Fouts's (1975b, p. 143) subjects combined signs much earlier than Washoe had, probably because they were taught ASL earlier than she had been.

Fouts (1975a,b) and his colleagues (Fouts and Couch, 1976; Fouts et al., 1976) also demonstrated that a 3 year old male chimpanzee (Ally) could transfer object names in English and their ASL signs in order to refer to their physical referents. Chimpanzees readily learn to understand verbal object names and simple commands. First Ally learned the English names for 10 objects. Then, without reference to the objects, he learned ASL signs for the words. Later when trainers showed the objects to Ally and signed "what that," Ally could sign correctly for the 10 referents. Thus, Ally exhibited notable cross-modal transfer, as well as mnemonic and cognitive skills.

Fouts bred Washoe and Ally in order to study the transmission of signing from mother to child. While pregnant, Washoe vomited a meal of yogurt and raisins. When asked "what that," she signed "berry lotion." As her pregnancy advanced she was asked "what in your stomach"; she repeatedly signed "baby" (Fouts et al., 1982).

Washoe's infant died 2 months after birth. In 1979, Fouts acquired a 10-month old male (Loulis) for her to raise and, in 1980, moved them with Moja to Ellensburg, Washington. By 1982, Washoe had an ASL vocabulary of 180 signs (Fouts et al., 1982). Loulis's vocabulary at 2.5 years was 17 signs, only one of which was used by humans in his presence. Loulis also could learn signs from the 8-year old signing female chimpanzee, Moja. Loulis manually babbled before producing full-fledged signs like "tickle" and "drink" (Fouts et al., 1982).

People use only 7 signs (mostly Wh-words) in Loulis's presence (Fouts,

1983). Washoe understands English so they need not sign in order to communicate with her when she is with Loulis. Thus, apparently Loulis has acquired most of his signs by imitating Washoe and Moja. Further, Washoe has actively taught Loulis to sign. She molded his hand into the configuration for food and touched his mouth with it. He now uses this sign. She placed a chair in front of him and signed "chair" (meaning sit down) 5 times. He has not been seen to use the sign. Fouts et al. (1982) concluded that signing and various examples of object manipulation show that chimpanzees acquire at least some behaviors via cultural transmission.

It will be most interesting to learn whether infant chimpanzees imitate signing chimpanzee mothers more readily than human fosterers.

Projects Cody, Chantek and Princess

Emboldened by Furness's (1916) glimmer of success with *Pongo*, Laidler (1978) taught an infant male orangutan (Cody) to utter "kuh"; "puh"; "fuh"; and "thuh" for drinks in a mug; contact and comfort; solid food and food in a pan; and the continuance of brushing, respectively. The study was terminated when Cody was 15 months old. Laidler based Cody's tuition on a modern method for teaching autistic children. He did not attempt to teach him English words.

Fouts (1975a) worked briefly with an infant male orangutan. He acquired ASL signs and paired them sensibly. Further details about the study are not available.

In 1978, Miles (1983) obtained a 9-month old male orangutan, named Chantek, from the Yerkes Regional Primate Research Center. Chantek was raised and trained in a house trailer and adjoining playground in Chattanooga, Tennessee. His foster home was like Washoe's except that he was proffered more opportunities for arboreal and other suspensory behavior, including nesting in a hammock, and was exposed to a smaller group of caretaker/trainers. ASL training occurred mainly when Chantek seemed most attentive and socially motivated. He was not drilled or held to an acquisition schedule.

After one month of training, Chantek signed "food-eat" and "drink." Eighteen months later he had mastered 56 signs. During the second month of training, he spontaneously combined signs into sequences like "come food-eat." After 3 years of training he produced three- and four-word sequences—e.g. "key milk drink open," which was interpreted to mean, "Open the trailer door so that I can enter for a drink of milk." The mean length of Chantek's utterances is 1.9 signs with the longest consisting of 5 signs (Miles, 1983).

Chantek's signing was slower and "more articulate" than the signing of the chimpanzees that Miles (1983, p. 50, 1977, 1978) had studied in Oklahoma. Chantek signed at half the speed exhibited by Ally. Miles (1983, p. 56) concluded that Chantek's slower, more deliberate, and spontaneous signing and the distinctive "insightful" cognitive style of orangutans (Chapter 6) may indicate their superior linguistic and cognitive abilities relative to those of other apes.

Over a period of 19 months, Shapiro (1982) taught a 3.5 year old female Bornean orangutan (Princess) 37 ASL signs in Galdikas's field camp. Her rate of sign acquisition also compared favorably with those of Washoe and Koko.

Project Koko

This adventure began in 1972 at the San Francisco Zoo and continues today on a 7-acre wooded area at Woodside, California (Patterson, 1978a-d, 1980, 1981, 1983; Patterson and Linden, 1981). For 11 months, Koko was trained publicly at the zoo before moving to a Washoesque foster home with Patterson and her boyfriend. In 1976, they were joined by an infant male gorilla, Michael, which Patterson hopes will mate eventually with Koko (Patterson and Linden, 1981).

Koko was a year old when Patterson began to tutor her in ASL. Unlike the Gardners, Patterson and her assistants (including persons fluent in ASL) simultaneously spoke to the subject while signing. Koko quickly learned to sign but she did not talk back. After 18 months of training, Koko had mastered 22 signs, the first of which was for food. By 3.25 years of age she had repeated 236 signs and had mastered 78 of them according to Patterson's (1980, p. 506) criteria, viz. "spontaneous and appropriate use on at least half the days of a given month." Koko had mastered 161 ASL signs a year later and 246 signs as a 5.5 yer old. Her rate of development and mastery of ASL are quite on a par with the accomplishments of other great apes (Patterson and Linden, 1981; Patterson, 1978a).

Patterson (1978a,b, 1980; Patterson and Linden, 1981) also claimed that Koko spontaneously combined signs into meaningful and often novel statements comprised of up to 11 signs (though mean length is 2.7 signs) and that she used ASL to talk to herself, joke, lie, complete nursery rhymes, argue, threaten, show that she knows pig Latin, express her feelings and desires, insult offenders, and discuss where she will go when she dies. On a program in the *Nova* series, Patterson even narrated that Michael used ASL to relate to her the horror of his bloody capture in western Africa.

Now we turn to 4 projects in which experimental control was at the forefront.

Project Sarah et al.

In 1966, Premack and Premack inaugurated a study of two-way communication with a 5-year old wild-born female chimpanzee, Sarah, and 4 other young subjects (Premack, 1970, 1971a,b,c, 1976, 1976a,b,c, 1977, 1978; Premack and Premack, 1972, 1975, 1983; Premack and Schwartz, 1966).

They adhered to strict laboratory procedures instead of having freer, more homey social interactions like those that characterized the ape ASL projects. Nevertheless Sarah had close contact with humans until she severely bit a trainer and had to be kept locked up.

The Premacks did not aim for Sarah to imitate human language. Instead they were interested in Sarah's capacity to employ a relatively small vocabulary productively and syntactically. The "words" were metal-backed pieces of plastic in various colors and random (non-iconic) shapes. Sarah affixed them to a magnetized board in vertical series. Within 6 years, Sarah had a vocabulary of 130 units. She arranged them in series with up to 8 units. Peony, Elizabeth, and especially, Walnut and Gussie, were less impressive students (Premack, 1976a).

Premack and Premack (1983, pp. 115-116) ultimately concluded that while apes recognize equivalence between their world and the plastic language, they cannot acquire even the weak grammatical system of young human children. They agreed with Terrace et al. (1979) that mere word order is not the equivalent of syntax. Sarah's strings of plastic signs were constructions, not sentences. Sarah's plastic constructions were simple one-to-one correspondences between words and the actual items they refer to. Human sentences can convey much more abstract concepts and are infinitely variable.

Sarah learned to designate sameness or difference and mastered the concept "name-of" which greatly facilitated her acquisition of new names. She seemed to learn interrogatives (who, what, why), the conditional "if-then," and prepostional concepts like on, under, and to the side of. Sarah could describe features of objects in the absence of the objects when presented with their plastic signs. For instance, when shown the plastic sign for apple, she chose plastic signs for red versus green, round vs. square, and having a stem. Thus she seemed to exhibit metalinguistic phenomena, i.e. the use of symbols to discuss other symbols (Premack, 1970, 1976a).

Project Lana

In 1972, Rumbaugh and a multidisciplinary team of co-workers commenced an experiment on language learning and communication with an infant female chimpanzee, Lana, at Yerkes Regional Primate Research Center in Atlanta, Georgia. They particularly wanted to establish whether apes have the capacity for linguistic productivity, i.e., can they create and understand new messages? Rumbaugh and linguistic colleagues devised a series of artificial lexigrams and created a new language, Yerkish, based on them and simple grammatical rules. The Yerkish grammar was designed after correlational grammar, in which there is no distinction between syntax and semantics. The lexigrams were color coded according to 7 gross semantic categories: autonomous actors; spatial objects and concepts; ingestibles; body parts; states and conditions; activities; and prepositions, determiners and particles. Two additional colors signalled sentential modifiers that had to be placed in the initial position of messages.

After 9 months of testing with lever pressing and matching-to-sample tasks in which hand-drawn non-iconic geometric lexigrams were employed, the 2.25 year old Lana was placed in the language training area. It had a keyboard for her (Figure 20), another for the experimenter, and panels on which they could read what was typed. All communications were recorded automatically by a computer. Lana could request a variety of foods, toys, contact with the trainers, music, movies, a romp outside, or a view of the real world through a window. She began with please and had to end with a period. Mistakes could be erased with the period key. Periodically, the lexigrams were shuffled on Lana's keyboard. Low back-lighting indicated that a key could be activated and no back-lighting meant that the key was inactive. When active keys were pressed, they lit up more brightly. Lana generally had 50 keys back-lit out of a total of 75 keys on the board (Rumbaugh, 1974, 1977a,b,c, 1978, 1981b; Rumbaugh and Gill, 1975, 1976a,b, 1977; Rumbaugh, Gill, Brown et al., 1973; Rumbaugh, Gill and von Glasersfeld, 1973, 1974; Rumbaugh, Gill, von Glasersfeld et al., 1975; Rumbaugh, von Glasersfeld, Gill et al., 1975; Rumbaugh, von Glasersfeld, Warner et al., 1973, 1974; Rumbaugh and Savage-Rumbaugh, 1978; Rumbaugh et al., 1978; Savage-Rumbaugh, Rumbaugh and Boysen, 1980; Buchanan et al., 1981; Dendy, 1973; Essock, 1978; Essock et al., 1977; Gill, 1977, 1978; Gill and Rumbaugh, 1974a,b; von Glasersfeld, 1974, 1976, 1978; von Glasersfeld et al., 1973; Pate and Rumbaugh, 1983; Warner et al., 1976).

Lana readily mastered the system and used it to obtain what she wanted and needed in order to subsist. The speed with which she typed her requests is quite impressive. Within 30 months of training and practice Lana had mastered 75 of the 125 lexigrams to which Rumbaugh and Gill (1975) aimed to introduce her.

Lana spontaneously learned to read displays that were projected above the keyboard and to hit the period key when she had erred while typing instead of completing a series that would not be rewarded. Once she had grasped the basic mechanism for using names, she rapidly learned new ones and began to request the names of objects for which she had no lexigram. She combined available lexigrams in novel ways in order to communicate about things for which she had not been provided lexigrams. She used the lexigram for "this" to refer to specific incentives (Rumbaugh and Gill, 1976a).

Lana extended "no" from simple negatives to protests about malfunctioning food dispensers and problems with trainers. And she began to compose novel sequences of lexigrams in addition to using the standard chains of words that characterized her training period. In the 26th month of the study she first initiated a two-way communicative exchange with a

Figure 20: Lana's computer-controlled language training situation. The overhead bar must be pulled for the console to be activated. As the keys are pressed lexigrams on the surfaces of the word-keys are projected from left to right above the keyboard. Juice and water dispensers are immediately below the console; banana and M&M dispensers are located in the lower right of the picture. Additional keys (and lexigrams) were added as the study progressed. (Photo courtesy of D.M. Rumbaugh.)

trainer, who was outside her chamber, anent his Coca-Cola: "?Lana drink Coke out-of room." (Rumbaugh and Gill, 1975). Subsequently, other complex Yerkish exchanges occurred when Lana wanted something exceptional or when a practical problem arose within the system (Rumbaugh, 1977b).

Aiming to establish a Lana-like project, Asano et al. (1982) taught 2 female and 1 male wild-born chimpanzees non-iconic geometric lexigramic names for 8 objects and 5 colors. The subjects were approximately 2 years old when training began in 1978 at the Primate Research Institute of Kyoto University. All lexigrams were white on a black background and were shuffled periodically on the keyboard and backlit to indicate computer activation. The project was discontinued when it appeared that it would not be productive as a language study (Sebeok, 1982).

Project Sherman and Austin

After Project Lana, Rumbaugh and Savage-Rumbaugh, who had worked with the Oklahoma and Gombe projects, inaugurated an elegant study of symbolic communication between chimpanzees via the computerized lexigraphic apparatus (Savage-Rumbaugh, 1979, 1981, 1984b; Savage-Rumbaugh and Rumbaugh, 1978, 1980, 1982; Savage-Rumbaugh et al., 1978a,b, 1980, 1983; Rumbaugh et al., 1982; Bassett, 1983). They induced two arduously language-trained, laboratory-born male chimpanzees to communicate about and share food. Austin, the 3.5 year-old, was an admirable altruist, while Sherman, the 4.5 year-old, was a bit stingy. Once the subjects had learned to communicate via the keyboard with each other regarding the identity of one of 11 possible foods or drinks in a container, they were tested to see whether they could convey the information by natural gestures and vocalizations. They failed to do so (Savage-Rumbaugh et al., 1978a).

If they made a mistake using the keyboard, it was always another food or drink that was named instead of inedible objects that also were represented by lexigrams on the keyboard. At the culmination of the experimental series, one chimpanzee was given an array of foods, which could be viewed by the partner that had access to the keyboard. The possessor could see projected requests of the cadger and usually read them accurately and handed correct tidbits through a small window in the transparent plastic partition that kept them physically apart (Savage-Rumbaugh et al., 1978a).

Savage-Rumbaugh and her associates (1980) then tested the capacities of the chimpanzees to categorize objects as food or tools and to comprehend the representational function of lexigraphic symbols. After learning the lexigrams for food and tools, Sherman and Austin were able to categorize photographs and lexigrams for specific foods and tools correctly. Savage-Rumbaugh et al. (1980) cogently argued that Washoe, Sarah, Lana, Nim and other chimpanzees that used symbols in ways that emulate human usage did not comprehend their representational functions. Whereas Sherman and Austin could immediately label novel objects as food or tools, Lana could not. Sherman's only "error" was labelling sponge as food; but he ate part of it.

Lana could sort foods from tools but she could not encode the perceived functional relationships between the foods and the tools symbolically. Thus, Lana, Washoe, et al. appear to have learned contextually appropriate usages for their lexigrams or ASL signs but really did not know what they represented. This is because of the training regimes that were used. In short, they were probably problem-solving (Chapter 6) instead of using language with the intent of conveying meaning. Sherman and Austin evidence that chimpanzees can comprehend the meaning of lexigrams and presumably other sorts of symbols if appropriate training procedures are employed (Savage-Rumbaugh et al., 1980).

Sherman and Austin also learned to use the lexigramic system to work cooperatively for food that had to be obtained with tools (Figure 21). They were alternately placed in rooms containing manually inaccessible food and a kit with 7 kinds of tools. The possessor of food or drink had to request an appropriate tool from the hardware chimpanzee. He would then share his reward with the tool supplier (Savage-Rumbaugh et al., 1978b).

Project Nim

Nim Chimpsky was to other ape language projects what the U.S.A.-bomb was to Nagasaki.

In 1973, Terrace, a psychologist at Columbia University, obtained a 2-week old institution born male chimpanzee and dubbed him Neam ("Nim") Chimpsky, loosely after Noam Chomsky (Terrace, 1979a,b, 1981, 1982, 1983; Terrace and Bever, 1976; Terrace et al., 1979, 1980). Baby Nim was raised initially in a hectic New York townhome by a family of friends of Terrace and was taught ASL by molding and guidance. He was to be highly socialized in order to maximize his chance to develop a linguistic concept of self (Terrace and Bever, 1976). There was fequent turnover of his 57 playmate-teachers (Terrace, 1979a, pp. 256-261) and shifts of study location and living quarters during the four-year study period. By 1977, Nim had acquired 125 signs (Terrace, 1979a).

Unlike previous workers, Terrace routinely made videotapes of Nim's signing sessions. Analysis of them revealed that Nim's signs were often prompted by his teacher's prior signs. Commonly he merely repeated the instructor's signs, perhaps with additions like Nim, me, you, hug or eat. Terrace (1979a) detected similar inadvertent prompting on the films of Washoe and Koko and their trainers. This is similar to the Clever Hans phenomenon wherein a performing horse reacted to subtle cues of his trainer instead of truly counting or otherwise comprehending human symbols.

a

b

c

d

e

f

Figure 21: Sherman and Austin use a computer-controlled language to work jointly with a tool and obtain a food reward. (a) Austin watches as food is placed in a long, narrow tube. He will need a stick to push it out. Sherman's view of the proceedings is blocked so initially he is ignorant of the nature of the problem. (b) Austin requests "stick" via the keyboard. (c) Austin looks to see whether Sherman is attending. (d) Sherman (right) selects the stick from the tool kit and Austin (left) approaches the window. (e,f) Sherman passes the stick through the window to Austin.   (continued)

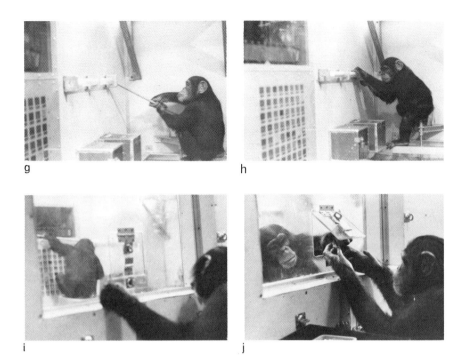

Figure 21 (continued): (g) Austin begins to insert the stick into the food tube. (h) Austin pushes the stick through the tube. (i) Sherman keeps close watch over Austin's food quest. (j) Austin gives Sherman a portion of the food that he obtained with the stick. (Photos courtesy of E. Sue Savage-Rumbaugh.)

Analysis of strings of signs produced by Nim revealed that they could be viewed simply as unstructured combinations of signs, in which each one is separately appropriate to the situation at hand. Although the mean length of Nim's utterances was only 1.6 signs, he produced strings of up to 16 signs in order to stress his wants. The longer strings were riddled with redundancy as shown by his most sustained utterance: "give orange me give eat orange me eat orange give me eat orange give me you." While in written form this is reminiscent of poetry by e.e. cummings (1972, p. 614), the two really are worlds apart. As the mean length of human utterances increases, their complexity also progressively increases (Terrace et al., 1979a).

Overview of Ape Language Projects

The cloud that Terrace's revelations cast over the ape language studies had harbingers in the criticisms of other scientists (e.g. Mistler-Lachman and Lachman, 1974), including the commentaries of the ape language researchers on one another's projects. Two camps are obvious—those who fostered their subjects (the Gardners, Fouts, Miles, and Patterson) and those who elected more traditional laboratory settings for their studies (the Premacks and the Rumbaughs). While Terrace chose the former approach and at first leveled his strongest criticisms at the laboratory projects, he and his associates (Petitto and Seidenberg, 1979; Seidenberg and Petitto, 1979, 1981) eventually turned also on the interpretations from home-rearing studies.

They were answered promptly by Gardner (1981), Van Cantfort and Rimpau (1982), and others, who leveled incisive criticisms at the methodologies that had been employed in Project Nim and for interpreting videotapes and other data on the signing apes.

Other critics joined the fray (Epstein et al., 1980; Brown, 1981; Michael, 1984; Muncer, 1983b; Muncer and Ettlinger, 1981; Ristau, 1983a,b; Ristau and Robbins, 1982a; Sanders, 1985; Thompson and Church, 1980), the most hard-boiled and unrelenting of whom were the Sebeoks. They organized symposia and published anthologies that focus on problems with the experiments and the interpretation of results therefrom (Sebeok and Umiker-Sebeok, 1980; Sebeok and Rosenthal, 1981; Umiker-Sebeok and Sebeok, 1982; Sebeok, 1982). Ristau and Robbins (1982b) presented a most far-reaching, thoughtful and fairminded critical review of the great ape language studies. It would be long-winded to attempt to repeat or supersede them here.

The upshot of reactions to the past 18 years of ape language research is that the great apes are generally thought not to have mastered syntax of the kind that allows humans to produce sentences creatively. Embedding has barely been explored and it is unlikely to occur until simpler sentence-like constructions are generated spontaneously by the subjects.

As investigators of two-way communication with apes realized the elusiveness of syntax, they turned to problems of semantics, just as linguists have shifted from a heavy concentration on syntax to questions of semantics in human speech. But it is much easier to demonstrate that great apes "know how" than that they "know that" (Ristau and Robbins, 1982b). Clearly apes can label objects with arbitrary and iconic symbols and remember a remarkable number of arbitrary signs. However, the skeptics argue cogently that they have not exhibited comprehension of meaning such that the English glosses that are applied by investigators are the unequivocal equivalents of human words. Thus, although apes can name and symbolize, we still do not know precisely what they know or that their use of symbols is homologous with our own symbolic capabilities. A productive direction for future studies might involve more emphasis on symbolic communication in naturalistic social relations instead of focusing on food and paraphernalia of the Yankee bourgeois child's world (Ristau and Robbins, 1982b).

NATURAL COMMUNICATION

Because most long-term field studies of apes have been focused on social and ranging behavior and because communication is recognized as an imperative of sociality, considerable effort has been expended to catalogue their vocalizations, communicative postures, facial expressions, and athletic displays.

Hylobatid Apes

Carpenter (1940) was the first scientist to attempt a systematic study of ape vocalizations in the field. As usual, he was wholly candid about the limitations of his recording methods and crude verbal descriptions of Chiengmai white-handed gibbon signals. He urged that especially equipped laboratory studies be conducted to complement the field studies. He noted that whereas the functions of high volume intergroup calls could only be discerned in the wild, subtler intragroup vocalizations would have to be recorded in a laboratory. He listed 9 types of vocalizations and their probable functions. But he acknowledged that verbal descriptions failed to convey their richness and complexity and sometimes border on the absurd (Carpenter, 1940).

A wide range of stimuli could evoke the particular vocal responses of *Hylobates lar carpenteri* and there was considerable variation in individual responses thereto. But because specific vocalizations persisted in captive assemblages of mixed species, Carpenter (1940) suspected that there are notable genetic factors underpinning hylobatid vocalizations.

The loud choral morning song bouts of Chiengmai gibbon groups served to maintain the integrity of their territories in lieu of daily combat. Morning songs peaked between 0730 h and 0830 h during the period when groups were travelling most. Less dramatic chirps, squeals, chatters, clucks and facial expressions seemed to facilitate group cohesion and harmony and to coordinate group progressions through the canopy (Carpenter, 1940).

Ellefson's (1968, 1974) study of *Hylobates lar lar* securely established the sexual divocalism of white-handed gibbon territorial calls. He documented that the distinctive morning song bout consists of an elaborate female great-call and duetted male hoot series coda. The great-call sets off bouts of vigorous swinging by the adults and chases by males if another group is in view (Ellefson, 1968, p. 193). Ellefson agreed with Carpenter that the morning songs probably announce the locations of lar groups in a region. The distinction between male and female vocalizations appears very early in childhood (Ellefson, 1974, p. 78).

Ellefson also noted that contesting males conflict-hoo and engage in acrobatic brachiating displays to dissipate agonistic tensions. Thus, they often avoid actual fighting, particularly if they are away from preferred food sources.

Most of the other vocalizations, gestures and facial expressions in Ellefson's appendix (1974, pp. 127-132) are produced by both sexes in intragroup contexts. While designating them with his own labels, Ellefson noted correspondences with Carpenter's "types" wherever possible. The fact that Ellefson's study was much longer and focussed on a few groups whose members were identified and followed regularly allowed him to augment Carpenter's gibbon glossary.

Baldwin and Teleki (1976) further expanded the list of communicative signals by white-handed gibbons from a study on 6 young free-ranging captives. They identified 12 facial expressions; 19 expressive postures, gestures and movements; 9 types of vocalizations; and variants thereof. Most of the visual signals are clearly documented photographically. Although the social composition of the Hall Island colony is atypical of *Hylobates* and some of the subjects had been injured during invasive neurological experiments, the results correspond closely with those of Carpenter (1940), Ellefson (1974) and Chivers (1972, 1974) and provide a solid base for future comparative studies on hylobatid ethograms.

In Tenasserim lar gibbons, unmated subadult males in their parents' territories, and mated adult males sing solos over spans ranging between a few minutes and up to 4 hours before or during sunrise from their lodge trees or elsewhere later in the morning. Subadults call more frequently and, on average, for longer periods. Further, while males engage in chasing matches between their territories, they utter distinctive calls (Raemaekers and Raemaekers, 1985).

Females solo less often than males do (Raemaekers and Raemaekers, 1985). Their single great-calls may induce duets in which they deliver complete arias which culminate in great-calls to which the male appends a coda. Raemaekers and Raemaekers (1985) found that the great-calls of known females were individualistic. Young females sometimes accompanied the great-calls of adults.

Most Tenasserim duetting ($\bar{x}$ duration $= 12$ min) occurs during midmorning, peaking around 0830 h, from anywhere in the group's territory. Sometimes duets contain rhythmic bursts of ooaa calls, the significance of which was not elaborated by Raemaekers and Raemaekers (1985).

Raemaekers and Raemaekers (1985) conducted playback experiments over a 10-month span in the Khao Yai National Park, Thailand, in order to test the reactions of Tenasserim lar gibbons (*Hylobates lar entelloides*; n = 8 groups) to loud calls and song bouts in their territories. All target subjects in the playback experiments approached stimulus duets in the centers of their territories (n = 8 tests) but only 2 out of 8 target groups approached the sources of duets on the borders of their ranges. However, target subjects called more often (50% of 8 trails) in response to stimulus duets on the border than at the core (20% of 5 trials).

Adult female Tenasserim lar gibbons were among the leaders of group approaches to played back conspecific female solos in the centers of their ranges significantly more often (88% of 8 trials) than they led approaches to male solos (none of 8 trials) or duets (14% of 7 tests) at the cores. Adult males were generally among the leaders of approaches to the stimulus duets (all of 8 tests) and male solos (88% of 8 trials). They were among the leaders toward only a quarter of 8 group approaches to female solos. Only centered playbacks of female solos consistently induced calling, including duets in which the target female was especially vigorous vocally (Raemaekers and Raemaekers, 1985).

This data suggests that there is strong intrasexual competition in Tenasserim lar gibbons. While resident males might tolerate alien females, the resident female would evict them. The resident male's antagonism toward other males induces him to chase off alien males and their cohorts (Raemaekers and Raemaekers, 1985).

Playbacks of solos by female capped gibbons (*Hylobates pileatus*) in the centers of Tenasserim gibbon ranges induced male led approaches in 62% of trials. The target group called during only one of the 8 tests. This indicates that female lar gibbons cannot tell the sex of capped gibbon callers (Raemaekers and Raemaekers, 1985). Since male lars approach them and female lars might not react aggressively to them, hybridization could occur where the 2 species meet. Indeed there are bigamous male gibbons in the Khao Yai National Park. One of the male's mates is always a conspecific and the second female is of the other species or a hybrid between them (Raemaekers and Raemaekers, 1985; Brockelman and Srikosamatara, 1984; Chapter 1).

The morning calls of agile gibbons (*Hylobates agilis agilis*) are similar to, and yet readily distinguishable from, those of closely related white-handed gibbons. The agile female's great-call consists of shorter notes, is more ornamented, and has a more sustained decrescendo. The short phrases of the male are quite different (Marshall et al., 1972) and the agile male's coda sometimes overlaps the female's finale. Whereas the entire agile family may join the coda, only the subadult male white-handed lar sings it (Marshall, 1981). Sumatran agile gibbon vocalizations are like those of Peninsular Malaysian conspecifics (Marshall et al., 1972).

Gittins (1978b, 1984b) noted that male agile gibbons sing solos for long spans while females rarely sing alone. Males begin the morning song bouts with hoot series and are later augmented by the female great-calls. Gittins speculated that male singing advertises their territories and females signal potential rivals that they are already mated. He observed that young males are discouraged from joining the adult song but young females are allowed to duet great-calls with their mothers.

The female parts of Peninsular Malaysian agile gibbon duets are highly individualistic as revealed by statistical analyses of field recordings on 8 subjects (Haimoff and Gittins, 1985). This may allow them to identify one another while singing contemporaneously.

Gittins's (1984b) focal study group at Sungai Dal, Peninsular Malaysia, sang on most (88% of) days. On average, they sang 2 bouts (r = 0-5), commonly one at dawn followed by another later in the morning. Overall they sang 35 min per day (Gittins, 1984b).

Like white-handed gibbons, agile gibbon males conflict-hoo, move about on branches, chase, and hang stiffly in front of their opponents while the females and young keep away from the fracas. In addition to "hoos," agile males emit low "whistles" during their display bouts (Gittins, 1980).

The morning song bouts of siamang have been studied intensively at several localities in Peninsular Malaysia (Chivers, 1971, 1972, 1974, 1976; Chivers and MacKinnon, 1977; Chivers et al., 1976; Kawabe, 1970; McClure, 1964; Papaioannou, 1973) and in numerous laboratory pairs (Lamprecht, 1970; Tembrock, 1974; Haimoff, 1981, 1983; Haraway et al., 1981). Detailed analyses of sonograms revealed no distinctions between the morning songs of wild Sumatran and Peninsular Malaysian siamang (Haimoff, 1983).

Inflation of the laryngeal (Figure 22) sac produces a "boom" which siamang pairs use to coordinate and synchronize their duets (Haimoff, 1981, 1983). The resonant effect of the inflated sacs produce louder calls than those of white-handed gibbons (Chivers et al., 1975).

Figure 22: Siamang singing. Note inflated laryngeal sacs.

Typically, duets begin when stationary adults are close to one another in an emergent tree and exchange single barks. These escalate to variable sets of multi-unit barks after which the male moves away and barks rapidly. Then, as the female continues to execute bark series, the male periodically delivers bitonal and ululatory screams. During the former, he moves back toward the female. After the ululatory scream he moves vigorously about the tree while screaming loudly. Concurrently, the barking female also moves athletically. Maturing family members may chorus with their parents but newly mated pairs still seem to require notable spans to develop full-fledged, well synchronized morning calls (Chivers, 1972, 1974, 1976; Haimoff, 1981).

Unlike white-handed gibbons, male siamang are mute during territorial chases, branch shaking and staring matches (Chivers, 1976). However mum they may be, siamang still mean business. After Chivers and MacKinnon (1977) played back calls centrally in a siamang territory, the occupants reacted with violent visual displays, repeated returns to the transgressed sites, and stepped up patrols of the area. Chivers and MacKinnon (1977) concluded that siamang territorial calling is largely innate.

Without providing an enumeration of siamang signals, Chivers (1976) proposed that their intragroup repertoire is smaller and less often employed than that of lar gibbons. Perhaps their closeness and lifelong bonding obviate the need for a larger repertoire of conspicuous signals.

Studies on sympatric hylobatid species show that siamang usually sing later in the morning and less often than white-handed and agile gibbons do. Gittins and Raemaekers (1980, p. 75) concluded that siamang tend to sing every third day, white-handed gibbons once per day, and agile gibbons twice in the morning. Most Peninsular Malaysian siamang song bouts occur between 0800 h and 1100 h while the gibbons sing mostly between 0600 h and 0900 h. They sing little after noon (Chivers et al., 1975). Sumatran siamang sing mostly between 1030 h and 1130 h while sympatric white-handed gibbons generally sing between 0830 h and 1030 h (Rijksen, 1978, pp. 117-118).

Like siamang, hoolock gibbons sing elaborate bisexual duets and do not engage in unisexual choruses or countersinging bouts (Marshall and Marshall, 1976; Marler and Tenaza, 1977; Tilson, 1979). Youngsters of both sexes may join the song. Males begin with a short series of high notes, which they repeat until the female joins in. Then both execute accelerated lines of alternating high and low notes. There is no clear-cut sexual divocalism in the song bout. Assamese *Hylobates hoolock* usually duetted once a day; the mean times are 0901 h in summer and 1004 h in winter. Occasionally in summer they also sing in the afternoon (1400-1600 h). Like siamang, hoolock neighbors tend to wait until one has finished its song before beginning theirs. Hence the duets are heard sequentially through the forest (Tilson, 1979).

Hamilton and Arrowood (1978) made sonograms of captive hoolock copulatory sounds. Like humans and baboons, humping hoolock female vocalizations begin earlier and are more complex than the grunting male's are. Both are elaborations of heavy breathing but their functions remain obscure.

Pileated gibbon duets are like those of siamang and hoolock gibbons in that the male and female contributions overlap instead of alternating as in other gibbons. However, like white-handed gibbons, pileated gibbon males end the song with a solo coda (Marshall and Marshall, 1976; Marler and Tenaza, 1977). Brockelman (1975) described the female great-call as a series of hoots which rise steeply in pitch and accelerate into a long bubbly trill. The male calls "Oh-Ah" several times on notes nearly an octave apart and often follows with a brief low bubbly trill.

Geissmann (1983) recorded a captive female *Hylobates pileatus* singing the full male section of the duet, presumably to instruct or inspire a "timid" male with which she was newly paired. Before cohabitation, each had sung solos that characterize its own sex. When they were introduced into the same cage she uttered a complete male solo and they took turns repeating male phrases. Later she returned to female great-calls and the male properly duetted with her. One year later they engaged in another bout of alternately singing male phrases. Geissmann (1983) could not explain these events.

The female great-call of Müller's Bornean gibbon (*Hylobates agilis muelleri*) is like that of pileated gibbons (Marshall and Marshall, 1976). Marler and Tenaza (1977) reported that instead of singing with the female, the male simply adds a brief coda to her finale. And Marshall and Marshall (1976) stated that instead of duetting, males solo before dawn while females call later in the morning. Per contra, Haimoff (1985) and Mitani (1984, 1985a,b) found that although the female dominates the duet, the male participates complexly early and late in the song bout. They confirmed that males also sing solos. The mated Kutai males usually soloed for up to an hour before dawn (0400-0600 h) about once per week. Solitary males sang more sporadically. Kutai female morning solos were rarer and consisted of their part of the duet (Mitani, 1984).

In the Kutai Game Reserve, East Kalimantan, Indonesia, pairs of Müller's gibbons usually duetted high in the canopy on dry mornings between 0600 h and 1000 h (Mitani, 1985a). Ninety-five percent of duets were performed before 0900 h (Mitani, 1985b). On average, duetting sessions lasted about 15 minutes. Countersung duets ($\bar{x} = 18.35 \pm 8.81$ min) were significantly longer than duets that pairs sang alone ($\bar{x} = 13.76 \pm 7.90$ min). Approximately 30% of 113 duetting bouts involved interactive singing among neighbors (Mitani, 1985b).

Mitani (1985a) could distinguish the different Kutai groups by the individualistic temporal patterns and acoustic features of their morning song bouts. The gibbon pairs tended to sing near the boundaries of their ranges (Mitani, 1985b).

Mitani (1984, 1985a,b) conducted playback experiments in order to discern possible functions of the gibbon song bouts. In >1,500 h of observation, his focal study group came within 100 m of another group only 7 times, all of which elicited duetting. The focal group countersung 5 times, engaged in one chasing match, and avoided contact with the other group by unilateral or mutual withdrawal 6 times (Mitani, 1985b).

According to the location of recordings in their range of about 40 hectares, Kutai gibbons reacted to playbacks of duets by some or all of the following: countersinging, display-swinging, uttering alarm "hoos," and approaching the stimulus. They did not seem to distinguish recordings of their own duets from those of other groups (Mitani, 1985a,b). They reacted most dramatically to calls that were central in their range, not only by engaging in all of the behaviors mentioned hereabove but also by remaining in the playback area for extended periods and increasing the duration of their song bouts subsequent to the disturbance (Mitani, 1985b).

Duet playback near the boundary of their range consistently elicited only countersinging from the residents, though they might also approach or move away from the stimulus thereafter. Duets within a neighboring range rarely stimulated countersinging or approaches by the target group (Mitani, 1985b).

The female of a Kutai couple commenced the duet, usually alarm hooed, and led the group movement toward or away from duet playbacks in their territory. Playbacks of female solos elicited the same response by the target group as the duets did. However, following playbacks of male solos, the male of the target group silently led them toward the stimulus. Mitani (1984, 1985b) concluded that adults of each sex defend the group range against like-sex intruders albeit with backup from her or his mate (Chapter 8).

Until recently, Javan gibbon calls were poorly known. Marshall and Marshall (1976) reported that most morning song bouts by *Hylobates moloch* consist of female great-calls which end rallentando. They also heard male hoots between female great-calls early in the bouts. But no full-fledged duets were recorded.

Kappeler (1984b) described 4 types of loud vocal bouts based on field studies and recordings of vocalizations by moloch gibbons at Turalak in the Ujong Kulon-Gunung Honje Nature Reserve, West Java, Indonesia. As in Müller's gibbons, adult females are the lead singers in moloch territorial maintenance and defense ($\bar{x} = 6.4$ hectares; $r = 11.2$-20.9 ha; $n = 6$ groups). In fact, male molochs rarely sing. Kappeler (1984b) detected no singing by 5 mated males during his 18-month study. An unmated male sang once, perhaps to attract a mate.

Moloch great calls are quite individualistic, so much so that Kappeler

(1984b) speculated that this allows the females to recognize one another. When a female commenced a song bout in her territory, the females of neighboring territories often began to sing immediately. Sometimes this produced a chorus. Subadult females commonly joined the prima donnas in song bouts.

Turalak females rarely (3% of 392 song bouts) sang before sunrise. Most (87%) song bouts occurred within the first 3 hours after dawn and they were rare (10%) thereafter. Kappeler (1984b) recorded no song bouts after about 1500 h.

The soloist ascends to the crown of one of several emergent trees in her territory. When she sings more than once per day, a new song tree is used for each encore. During the cadenzas of her great calls, she brachiates rapidly through the crown of the song tree. Sometimes this sends dead branches crashing to the forest floor (Kappeler, 1984b).

The songs of alien females inside or at the periphery of a resident female's territory prompt her to climb the nearest song tree and respond in kind. In 3 cases, the intruders then became silent and fled with the resident males in mum pursuit of the trespassing pairs. Playbacks of alien great calls also elicited song bouts by the resident female in the territory (Kappeler, 1984b).

Kappeler (1984b) observed 10 encounters between Turalak gibbon groups. In 4 of them the females countersang and parted with no overt aggression between any members of the groups. In 5 episodes, upon mutual awareness, all members of both groups screamed loudly. The males brachiated rapidly toward one another and faced off while the screaming continued. Occasionally the females uttered great calls during which younger group members fell silent. This seemed to prompt vigorous bouts of chasing over all stories of the forest between the adult males, which had been sitting opposite one another. Other group members stay back from the fracas and scream like cheer leaders who would be a credit to any champion. The contests usually ended abruptly after a vigorous chase when the mum males wandered off instead of sitting to face off again (Kappeler, 1984b).

In the shortest conflict that Kappeler (1984b) witnessed, only the males screamed and chased while other group members remained silent.

Kappeler (1984b) also observed harassing call bouts, in which all group members uttered loud ad libitum staccato screams and moved about agitatedly in the presence of a leopard (n = 3) and a human (n = 1).

Concolor gibbon calls have been studied more thoroughly in captivity than in the field. Goustard (1979a) noted subspecific differences between songs of male *Hylobates concolor leucogenys*, *H. c. gabriellae*, and *H. c. hainanus*. For instance, although the topography of segments are the same in Northern white-cheeked and red-cheeked gibbons, there are differences between frequencies of emission and the duration of components in their segments (Goustard, 1965, 1976). Differences between songs of female *H.*

c. leucogenys and *H. c. gabriellae* are less marked than those of the males.

Marshall et al. (1972) noted the high pitch of concolor calls. Marshall and Marshall (1976) remarked that the climax of the female great-call is at the highest pitch uttered by any gibbon. Males begin the duet with staccato "eek" phrases that lead to a variable number of inflected shrieks. The female enters with low growls that develop into a whine. Then she soars in pitch, intensity and speed, abruptly breaks off, and finishes with a descending series of chirps. The male responds with a coda that is more elaborate than his prelude (Marshall et al., 1972; Marshall and Marshall, 1976; Goustard, 1979a). The utterance of great-calls appears rather suddenly in captive pubescent males (DeMars et al., 1978).

Wild Laotian white-cheeked gibbons duet just before sunrise and sometimes in the late afternoon. Goustard (1979a) stated that duetting is more extensive on clear than cloudy days. On the basis of experiments with different groupings of subjects Goustard (1979a,c, 1982, 1984b) concluded that the great-calls of male, female and young concolor gibbons chiefly serve to reinforce their social bonds. Schilling (1984) added that in the wild they probably advertise territorial claims.

Tenaza and Tilson (1977) list 10 types of Kloss gibbon vocalizations; Marler and Tenaza (1977) list 11. Two of them, "sirening" and "alarm trills," function as long-distance alarm calls. They are emitted by males and females when humans are detected on the forest floor. Humans have hunted Kloss gibbons for centuries in the Teitei Peleigei Forest of eastern Siberut Island. They abandon the chase if detected by the prey because their arrows are no match for alert, rapidly fleeing gibbons in the canopy.

Tenaza and Hamilton (1971) confirmed the observation of Chasen and Kloss (1927) that the morning song bouts of *Hylobates klossii* are quite distinct from those of siamang and instead are closer to those of other gibbons. Marshall and Marshall (1976, p. 237) proclaimed the Kloss gibbon great-call to be "probably the finest music uttered by a land mammal." Neighbors may be less appreciative of their vocal accomplishments as males are inclined to sing from their lodge trees during the 4 hours before sunrise (Tenaza, 1975). At the onset of the fruiting season some male singing shifted to postdawn. Females always sang during the 4 hours after dawn and after departure from lodge trees. Their mates remain silent or softly "whistle" during breaks in their arias.

On average, adult males sing every 4.4 days and females sing every 4.2 days. Subadult males sing once every 2.5 days; their song bouts last 63.1 min versus 57.9 min for adult males (Tilson, 1981). Generally males respond to singing by other males, and females react to other females. Because of overlaps these bouts usually develop into unisexual neighborhood choruses. The countersinging males appear to be relaxed and they tend to take turns while countersinging females engage in vigorous locomotor displays and sing simultaneously (Tenaza, 1975, 1976). The close proximity of adult males seems to suppress the vocal efforts of younger males. Male songs are individualistic enough that Tenaza (1976) could identify the singers.

Tenaza (1976) concluded that male singing functions to maintain exclusive use of their emergent lodge trees which are scarce due to the high density of gibbons (Chapter 4). The louder songs and athletic displays of females, which are executed along their territorial boundaries, indicate readiness to chase out trespassers. The lack of bisexual duetting may be a response to human predation. The quiet males, which stay below the calling females, can serve as lookouts while their mates defend the territorial border. Chorusing may reduce the risk of individual victimization from terrestrial predators, viz. humans (Tenaza, 1976).

Over a 7-month span in central Siberut, Whitten (1984a) monitored the songs of a mated male *Hylobates klossii*. Overall he sang once every 1.7 days (on average, before dawn every 2.0 days and after sunrise every 9.8 days; n = 89 days). On 2 days he sang once before dawn and twice thereafter. Nocturnal rains and low temperatures seemed to surpressd predawn singing.

There was no significant difference between the durations of his predawn (median = 46 min) and post-dawn (median = 40 min) song bouts. Starting with a single piping note, the song progressed through whoos and whoops to a cadenza of complicated trilling phrases. The length and elaborations of male Kloss songs, particularly by comparison with those of Kloss females, led Whitten (1984a) to infer that mere territorial advertisement is not adequate to explain Kloss male song bouts. He proffered that they may also provide information about the fitness of potential rivals.

The capacity of hylobatid apes to hybridize offers intriguing possibilities for determining the extent of genetic specificity for their vocalizations (Tenaza, 1985). Marler and Tenaza (1977; Tenaza, 1985) found that the song of a hybrid female was intermediate between those of her white-handed father and Müller's gibbon mother. The song and duetting behavior of her full brother basically resembled those of white-handed males but also contained an element like that of Müller's gibbon male songs. It will be especially interesting to follow the development of siabon vocalizations since they can lack laryngeal sacs (Wolkin and Myers, 1980). Shafer et al. (1984) reported that a female siabon progressively developed siamang-like vocal features despite the continued absence of inflatable laryngeal sacs.

Orangutans

Most field researchers concur with Yerkes's supposition that orangutans are probably the least vocal of the great apes (Davenport, 1967; Horr, 1975; MacKinnon, 1971; Galdikas, 1983). However, MacKinnon (1974a) catalogued 16 vocalizations made by free and captive subjects. His information was derived mostly from free-ranging Bornean beasts. Rijksen (1978) described 17 sounds that are uttered by wild, captive, and rehabilitant Sumatran orangutans. Maple (1980, pp. 80-81) compiled a useful comparative table of their listings; it shows only 12 that are the same. Additional field studies are needed to detail the functions of many orangutan vocalizations, perhaps to complete the repertoire, and to discern whether there are differences between the Bornean and Sumatran subspecies.

By far the most intensively studied vocalization of *Pongo* is the dramatic long call (MacKinnon, 1971), which is only given by prime adult males (Horr, 1972, 1975, 1977; Rodman, 1973; MacKinnon, 1971, 1974a, 1979; Galdikas, 1978, 1979, 1982a, 1983; Rijksen, 1978; Mitani, 1985d). Galdikas (1979, p. 212) described it as "the most impressive and intimidating sound to be heard in the Kalimantan forest." The laryngeal sacs are voluminously inflated and contract heavily and irregularly during the call, which begins as a low soft grumble with vibrato, increases to a leonine roar, and terminates decrescendo with soft grumbles and sighs. During the roar, the subject protrudes and parts his lips and extends his neck. The lips are closed during the grumbling finale (Hofer, 1972; MacKinnon, 1971). Short (1981) suggested that the cheek pads (Figure 23) act like parabolic reflectors to locate callers.

Figure 23: Adult male Sumatran orangutan. Note cheek pads and 'tense mouth face'. (Photo courtesy of H.D. Rijksen.)

Pongid long calls can be heard both day and night in Bornean and Sumatran forests. They are audible up to 2 km from the source (Galdikas, 1982a). Rodman (1973) heard long calls on average twice per month in the Kutai Reserve, East Kalimantan (Borneo). But there were only two adult males in the study area and one of them did not call, perhaps because he was past prime.

During a 16-month follow-up in the Kutai Game Reserve, Mitani (1985c,d) had more male subjects to observe and he tested their reactions to played back long calls in their ranges. The resident population included an adult male, 2 subadult males and 3 females with youngsters. In addition, 6 adult males, >2 subadult males and a solitary adult female visited the study area (Mitani, 1985c,d).

The resident adult male, which was dominant over all others, called 3 or 4 times per day. Other adult Kutai males called significantly less often. The head caller and his subordinates performed, usually in the lower story of the canopy, at all times of the day and occasionally at night (Mitani, 1985d).

Most (82%) of the dominant male's long calls (n = 157) appeared to be spontaneous, 8 followed loud branch or tree falls, 6 immediately preceded 67% of 9 copulations, 6 overlapped played back calls of another male, 1 occurred as he chased another adult male, and 3 immediately succeeded similar chases. He called significantly more often on days after he was with females than on days that preceded, and during, bisexual associations and when he was alone (Mitani, 1985d).

Subordinate male and adult female Kutai orangutans typically oriented toward the resident male's long calls (n = 223) when they were <400 m away; farther ones were generally ignored. Two adult males kiss-squeaked after 2 (18%) of 11 calls. Other subjects kept quiet. Adult males usually increased distance between themselves and the calling site. Adult females did not approach the caller, even when they were accompanied by subadult male tagalongs that had engaged them in forced copulations and which probably would have to leave them if they were to approach the caller (Mitani, 1985c,d).

Mitani's (1985d) playback experiments supported his naturalistic observations that responses to long calls vary as a function of male dominance relationships. Whereas the resident adult tended to countercall and approach the stimulus, subordinate adults and subadults moved away. Further, they tended to beat a hasty retreat and not to call for a while. A few subordinate targets expressed their agitation by bubbling, kiss-squeaking or shaking branches but these displays are not much to shout about. The 3 females tended to eschew the taped calls just as they seemed to shun live recitals (Mitani, 1985d).

Horr (1975) reported that in the Lokan area of Sabah, East Malaysia, orangutans normally called once a day or less frequently. Most calls were heard early in the morning or late in the afternoon. Occasionally, they called at night and more than once per day (Horr, 1975).

MacKinnon (1971) noted long calls 241 times during 394 days in the Ulu Segama Reserve, Sabah. Several were given on some days and none over spans of several days. Long calling peaked about dawn and there was a second, larger peak in late morning at Segama.

Most of the 33 calls that MacKinnon (1974a) recorded in the Ranun River area of Sumatra occurred at night. Further, the Ranun calls had fewer call-units (rarely >25) than the Segama calls did (sometimes up to 50 units). And individual units were longer at Segama than at Ranun. In contrast with MacKinnon's observations in Sumatra, Rijksen (1978, p. 234) heard the 3 adult Sumatran males of the Ketambe area call mainly late in the morning.

The most extensive study of long calls was conducted by Galdikas (1983) on Bornean orangutans in the Tanjung Puting Reserve of southwestern Kalimantan Tengah. Her subjects called more frequently than the orangutans in other areas. Focal males uttered long calls 907 times over a four year span. Overall, Galdikas recorded a mean of 1.5 calls per day. Dominant males called more frequently than less dominant ones, e.g., one of the former called on average 4 times per day. They pushed over snags, causing loud crashes, as preludes to or during 3% of observed long calls. In this behavior Tanjung Puting orangutans are unique.

The calling frequency of males that were followed all day peaked between 0900 h and 1100 h. The calls of distant subjects were distributed widely during the day and night with 31% of them being heard between 0500 h and 0800 h. Afternoon calls were most often heard between 1600 h and 1700 h. Twenty-nine percent of distant long calls were noted between 1700 h and 0500 h (Galdikas, 1983).

Researchers have suggested several functions for long calls, all of which relate broadly to the reproductive behavior of the species. MacKinnon (1971, p. 170) initially proposed that the long calls maintain spacing among adult males and do not serve to attract females to them. Whereas Segama females typically showed no reaction to a call and occasionally even hid or moved away upon hearing it, adult males responded by calling themselves, approaching, and perhaps chasing or displaying toward the caller. But sometimes the male receiver remained unusually quiet and once a female seemed to hurry toward a caller (MacKinnon, 1974a). Calls were commonly uttered spontaneously and immediately after loud percussive noises, like falling trees and breaking branches. Overcrowding due to deforestation may have caused an increased frequency of long calls in one part of the Segama area (MacKinnon, 1971, 1974a, 1979).

Horr (1972, 1975), Rodman (1973), Galdikas (1979, 1983), and Rijksen (1978) agreed that long calls serve not only to space adult males in the forest but to attract sexually receptive females to the caller. It pays to advertise. Mitani's (1985c,d) observations cast doubt on the latter function. But he acknowledged that parity might influence whether a female will pursue a distantly calling male. The 3 Kutai females were parous and perhaps anestrous. Mitani (1985d) reasonably suggested that we must await further research, focussed on a spectrum of estrous females, before drawing conclusions about the orangutan long call as a sex attractant.

Horr (1977) and Rijksen (1978) suggested that variants in the long calls may serve to identify the callers to other orangutans. Humans at Tanjung Puting could distinguish the calls of 4 males when they were within 400 m (Galdikas, 1983).

Galdikas (1983) provided the fullest quantitative data on the contexts of long calls. She found that only 6.7% of calls by focal males followed sudden sounds such as falling snags and trees or animal noises. They were much more responsive to nearby crashes than to distant ones. The remaining 93.3% of the calls by focal males were spontaneous, though the possibility that a number of them were stimulated by the presence of humans cannot be ruled out. Ten percent of the spontaneous calls began while the subjects were on the ground and about one-third of these were completed in trees. Eleven percent of the spontaneous calls were uttered from night nests and another 10% were in or while entering food trees.

Galdikas (1983) cited two particular contexts which commonly elicited long calls, When two males were visible to one another invariably one or both of them called. They also called on 31.5% of occasions when they met other orangutans that they had not seen for a while. By observing the reactions of known males to the individualistic long calls of others, Galdikas (1983) worked out a dominance hierarchy among 4 males at Tanjung Puting. Less dominant males often fled terrestrially or arboreally upon hearing the long calls of more dominant males, the foremost of which never moved away from the caller. Indeed the more dominant males frequently crashed snags, called, and moved toward less dominant close callers. It pains to advertise.

The second context of frequent long calls or parts thereof is copulation. Males called or uttered the grumbling segment of the long call during 87% of 30 copulations. Galdikas (1983) did not mention any peculiarities of these vocalizations that might broadcast that the male had scored.

Galdikas (1979, pp. 208-209) stated that females uttered low, hoarse grunts and groans while resisting copulation but later quieted down when apparently accepting the male's efforts. MacKinnon (1971, 1974a) noted "mating cries" by females during some copulations and remarked that they are essentially like fear screams except that the former are more rhythmic and continuous. Rijksen (1978, p. 230) termed them "mating squeaks." Like humans, some Sumatran female orangutans habitually vocalized while others remained silent during copulation. The coupled female in Figure 24 looks more ecstatic than fearful to me; but her mother might view it differently.

Because parturition is even less commonly witnessed than copulation we have minimal evidence about the extent to which it is signalled vocally. DeSilva (1972) reported that a rehabilitant primiparous Bornean female

Figure 24: Ventroventral copulation by suspended Sumatran orangutans. The female is squealing. (Photo courtesy of H.D. Rijksen, 1978.)

whimpered and cried tearfully as she gave birth in an open cage. Galdikas's (1982a) accounts of 2 births in tree nests suggest that parturition is basically a nonvocal event for multiparous and primiparous females in the forest, where it is probably unwise to issue birth announcements.

Youngsters and, to a lesser extent, adults utter a variety of squeaks, grunts, moans, barks, and screams during playful and fearful situations (MacKinnon, 1974a; Neimitz and Kok, 1976; Rijksen, 1978; Maple, 1980). The noisy tantrums of youngsters are especially audible in the forest (Horr, 1977; Rijksen, 1978). Frightened rehabilitants and captives sometimes grind their teeth audibly (Rijksen, 1978).

Alarmed and annoyed adults produce several peculiar sounds which may be accompanied by branch breaking and other physical displays. In Sarawak, East Malaysia, 3 Bornean orangutans that were presumably annoyed by Schaller (1961) interchangeably produced (a) "kissing" sounds by sucking air through their puckered lips and occasionally kissing the backs of their flexed fingers loudly; (b) "gluck-gluck-gluck" noises by gulping several times with the lips pursed; and (c) bitonal burps (from a low note to a high note). The first two sounds have been confirmed by MacKinnon (1974a) and Rijksen (1978). They termed the second (gulping) sound "grumph." MacKinnon (1974a) noted that Sumatran orangutans kissed the backs of their hands as well as their knuckles. Neither observer described bitonal burps though they may be counterparts of MacKinnon's "gorkums." MacKinnon (1974a) reported that kiss sounds were the most common and grumphs the second most common vocalizations by his subjects, probably because they are reactions to human intrusions.

Very annoyed females and subadult males uttered lork calls and displayed violently. Rijksen (1978; p. 233) suggested that the lork call also serves to advertise an adult female much as the long call serves adult males. Kisses, grumphs, gorkums and lorks seem to indicate increasing levels of annoyance. Instead of making kissing sounds, some subadults blow raspberries. Occasionally during intimidation displays adult males produced "ahoor calls" by gasping sharply and then uttering an explosive grunt (MacKinnon, 1974a).

The communicative postures and gestures of Sumatran orangutans are much better documented than those of Bornean orangutans. Compared with Rijksen's (1978, pp. 177-238) ethogram, based heavily on rehabilitants, the accounts of other workers are relatively impoverished.

Males use their bulk and other secondary sexual characteristics to good advantage during intimidation displays. The inflated laryngeal sacs and piloerection on their shoulders and arms make them look even larger than usual (MacKinnon, 1974a; Rijksen, 1978). Further they adopt a number of extended suspensory and rigid bipedal and quadrupedal postures and move their bodies in ways that sport their inflated shapes. Long hair on the backs and upper limbs of Sumatran males is most impressive as they sway it back and forth (MacKinnon, 1974a; Rijksen, 1978).

Although snag crashing seems to be restricted to orangutans of Tanjung Puting (Galdikas, 1983) others commonly wave, shake, break and drop branches at annoyances. Some individuals elaborate on branch shaking by executing dramatic dives and lunges wherein they release some grips and dangle headfirst below footholds or sidelong from an ipsilateral hand and foot (Davenport, 1967; MacKinnon, 1974a; Rijksen, 1978). Snag crashing also might have developed from branch shaking.

Branch busting orangutans commonly stared and exhibited "tense mouth" facial expressions in which the lips were tightly pressed together (Rijksen, 1978; Figure 23). Threatening individuals also gape and yawn widely thereby displaying their teeth (MacKinnon, 1974a, p. 61).

The teeth are exposed when fearful orangutans grimace and playful youngsters draw back the corners of their rather widely open mouths (Rijksen, 1978, pp. 220-222). Orangutans sometimes protrude their prehensile lips in contexts where they seem to be mildly fearful, e.g. when trying to appease or to approach another animal. This may accompany whimpers and moans, especially in youngsters (MacKinnon, 1974a; Rijksen, 1978).

Both MacKinnon (1974a) and Rijksen (1978) noted many similarities between orangutans and common chimpanzees in their facial expressions and apparent meanings. Indeed *Pongo pygmaeus* has a remarkable repertoire of visual signals for so separate an ape.

Common Chimpanzees

The common chimpanzee has long been acknowledged as the noisiest and most demonstrative of apes. They could hold their own at any rock concert. Despite experience with clamoring captives, Nissen (1931, p. 89) was astounded by the cacophony of his wild Guinean subjects. The din that they produce in the forest must make many a predator salivate and perhaps hesitate, as there are no documented instances of nonhuman predation on boisterous chimpanzees. Reynolds (1964, p. 165) speculated that their outsized ears might be related to their vociferousness.

Chimpanzees have a diverse repertoire of graded vocal signals which are variably combined with subtle and striking facial expressions; body postures; and athletic displays, during which they brutalize their surroundings. Largely through the efforts of Goodall (1963b, 1965; Lawick-Goodall, 1967, 1968a,b, 1971) and Marler (1965, 1969, 1976; Marler and Hobbett, 1975; Marler and Tenaza, 1977) the communication of provisioned Gombe chimpanzees has been most thoroughly described and, to a lesser extent; interpreted functionally. On the basis of sound spectrographic and cine film analyses Marler (1976; Marler and Tenaza, 1977) reduced the number of classes of chimpanzee vocalizations from 24 (Lawick-Goodall, 1968a,b) to 13. He was hampered in this exercise because many calls grade into one another and sometimes only what seemed like parts of series were uttered.

Pant-hoots were the most frequently heard chimpanzee vocalizations at Gombe (Marler, 1976). They are uttered most often by males. In both sexes they are individually distinctive to human and presumably also chimpanzee receivers (Lawick-Goodall, 1968a; Marler and Hobbett, 1975; Bauer and Philip, 1983). Pant-hoots are usually voiced on exhalation and inhalation. Complete male pant-hoots increase steadily in volume to a roaring or shrieking climax at the end of the crescendo. Female pant-hoots lack such dramatic climaxes and tend to begin with deeper pitches (Marler and Hobbett, 1975). Thus the sex of a caller should be apparent to listeners even if they do not know him or her personally. Reynolds (1964) noted that chimpanzee "choruses" are audible over more than 3 km.

Pant-hoots occur in a number of contexts. Chimpanzees seem to listen carefully to distant pant-hoots and then answer them. They also pant-hoot spontaneously while feeding, abed at night, meeting other chimpanzees, subdividing into smaller foraging parties, eating mammalian prey, and otherwise generally aroused (Reynolds and Reynolds, 1965; Izawa and Itani, 1966; Lawick-Goodall, 1968a,b; Marler, 1976; Teleki, 1973a; Ghiglieri, 1984a,b). Pant-hoots probably serve to announce the locations of individuals so that friends can meet and foes can be discreet. It is also likely that pant-hoots serve other functions that might be diagnosed from further fieldwork, hopefully including non-provisioned subjects. For instance, Ghiglieri (1984b) found that the arrival of later parties of Ngogo chimpanzees was significantly associated with pant-hooting at fruit trees.

In addition to pant-hooting, highly excited male chimpanzees rhythmically slap, stamp and pound resonant tree buttresses, trunks and logs, thereby producing sounds that are audible several kilometers away (Nissen, 1931; Kortlandt, 1962; Reynolds, 1965; Reynolds and Reynolds, 1965; Goodall, 1963b; Lawick-Goodall, 1968a,b, 1971; Izawa and Itani, 1966; Jones and Sabater Pi, 1971; Sugiyama, 1973b; Nishida, 1979a; Mori, 1982; Ghiglieri, 1984b). Such drumming is usually accompanied by vocalizations and sometimes also by other sorts of vegetal assaults.

Nissen (1931) reported that drumming occurred frequently during the day, quite rarely at dusk, and not at night in Guinea. Like Nissen, Jones and Sabater Pi (1971) noted "frequent" drumming in Equatorial Guinea, with 93% of sessions heard during the first 6 daylight hours and little during the rest of the day or first hour after dawn. The peak of drumming was between 0730 h and 0900 h.

Reynolds and Reynolds (1965) observed that 8 times irregularly over a 9-month span Budongo chimpanzees drummed and called continuously for several hours. However, most drumming lasted only a few minutes. On one day 18 outbursts of drumming were heard between 0730 h and 1630 h. The peak was between 1300 h and 1330 h and 83% of them preceded 1330 h. Reynolds and Reynolds (1965, p. 407) called this a "typical" day but provided no statistics that would allow this label to stick.

Lawick-Goodall (1968a,b) noted that, in Gombe, male fellow travelers took turns drumming on particular tree buttresses along the trails. Reynolds (1965) found that the buttresses of ironwood trees were favorites of drumming Budongo chimpanzees. The absence of nocturnal drumming (in contrast with calling) may be related to the fact that percussionists would have to leave their nests and venture near the ground.

The remaining 12 chimpanzee vocalizations (pant-grunt, laughter, squeak, scream, whimper, bark, waa bark, rough grunt, pant, grunt, cough, and wraaa) appear to be directed at nearby listeners, though screams can carry over notable distances. Like orangutans, young chimpanzees throw noisy tantrums when frustrated, e.g. during weaning (Tomilin and Yerkes, 1935; Andrew, 1962; Lawick-Goodall, 1968a,b; Clark, 1977; Nicholson, 1977). Adults, including deposed alpha males, also exhibit tantrums (de Waal, 1982).

Gombe females either ran off screaming after copulation or remained quite calm about what had just transpired. Often during copulation they squeaked and grinned. Sugiyama (1969, p. 211) observed similar behavior by Budongo chimpanzees. Several Gombe males invariably panted at the culmination of copulation and sometimes slowly smacked their lips before the big moment (Lawick-Goodall, 1968a,b). Both sexes pant quietly during grooming sessions, peaceful reunions and mutual greetings (Marler and Tenaza, 1977).

Frolicking chimpanzees, and especially tickled youngsters, produce sounds that vaguely recall human laughter (Reynolds and Reynolds, 1965; Lawick-Goodall, 1968a,b; Marler, 1976; van Hooff, 1981).

Chimpanzees threaten nuisances mildly with coughs and passionately with barks. Subordinate animals pant-grunt as distance is closed between them and more dominant individuals. Pant-grunts have been invaluable for determinging the relative statuses of individual Gombe chimpanzees (Chapter 8). They are species-specific, that is to say, unlike other chimpanzee signals, they were directed only at other chimpanzees and not toward the baboons that also interacted with them in the Gombe camp (Bygott, 1979, p. 409).

Rough grunts occur when they approach and eat preferred foods (Marler and Tenaza, 1977). Marler concluded that rough grunts, and probably other vocalizations, are not learned because Viki, which had been isolated from other chimpanzees (Hayes, 1951), uttered them. Similarly, Temerlin (1975, pp. 113-114) remarked that a female chimpanzee, which grew up in his home, made many of the sounds and facial expressions of Gombe chimpanzees despite the fact that she had not had opportunity learn them from fellow chimpanzees. She also produced some unique vocalizations.

Adult females and adult and adolescent males utter wraaa calls in response to human intruders, predators, and other beasts that alarm them. They sometimes also slap or pedally thump the ground or a tree sharply. These percussions and wraaa calls not only threaten the stimulator but also warn companions of potential danger (Nissen, 1931; Reynolds and Reynolds, 1965; Izawa and Itani, 1966; Lawick-Goodall, 1968a,b; Marler and Tenaza, 1977). Wraaa calls were especially conspicuous among the reactions of Gombe chimpanzees following the death of an adult male which fell from a tree (Teleki, 1973d).

The visual signals, and especially postures and facial expressions, of captive chimpanzees have been studied intensively by Ladygina-Kots (1935), van Hooff (1962, 1963, 1967, 1971, 1972, 1973, 1976), Reynolds and Luscombe (1969a-c, 1976), Chevalier-Skolnikoff (1973), Berdecio and Nash (1981), de Waal (1982; de Waal and van Hooff, 1981) and Tomasello et al. (1985). These complement the field observations of Lawick-Goodall (1968a,b), Reynolds and Reynolds (1965), Marler (1976; Marler and Tenaza, 1977), and Plooij (1978, 1979), and provide a solid ethogram for the species. A number of them accompany particular vocalizations and indeed might be necessary for the production of certain sounds (Andrew, 1963a-c, 1964, 1965; van Hooff, 1981).

Like their vocalizations, chimpanzee visual signals tend to grade into one another. Marler and Tenaza (1977, p. 1011) classified chimpanzee facial expressions into 6 categories (tense-mouth face; open-mouth threat face;

grin; pout face; play face; and lip-smacking face). They omitted facial configurations that are associated with the physical production of vocalizations but which have little apparent valence as visual signals.

Before chasing or attacking a subordinate animal and before copulation, a male chimpanzee may glare fixedly at his target and exhibit a tense-mouth face in which his lips are tightly compressed. His upper lip may bulge as though it holds a pocket of air. In less tense situations, particularly if the would-be intimidator is not unduly afraid, he stares at the target with his mouth open a variable amount and his eyebrows depressed.

The frightened or squelched chimpanzee grins toward conspecifics with its mouth closed or open slightly. If contacts are nonaggressive chimpanzees grin with their mouths open widely and eyebrows elevated. Open-mouth grins also occur when subordinates threaten dominant group members and other species of which they are fearful.

Whimpering and mildly fearful, frustrated, and curious youngsters and adults protrude their lips, forming pout faces. Contentedly playing individuals allow their mandibles and lower lips to droop thereby producing play faces. Groomers stare closely at their tiny targets and slowly move their jaws in a lip-smacking face.

Hell knows no fury like a riled chimpanzee. If his glaring tense-mouth face doesn't quell a disturbance or keep rivals apart, a dominant male may rise with the hair on his arms and shoulders piloerected and charge quadrupedally. If he catches an offender he may grab its hair and stamp bipedally on its back. If the victim is small it may receive one or more body slams or be dragged by a limb or hair. The attacker commonly mouths, and less frequently bites, the unfortunate. Handfuls of hair may mark the scene. Milder attacks, particularly those of females, may involve slapping and scratching. Lawick-Goodall (1968a) and other humans and competitive baboons have received noteworthy knocks from disturbed chimpanzees.

Displays ranging from tenuous to athletic frequently occur instead of attacks. The disturbed animal may bark and jerk its head up and back by extending the neck. It may also rapidly fling a forelimb palmward or lunge and make hitting movements with the dorsum of the hand facing the nuisance. Downward swats and lunges may precede slapping bouts and follow attacks when the vanquished has been emboldened by a reassuring third party. Threateners vigorously jerk and shake attached branches and wave free objects or hurl missiles toward disturbers (Lawick-Goodall, 1968a,b). Males may "bluff-over" subordinates, i.e. leap or swing a hindlimb over the crouching target (de Waal, 1982). They also take to the trees where they dash about and brachiate (Sugiyama, 1969, p. 201).

Among the most striking displays of chimpanzees are bipedal runs and sway walks (=swaggers), especially when they are piloerected and waving their forelimbs and perhaps branches overhead. Males perform sway walks more commonly than females do. Youngsters amusingly mimic them (Lawick-Goodall, 1968a,b; van Hooff, 1973).

Victims caught in attacks have little recourse but to crouch, scream and hope for the best, especially when they are outweighed or outnumbered. But in a pinch, females and subordinate males will fight back and can score telling bites on overbearing males and other adversaries (de Waal, 1982). However, most altercations are avoided or resolved through a rich repertoire of submissive, appeasement, and reassuring postures and gestures.

Flight can carry one away from a fight or fright. Occasionally one can hide or more commonly move near the protection of others or away from the source of disturbance. A threatened individual may also avoid attack by crouching close to the ground, usually with its rump toward the aggressor. After attacks, victims commonly crouch and may back toward their tormentors. Rumpward posturing is called presenting. Presentations are adopted also by estrous females before copulation. Various grins, screams and squeaks accompany presentations, depending upon the intensity of the interaction. Subordinates may present to dominant individuals as they pass near them or are approached by them. These low key presentations may be straight faced. At Gombe, females presented most often to males. Males frequently received presentations and seldom executed them. Youngsters also presented much more often than they received presentations (Lawick-Goodall, 1968a,b).

When aproaching dominant animals face first, especially after they have displayed, subordinates may bow quadrupedally, flexing their elbows more than their knees. They also may bob one or more times when facing higher ranking chimpanzees (Lawick-Goodall, 1968a,b).

A bowing or crouching chimpanzee may press its lips or teeth to the face or body, usually the groin, of the dominant animal. If distance is not closed, it may reach toward the higher ranking individual or touch some part of its body, usually the head, back or rump. The submissive one may also mount and lightly embrace its attacker. It may execute a few pelvic thrusts and gingerly touch or hold the scrotum of the dominant male with its foot (Lawick-Goodall, 1968a,b).

Peace is signalled by the dominant animal touching the extended hand or proffered bottom or kissing or patting some part of the suppliant. Passing subordinates may also be touched briefly with a hand or foot. The genitalia of females may be more thorougly examined by males. The dominant animal may mount or embrace the subordinate in manners similar to reciprocals described hereabove. Occasionally, dominant chimpanzees present to subordinates and may receive scrotal manipulations from them (Lawick-Goodall, 1968a,b).

Reassurance gestures are sought and given not only between attackers and victims but also among other group members that are disturbed by intragroup altercations or extragroup scares. Many of them occur when subgroups merge and individuals greet one another (Lawick-Goodall, 1968a,b; de Waal, 1982).

Libidinous males may stand bipedally and rock sideways from foot to foot, shake branches, and put down Tarzan with arboreal swings and leaps (Lawick-Goodall, 1968b). Some flash their erect attenuated pink penes toward prospective mates by sitting with their thighs widely abducted and their pelves thrust forward (Tutin and McGrew, 1973; Tutin and McGinnis, 1981; de Waal, 1982). They can flip their phalluses up and down. A male may have to grunt to get a female to regard his prize and then beckon her to approach with his upper limb. If receptive to his advances, she crouches over his lap and the event is quickly consummated (de Waal, 1982).

Libidinous females sometimes pursue males and attempt to stimulate them with acts that are probably too shocking to repeat before older readers (de Waal, 1982, p. 161).

Bonobos

Preliminary accounts from the field (Mori, 1983, 1984; Badrian and Badrian, 1984a; Kano, 1980; Kano and Mulavwa, 1984; Kuroda, 1980; Chapter 8) and laboratory (Savage-Rumbaugh, 1984a; Savage-Rumbaugh et al., 1977) evidence that bonobos have a number of unique vocalizations, gestures and postures.

MacKinnon (1976) reported that an alarmed arboreally fleeing female bonobo uttered eerie, high-pitched screams that were similar to the calls of gulls. He concluded that they are generally quiet and appear not to have long-range contact calls or territorial signals. Although Badrian and Badrian (1977) agreed that Lomako bonobos are quieter than Gombe chimpanzees, they mentioned short calling bouts by them. Kano (1979) portrayed bonobos at other Zairean localities as much noisier than those at Lomako, stating that their incessant screaming made them easy to locate. He agreed that common chimpanzees are much noisier still and that bonobos have a distinct high-pitched, metallic alarm call, which both males and females utter.

Based on a 9-month study of the Wamba E-group, mostly at the feeding station, Mori (1984) identified 14 vocalizations and 47 nonvocal behavior patterns in the communicative repertoire of *Pan paniscus*. She compared them with data from her prior 6-month study on the provisioned Mahale K-group of *Pan troglodytes* and a one-month study on 2 groups of *Pan gorilla graueri* in the Parc National du Kahuzi-Biega, Zaire (Mori, 1983).

Mori (1983) noted that many behavior patterns are similar between bonobos and common chimpanzees. But she also found that some behaviors, like female homosexual hunching (Chapter 8), were specific to bonobos versus common chimpanzees. Further, unlike displaying common chimpanzees, terrestrially charging bonobos seldom ran bipedally.

Bonobo vocal repertoires were also different from those of other African apes. For instance, they hissed in tense situations. It appears from Mori's (1984, p. 259) brief descriptions of other vocalizations by *Pan paniscus* that they are distinctive. She specifically mentioned that bonobo pant-hoots, threat calls and wraahs are similar to those of eastern chimpanzees, though even here there are some departures from the chimpanzee model (Mori, 1983, pp. 490-491). Now we need fuller accounts of bonobo vocalizations and a pictorial record of their facial expressions and displays in the mode of those by Lawick-Goodall (1968a) for *Pan troglodytes* and by Fossey (1972, 1983) for *Pan gorilla*.

Savage-Rumbaugh et al. (1977) and Rumbaugh et al. (1978) described a number of spontaneous forelimb gestures which orchestrated copulatory positions between captive bonobos. These included positioning movements in which one individual touched another, thereby indicating how it was to posture its limbs, and various iconic hand motions that showed its partner what to do. Though falling short of the *Kama Sutra*, the remarkable variety of copulatory positions that were enjoyed by the bonobos seemed to require a flexible system of communication.

Gorillas

The thoroughness of studies on gorilla communication is second only to those of common chimpanzees among great apes. Fossey (1972) notably augmented the substantial observations of Schaller (1963, 1965a) on the vocalizations of mountain gorillas. Except for the chest-beating display, the gestures, facial expressions and communicative postures of gorillas are not as fully documented, especially photographically, as those of common chimpanzees.

Schaller (1963, pp. 210-221, 380-381) listed and discussed 22 vocalizations of wild mountain and captive gorillas. He heard 4 of them only once and 7 others were noted less than 10 times each. Fossey (1972) lumped some of the calls, which, as Schaller admitted, intergrade, thereby reducing his list to 15 specific vocalizations, 3 of which she had not confirmed in her own subjects. Based on 2,225 hours of contact with wild groups and observations on 2 juveniles in her camp, Fossey (1972) described 16 vocalizations under 7 categories.

Aggressive gorillas roar, growl and utter pant series (Fossey, 1972). Schaller (1963, p. 218) described high intensity roars as "probably among the most explosive sounds in nature." They stopped him cold on the trail and routed his helpers. Fossey persevered. Silverbacks roar to threaten humans, other gorilla intruders, and buffaloes instead of immediately fleeing from them.

Silverbacks, and rarely blackbacks and females, utter low dog-like growls when they are annoyed by an approaching animal. The latter usually retreat and sometimes growl back. Females, and less commonly silverbacks and juveniles, utter low whispery pant series from the perimeters of disputes and skirmishes among other group members (Fossey, 1972, 1983).

Mildly alarmed and curious adults emit tritonal question barks in which the middle pitch is higher than the neighboring pitches. They usually spark other gorillas to seek the source of disturbing sounds. Hiccup barks, consisting of staccato disyllabic bursts of sound, occur under much the same conditions as question barks, e.g. when a human observer moves suddenly or out of view (Fossey, 1972, 1983).

Schaller (1963) and Fossey (1972) found that gorillas often fall silent and slip away as quietly as possible when they are extremely afraid, e.g. when they detect approaching poachers and fresh aliens. During intragroup altercations, gorillas of all ages and both sexes were inclined to scream shrilly and repeatedly. Screaming spreads contagiously through the group during fearful quarrels and contests. Silverbacks particularly the less dominant ones, are the main screamers.

Females may scream during and after copulation (Fossey, 1972; Schaller, 1963). Copulating silverbacks utter rapid series of staccato hoots that progressively merge into a long howl (Schaller, 1963; Fossey, 1972). Fossey (1972) heard these copulatory pants during 4 out of 8 copulations in mountain gorillas. Harcourt et al. (1981) described copulating mountain gorilla females whimpering rapidly and males panting and occasionally whimpering. Hess (1973) noted copulatory sounds in captive western gorillas. His descriptions are not sufficiently detailed for comparisons with wild mountain gorillas, except that the female western gorillas did not include screamers.

Wraaghs are loud, explosive monosyllabic alarm barks which are pitched between roars and screams. They are uttered by adults, mainly silverbacks, and rarely by juveniles, when startled by sudden, close contacts (e.g. with buffalo) in dense foliage or upon hearing loud noises like thunderclaps, branches breaking and rock slides. Fossey (1972) noted individualistic wraaghs by the silverbacks with which she was intimately acquainted. Following wraaghs, groups usually retreat from the sources of disturbance (Fossey, 1972).

Distressed infants emit a variety of screeches, sobs, wails and other noises which Fossey (1972) treated generically as cries. She mentioned that occasionally they build to a crescendo of tantrumesque shrieks. Apparently they were not performed by weanlings since Fossey's (1972) list of causes does not include a mother rejecting nursing attempts.

Prolonged puppy-like whines are uttered by youngsters and females that fear injury or abandonment (Schaller, 1963). Fossey (1972) also witnessed whining by a silverback and blackback but she did not state their circumstances.

Wrestling, chasing and otherwise playful infants chuckle, particularly when engaged with one another or with tickling humans. A pair of blackbacks chuckled during a low level tickling session (Fossey, 1972; Schaller, 1963).

Gorillas emit series of pig grunts more often when moving from areas than when stationary. Silverbacks are the principal grunters though the vocalization can spread contagiously to involve other adults. Silverbacks grunt while orchestrating group movements and during lustful pursuits of females. Others might grunt during quarrels over access to food and the silverback will grunt in order to quell them. A mother grunts at her wayward infant and it returns to her (Schaller, 1963; Fossey, 1972).

Moving and stationary gorillas also emit disyllabic hoot barks. Although silverbacks hoot barked the most, all categories of gorillas were inclined to give a hoot when mildly alarmed. Hoot bark alarms induced group members to cluster and to climb for a better view. Fossey (1972, p. 48) found that "Of all vocalizations, the hoot bark was the most effective in clustering group members."

To initiate group movements, a silverback may hoot bark and posture. This may be answered by a chorus of pig grunts from other adults which is joined by the silverback (Fossey, 1972).

Contented gorillas produce deep prolonged soft belching sounds. This too is contagious in a stationary group as they feed, assemble for night nesting, rouse from sunning themselves, play and groom. Eructating observers can induce responses from habituated gorillas (Fossey, 1972, 1983).

Fossey (1972) noted considerable variability in the belch vocalizations. During prolonged spans of belching, whinny-like sounds were produced. Isolated horsey whinnies were uttered twice by a blackback and more often by a sick silverback which produced no other vocalization. He whinnied in contexts wherein healthy silverbacks would hoot bark or wraagh. Because his autopsy revealed pneumonia and pleurisy (Fossey, 1972) we might suspect that the sound is caused by pathology of the respiratory tract.

Hoot series are given during intergroup contacts and encounters between groups and single silverbacks (Fossey, 1972). They are rarely sounded by any animal except silverbacks. The displayer begins quietly with a series of low pitched "hoos" which steadily rise in pitch until they are strained and sustained, rather like a dog's whine. A hoot series may end in a harsh growl, a lapse into silence, ground thumping, assaults on vegetation, running noisily through the foliage, or, most spectacularly, a chest-beating display (Fossey, 1972; Schaller, 1963; Figure 25).

Although chest-beating was noted by Du Chaillu in 1861 and was regularly mentioned in later accounts on gorillas, the complete display and its circumstances were not detailed until a century later (Emlen, 1962; Schaller, 1963, 1965a, 1970). Schaller (1963, p. 222) observed that full-fledged chest-beating displays of silverbacks include 9 more or less distinct acts.

First is a hoot series, sometimes during which the displayer plucks a leaf or sprig and places it between his lips. Just before the climax of the hoot series, he rises bipedally and often tears off and flings a handful of vegetation into the air. The hoots merge into a growl as he rapidly tattoos his inflated chest. The laryngeal sacs, which commonly extend into the pectoral region, resonate the percussions, which can be heard up to 1.6 km away. While slapping his chest with slightly cupped hands, the displayer may kick sideways with one limb and tear foliage with his foot. Then he runs a short distance sideways, often tearing and slapping at vegetation en route. He may also assault foliage at the end of his run. As a finale, he sharply smacks the ground with one or both palms. The entire display takes longer to write down than to perform, viz. no more than 30 sec and often much less.

Silverbacks may delete the hoot series or hoot only once before proceeding to a climax. Indeed Schaller (1963) found that only 10% of chest-beating displays included a notable hoot series. Hooting false starts are also common and would-be show-offs seem to be easily distracted during the overture, which signals other group members to clear the stage.

Females and youngsters also strike their chests but their slapping sounds lack the resonance of silverback thoracic percussions (Figure 25). Further, they do not perform the complete display. Sometimes group members rise and beat their chests as the silverback reaches his climax (Schaller, 1963). Often the silverback performs truncated displays and variably combines some of the middle acts in the 9-part sequence.

Schaller (1963, p. 231) concluded that chest beating is basically innate. The following situations can induce chest beating by mountain gorillas: the presence of humans, other gorilla groups and lone males; hooting, chest-beating and thumping by distant gorillas; displays by other group members; during play; and perhaps spontaneously (Schaller, 1963; Fossey, 1983; Carpenter, 1937).

In Equatorial Guinea, Jones and Sabater Pi (1971) heard most chest-beating in the morning and after the gorillas had nested for the night. They could not establish causality except in cases of human intrusion. Nocturnal chest-beating was either accompanied by vocalizations or was solo.

The full chest-beating display is unique to gorillas. Although excited chimpanzees hoot, stand and run bipedally, assault and throw vegetation, and thump the ground, they rarely beat their chests (Lawick-Goodall, 1978a, p. 273). Chimpanzees have never been reported to perform a complete silverback display or to comprise a thoracic percussion section after a chest-beating male's cadenza.

Communicative body postures and, to a lesser extent, facial expressions, of wild gorillas are not described fully enough to draw close comparisons with those of Gombe chimpanzees. Schaller (1963, pp. 208-210) sketched the facial expressions of his subjects under 8 emotional states, ranging from placidness to fear. He commented that many of them resemble those of captive chimpanzees as described by Foley (1935) and Yerkes and Yerkes (1929). Indeed we can recognize in Schaller's descriptions crude counter-

Figure 25:   Chest-beating display by male Visoke mountain gorilla. (Photos courtesy of David P. Watts.)

parts of the chimpanzee tense-mouth face, open-mouth threat face, grin, pout face, and play face.

Marler (1976; Marler and Tenaza, 1977) noted remarkable similarity between the vocal repertoires of gorillas and common chimpanzees. He matched all 13 chimpanzee vocalizations with calls from gorillas, leaving roars, growls, whines and whinnies as peculiar to gorillas among the African apes. Marler (1976, p. 246) was most confident of similarities between pant-hoots and hoot series; laughter and chuckles; screams; rough grunts and belches; and pants and copulatory pants; less confident of similarities between squeaks and whimpers and gorilla cries; waa barks, wraaas, and wraaghs; and grunts, coughs and pig grunts; and least confident in comparisons between pant-grunts and pant series and several barks.

Radically different field conditions frustrate quantitative comparisons between the calls of nonprovisioned gorillas in the process of habituation (Fossey, 1972) and the heavily provisioned, well habituated Gombe chimpanzees (Marler, 1976). Nevertheless, Marler reasonably related the outstanding vociferousness of silverbacks, including their singular roars, to their unique position as guardians of relatively close-knit groups versus the more flexible and fluid social groupings of common chimpanzees (Chapter 8).

Tattooing the thorax versus a tree buttress carries the obvious advantage of always having a drum readily at hand. Whether there are particular communicative differences between the two actions or chest-beating merely developed in areas where appropriate drumming trees were insufficient for early gorillas is unknown.

MENZEL'S CHIMPANZEES AND THE SOCIAL IMPERATIVE

In a break from the language oriented approach of comparative psychologists and the signal-oriented approach of ethologists, Menzel (1971a,b, 1973b, 1974, 1978, 1979; Menzel and Halperin, 1975) devised a group-oriented approach to chimpanzee communication. It produced intriguing results which underscore the need for innovative projects that test the natural communication of apes as social phenomena. Menzel (1979, p. 360) concluded that greater emphasis should be placed on "what they are communicating and how they do it" instead of their capacities for communication. The behavior of groups would be studied instead of compiling lists of individual vocalizations, facial expressions, and postures or trying to induce subjects to emulate features of human language.

Menzel's experiments were conducted over a span of 6 years with 5 female and 4 male juvenile wild-born chimpanzees in an enclosed, nearly one acre, field in southeastern Louisiana. Initially the field was treed. But the subjects destroyed much of the natural vegetation. Six of them had lived together in the field for one year before testing. They formed a cohesive and relatively compatible social group (Menzel, 1971b, p. 221, 1974).

While the rest of the group was caged out of view, Menzel (1971b) showed one chimpanzee, designated the leader, a cache of food or fearful object somewhere in the field. Five of the subjects served alternately as leaders; the others were too distressed by separation from the group to adopt the role of leader (Menzel, 1973b). After the leader had been reunited with the group they were released together into the field. If food was the incentive, the group followed the leader and sometimes even ran ahead of him. The followers seemed to orient more to *that thing out there* than to the leader. If the incentive was a fearful object, the group moved cautiously.

In trials wherein one leader had been shown hidden fruit, which is their preferred food, and another had been shown vegetables, the entire group tended to follow the fruitful leader. This indicated that there had been some sort of pooling of information in the holding cage. Further, the exchanges of information were quite subtle. Usually there were no conspicuous gestures, facial expressions or vocalizations, except that single alarm calls were sometimes uttered when leaders first saw novel frightful incentives. On 88% of 60 test trials, the group followed the leader to caches of food instead of detouring to retrieve single foods that were clearly visible in the field (Menzel, 1971b). When each of two agreeable subjects had seen a different quantity of hidden food, they often travelled together for the larger amount and then moved to the smaller cache. They generally ignored blank spots. On almost 80% of trials when the leaders split, the majority of the group followed the one that had seen the largest cache of food (Menzel, 1973b).

Exactly how the chimpanzees communicated information in the holding cage or momentarily after release is a mystery. Indeed the leader often wrestled with others or just sat by the door until they were freed. Isolation of the leader before the group was released did not affect their performance even though it was stressful for her or him. Menzel (1973b, p. 218) suggested that the "visual orientation and the locomotor postures and movements of the 'animal as a whole' contain sufficient information to account for the bulk of the communication about hidden objects."

Menzel (1971b) noted that his subjects succeeded so well because they were well acquainted with one another. The longer a pair had lived together before the tests the more closely interrelated were their movements, though the effects decreased markedly during adolescence when each of them became more independent (Menzel, 1973b, p. 196). No less should be expected of relatively stable wild populations of chimpanzees and perhaps other social mammals. Needless to say, the more subtly a social group communicates during routine foraging, nesting and other maintenance activities, the less attractive they will be to predators and opportunistic competitors. A major challenge now is to devise and conduct informative experiments on natural social groups.

8

Sociality and Sociobiology

"Functionally viewed, an act of fellow service may be more or less unselfish, but it is doubtful that any act, even in man, is purely altruistic." (Yerkes, 1939, p. 111)

"While it is now obvious that social relationships must be described in terms of social interactions and, in turn, social structure in terms of social relationships, the influence of higher levels on lower, social structure on relationships and relationships on interactions, is not so clear." (Harcourt and Stewart, 1983, p. 307)

Only within the past 25 years have scientists systematically collected extensive information about naturalistic social behavior in populations of gibbons, siamang, orangutans, common chimpanzees, bonobos and gorillas. In *The Great Apes* (1929) Yerkes and Yerkes cited anecdotes by travelers, adventurers, and hunter-naturalists and repeatedly bemoaned the inadequacy of such sources. Their synoptic table is riddled with errors. They concluded that gibbons were the most gregarious apes and extremely dependent upon one another. Yet they stated that the sexes were segregated at times, "if not regularly between breeding seasons" (p. 558). Purportedly gibbons were markedly nomadic, needed leaders because of their gregariousness, and had not been reported to have families as basic units of their "bands."

Orangutans were designated the least gregarious and sociable among the apes; males were dominant but the sexes were usually "segregated." Orangutans were less nomadic than other apes and had no stable groups (Yerkes and Yerkes, 1929, pp. 557-558).

Chimpanzees or gorillas were ranked undecidedly second in gregariousness; but Yerkes and Yerkes (1929) deemed chimpanzees first in sociability.

Chimpanzee and gorilla "bands" were led by dominant males during extensive nomadic travels, including seasonal migrations for food. Sexual segregation had not been reported for African apes. Instead they seemed to live in permanent families, consisting of "parents and offspring" (chimpanzees) or "male, females and young" (gorillas) (Yerkes and Yerkes, 1929, pp. 557-558).

Yerkes inspired Carpenter (1940) to inaugurate an intensive study of gibbons in Thailand. He was quite successful; but his monograph stood alone for approximately 25 years. Studies of Guinean chimpanzees (Nissen, 1931) and eastern gorillas (Bingham, 1932), also inspired by Yerkes, were far less informative than Carpenter's pilot project was, especially regarding the social behavior of the subjects. Thus, for many years there was no reliable base for comparisons between lar gibbons and other apes. Further, it was not known how applicable Carpenter's descriptions of lar populations were to general aspects of hylobatid behavior.

HYLOBATID APES

Carpenter (1940) established that Chiengmai lar gibbons characteristically live in monogamous family groups, consisting of an adult male, an adult female, and between 2 and 4 dependent youngsters and subadults. Among the 21 groups that he studied intensively, the modal size was 4 individuals. The smallest group was a childless adult pair.

The chronic bisexuality of lar gibbon groups had not been securely documented because their monomorphy (Chapter 1) makes sexual identification elusive for the casual observer in the forest. Carpenter had conducted daily behavioral observations over a 2.5 month span. Further, his team killed 6 families, which allowed thorough examinations of their reproductive organs.

Carpenter (1940) noted that the lar group was structured around an adult pair, which had strong bonds of attachment with one another. He detected no dominance of one sex over the other and denied that the males are more aggressive than the females. Both were involved in leading, coordinating and guarding their group. He could only speculate on how new groups were formed.

Carpenter (1940) indistinctly and incompletely observed 2 copulations. The subjects approached each other and embraced. Then the male explored the female's everted labia and mounted her dorsoventrally. Because youngsters in all stages of development were seen in the area and gunned females bore embryos and fetuses in various developmental stages Carpenter (1940) concluded that the Chiengmai gibbons did not have a breeding season. He believed that the possibility for copulation throughout the year might reinforce bonding of a mated pair.

Carpenter (1940) suggested that allogrooming also served to integrate the group socially, in addition to its obvious hygienic function. Although he stated that many instances of grooming were seen, he presented no quantification. Adults groomed reciprocally. Infants groomed others less often than they were groomed by adults.

Infants were carried by their mothers. Although males handled and sometimes played gently with infants, most youngsters engaged in isolated play or played with other young group members if they were available.

Carpenter (1940) witnessed no actual fights between Chiengmai gibbons. But he concluded that the groups were territorial on the basis of their regular song bouts and displays (Chapter 7), reactions to the sight and songs of other groups, and general restriction to a relatively small section of the forest. Hence they were not migratory. Carpenter (1940) concluded that when more or less evenly matched groups met and displayed, the winner would probably be the one nearest the focus of its own territory.

Ellefson (1974) confirmed most of Carpenter's basic observations on the social organization and ranging pattern of *Hylobates lar*. After a 5-month survey of 27 localities in Peninsular Malaysia, he spent 16 months at Tanjong Triang on the eastern coast where a small population had been isolated for a half century in a selectively logged forest. Modal group size was 3; Malaysian lar groups generally contained fewer youngsters than the Chiengmai groups had.

At Tanjong Triang the sex ratio was dramatically skewed in favor of males; there were 2 unmated males and 4 of the 5 youngsters in the 4 study groups were males. Ellefson (1974) fully habituated 2 groups, partially habituated a third group and the 2 single males, and failed to habituate the fourth group.

The gibbons of Tanjong Triang spent only 12% (70 min) of the average waking period engaged in social behavior. Half of this time was devoted to intergroup conflicts (i.e. territorial behavior). Adults groomed during 3% (15 min) of the waking period, usually in 2 separate bouts. Like the Chiengmai gibbons, adults groomed reciprocally and groomed youngsters more than the latter groomed them. Youngsters rarely groomed each other.

The youngsters played during less than 2% of their activity period. Play commonly occurred while the adults were grooming. Ellefson (1974, p. 5) concluded that gibbons are not highly social in the colloquial sense. But recall that they always moved, fed and slept as a unit (Chapters 3 and 4).

Ellefson (1974, p. 91) observed more sexual behavior than Carpenter did but he too concluded that it was rare. He witnessed 23 copulations on 18 days, about half in the morning and half in the afternoon. He noted that during the estrous period the adults were irritable toward the youngsters and squabbled between themselves. Both emitted "eee-eee whining-squeals" at the end of most copulations. Ellefson (1974, p. 93) speculated that this probably coincided with ejaculation.

In one group the male was consistently dominant over his young, nulliparous mate. In a second group no stable dominance relationship was evident between the adults (Ellefson, 1974).

The males of some groups were dominant over the males of adjacent groups as evidenced by the greater frequency of chasing and possession of food trees in contested areas (Ellefson, 1974).

Ellefson (1968; 1974) observed 126 intergroup conflicts, only one of which included fighting. Eighty-five percent of the contests occurred before 1030 h. Three groups were involved simultaneously in 22 of the conflicts. About 20% of the altercations ended prematurely because of Ellefson's presence. Whether he inadvertently deterred fighting in other contests is unknown.

The lion's share of territorial chasing, and presumably also fighting, is taken by the males. While the males are thus engaged the females and young can feed in the contested area. Ellefson (1974) estimated that there is about 20% overlap between the territories of adjacent groups.

Adults may groom during a fracas; the tense male combatants present to their mates, are groomed, and then return to the arena.

Ellefson (1974) failed to observe new group formation. So, like Carpenter, he was left to speculate, though with a fuller data base. A 4-year old male was threatened with stares, branch-shaking, lunges, and threatening approaches by the adults, particularly the male, when they moved into preferred food trees. Further the adolescent (4-6 year old) male did not join the morning song bouts while the juvenile (2-4 year old) did.

The 2 fully peripheralized adult males at Tanjong Triang called for long periods each morning, often before, during and after group song bouts had ceased. Because there were so few females at Tanjong Triang Ellefson (1974) could not elucidate the process whereby they were paired and whether they too suffered peripheralization by their parents. It seemed likely that nulliparous females would be attracted by lone calling males. But it was also possible that bold bachelors would usurp other males and take over their groups and territories. Subadults of both sexes could replace deceased parents in their natal groups (Carpenter, 1940; Ellefson, 1974).

The formation of a new lar group was first observed by MacKinnon and MacKinnon (1977) at Kuala Lompat, West Malaysia. Initially they noticed a male gibbon with a male siamang at fig laden trees. They often sang together in the morning and travelled together during part of the day (2-5 hours). One day, 25 minutes after they had called, a female gibbon appeared. The male gibbon attacked and briefly grappled with her and then followed her away from the siamang. The next morning the new pair sang together, including female climactic territorial great-calls (Chapter 7). They fed and travelled together for 3 hours. The honeymoon was abruptly busted up by a group of 3 gibbons that viciously attacked them and left with the female (MacKinnon and MacKinnon, 1977).

Both subjects were found alone the next day. The male called. But the fickle female began to travel with the lone siamang. She even sang with him. Three days after their separation, the 2 gibbons were reunited and resumed territorial morning song bouts. During their next battle with the homewreckers, the new pair held their own. The female was an active combatant, which is unusual among lar gibbons. The new group became firmly established in the area and produced an infant 32 months after their pairing (MacKinnon and MacKinnon, 1977; Chivers and Raemaekers, 1980; Gittins and Raemaekers, 1980).

MacKinnon and MacKinnon (1977) reasonably concluded that lone male calls attract prospective mates and are tolerated by neighboring groups. But when bachelorettes duet with bachelors other groups are threatened territorially and will attack them. Their speculation that hylobatid territories are more permanent than the families that occupy them is less defensible.

Caldecott and Haimoff (1983) noted that a widowed *Hylobates lar lar* in the Lima Belas Estate Forest Reserve, Peninsular Malaysia, continued to call for at least a year without attracting the only available male in the isolated region. Her songs contained almost the full repertoire of notes in the female part of a duet except the organizing sequence that precedes the great-call. She uttered only one or a few great-calls before quitting for the day. She had little to fear from other groups because the only other gibbons in the forest were another adult female and the unmated male (Caldecott and Haimoff, 1983).

Treesucon and Raemaekers (1984) reported a very different type of group formation by *Hylobates lar* in the Khao Yai National Park, Thailand. Indeed it contrasts with the MacKinnons' example like traditional versus modern patterns of family formation in the United States. A subadult male displaced the resident adult male of a group and was later joined by his 2 juvenile siblings after they had lost their mother.

As documented by Gittins's (1980; Gittins and Raemaekers, 1980) 2-year study at Sungai Dal, Peninsular Malaysia, the social behavior, activity budgets, territory sizes and ranging strategies (Chapter 3) of dark-handed gibbons (*Hylobates agilis agilis*) are basically similar to those of white-handed gibbons. Group sizes of 7 monogamous families at Sungai Dal ranged from 3 to 6 individuals. Allogrooming and play were very rare in the focal family of 4 that Gittins followed from roost to roost (or until lost) on 177 days during an 11-month period; these activities did not register at all in Gittins's percentages of daily activity period (Gittins and Raemaekers, 1980, p. 96). Most (95%) of their diurnal round was spent travelling, feeding and resting.

Adults of the group were codominant. When they moved single file, either one was likely to lead the group. The family moved and slept as a unit. But during the day the male and juvenile were more than 30 m away from the female (and infant) for 12% of the time and more than 10 m away from them for more than 50% of the time (Gittins and Raemaekers, 1980).

The agile gibbons of Sungai Dal spent 5% of their waking period calling (Gittins and Raemaekers, 1980) and the focal male devoted 13% of his activity period to territorial behavior (Gittins, 1980). The chief protagonists in disputes were adult males (76% of observations) though immature males (15%) and adult females (10%) sometimes also participated. The focal family averaged 0.69 territorial disputes per day. There were no disputes on 47% of days, one dispute on 40% of days, and 2 disputes on 11% of days. Once there were 4 disputes on the same day (Gittins, 1980).

Males seemed to be unable to space out disputes. They occurred randomly in response to chance meetings of neighbors near the boundary of their territories. The exact location of disputes seemed to depend upon the availability of suitable trees from which the males could display. Disputes were not correlated with time of day or particular seasons. They lasted between 1 and 147 min; the median duration was 28.5 min (Gittins, 1980).

Initially the combatants sat about 15 m apart in the upper canopy while other group members rested 10-20 m behind them. If the dispute lasted longer than 10 min, they moved further away from the males and resumed foraging (Gittins, 1980).

The males stare, face away, move about their stations, lunge toward one another, and begin to vocalize (Chapter 7). Then they hang in outstretched "star" postures to expose maximum surface area to their opponent. After 15-20 min of displaying they hotly chase one another back and forth over the boundary, while avoiding deep penetration into the other's territory. No physical contacts were seen during chases and none of the males bore facial scars from bites. Gittins (1980) detected no dominance of one adult male over another. Unlike Tanjong Triang, at Sungai Dal all contests were two-sided.

Gittins (1980) concluded that the extensive display behavior of agile gibbons enables them to assess one another's fitness conservatively and to break off the contest before someone gets hurt.

The pattern of siamang sociality was totally unknown to Yerkes and Yerkes (1929). Now it is among the best documented in apes largely because of the thorough long-term field studies and experiments of Chivers (1971a,b, 1972, 1974, 1976, 1977b, 1978) and his associates (Chivers and Chivers, 1975; Chivers and MacKinnon, 1977; Chivers and Raemaekers, 1980; Chivers et al., 1975; Aldrich-Blake and Chivers, 1973) in Peninsular Malaysia.

The monogamous families of siamang contain an adult pair and up to 4 immature individuals (Carpenter, 1940; McClure, 1964; Kawabe, 1970; Koyama, 1971; Chivers, 1971a et seq.; Papaioannou, 1973; Rijksen, 1978; Fox, 1972). They are notable among hylobatid apes for their cohesiveness (Figure 26) and the parental care which the male gives to weaned infants and juveniles (Chivers, 1971b et seq.).

Siamang families tend to feed, rest, groom and sleep in the same tree (40% of time) or in closely juxtaposed trees, which are usually connected

Figure 26: Malayan siamang family (n = 5) (a) grooming and (b) resting. (Photos Courtesy of D.J. Chivers.)

by climbers. A siamang is rarely more than 30 m from its group (Chivers, 1974, p. 183). One notable instance of separation was a female that gave birth while her group was out of sight (Chivers and Chivers, 1975).

Siamang travel single-file as a close-knit unit. In changing from one maintenance behavior to another, there are only brief intervals between first and last commencements by members of the group. On average, Chivers's (1974, p. 137) subjects started daily activities within 14 min of each other and retired within 19 min of each other. The entire group engaged in the same activity 73% of the time; synchronism decreased steadily from dawn (>80% of observations) until early afternoon (about 60% of observations) (Chivers, 1974, pp. 174-175). The adults stay closest to one another in food trees, with the infant nearby and older immatures further away. The subadult closes the distance during rest bouts and travel.

Chivers (1972) explained the marked cohesion of siamang as the product of social development in the young vis-à-vis patterns of parenting by the adults. Initially, the female is responsible for an infant. But when it is 8-10 months old, the male takes charge during the day. Hence the youngster acquires many of its feeding habits from close association with the male. At night the infant sleeps with the female. Generally, when she produces a new infant, its nearest sibling is a juvenile, which can feed by itself and negotiate all but the most challenging sections of their arboreal highways. At night the juvenile sleeps with the male (Chivers, 1974).

The subadult becomes intimately acquainted with group youngsters when it approaches the male for grooming. Then it may play briefly with his charge (Chivers, 1974).

Siamang groom one another during day rests (Figure 26) and when they settle for the night. The focal family at Kuala Lompat allogroomed on average 97 min (15%) of their waking period (Chivers, 1974, p. 199). Most bouts lasted less than 5 min. The adults and subadult groomed during the day and the adults groomed their young sleeping partners in the evening as they settled in the lodge tree. During the rest period the male reciprocally groomed the subadult male even though there was tension between them during feeding bouts. Overall the male groomed much more than he was groomed (Chivers, 1976). Youngsters commonly play or rest while the older animals groom.

Although the male is the focus of many group activities, the female often leads them about the day range. She is generally first to descend from and to enter a lodge tree and to shift from one activity to another (Chivers, 1974; 1976). She moves several meters and waits to be followed before moving onwards. If another member starts to travel first, she follows quickly and commonly takes the lead. Chivers (1974, p. 229) summarized the group's behavior as a compromise between following the female and remaining near the male.

Chivers (1978) concluded that breeding in Malaysian siamang is periodic, being timed so that pregnancy coincides with spans of increased food sources in the habitat. Females are receptive for several months every 2-3 years, i.e. after their current infant is fully weaned. At 2 different localities, Chivers (1974) observed 18 sexual bouts during 29 days (February through April, 1970) and 26 bouts in 49 days (January through May, 1970). He only witnessed dorsoventral copulations but other observers (Koyama, 1971; Aldrich-Blake and Chivers, 1973) have seen ventroventral couplings. Increased grooming frequencies between the female and male were correlated with copulations (Chivers, 1976).

Knowledge of siamang group formation is sketchy. However, a case reported by Aldrich-Blake and Chivers (1973; Chivers, 1974; Chivers and Raemaekers, 1980) suggests one mechanism. At Kuala Lompat a subadult male left his family after intensified episodes of aggression toward him by the adult male (Chivers, 1974, pp. 222-226). He took up residence in a previously siamangless area of the forest. There he was joined by a young female. Six months later they were joined by an older female. She stayed for about 2 months, left for 2 months, and then rejoined them. Whereas the male initially was most attentive to the young female, he ultimately shifted his interest to the older one. He played exclusively with the young female and copulated with the older female. The neighboring group, from which the male had been peripheralized, increased their singing as the new group formed but later lapsed back to previous levels. No fights were observed between the groups. Sometime during the succeeding 18 months the old female disappeared from the group (Aldrich-Blake and Chivers, 1973).

Female and male subadults receive much the same peripheralizing (and grooming) treatment from the adult male and avoidance by the adult female (Chivers, 1974, p. 230; Fox, 1972). The departure of subadults commonly coincides with the birth of infants into their home groups (Chivers, 1974, p. 231).

Chivers (1974) noted few aggressive interactions in the focal siamang family at Ulu Sempam and an average of 1.5 aggressive bouts per day at Kuala Lompat. He speculated that the lesser amount of intragroup aggression at Ulu Sempam was because the subadult male kept further away from the adult male. Further, the presence of neighboring groups may have reduced conflict within the focal group.

Siamang adults are generally more tolerant of their subadult group members than white-handed gibbons are. At Kuala Lompat the male was the most and the female the secondmost aggressive toward the subadult. The male was 4 times more aggressive than she was. The male juvenile was secondmost often the object of intragroup aggression. The male attacked the juvenile more often as he grew older. Male aggression increases toward the female during the mating period (Chivers, 1974).

Siamang exhibit lower levels of territorial behavior than lar gibbons do. Overt intergroup conflicts are infrequent. When siamang groups encounter one another the males display and chase back and forth over the territorial border while the females and youngsters hide (Chapter 7). Otherwise they maintain their relatively small territories (Chapter 3) by group morning song bouts and regularly ranging near the boundaries (Chivers, 1974, pp. 52-62).

Tilson (1979) confirmed McCann's (1933) observation that *Hylobates hoolock* live in monogamous family groups. The mean size of 24 groups in Assam was 3.2 ± 0.8 individuals; families contained between 0 and 4 youngsters. Tilson (1979) estimated birth intervals to be between 2 and 3 years. He also saw 2 lone young males and a single subadult female. All of them roamed across several family territories. The lone males were quiet and the female moved toward the distant calls of gibbon pairs. In one family, the male displayed toward the subadult when he entered their food tree; the female ignored him.

Tilson (1979) did not observe intergroup encounters. In the Hollongapar Forest each family regularly sunbathed during the winter months in a few sparsely leafed, emergent trees that were centrally located in their territories. Tilson (1979) did not mention whether they could be seen by their neighbors, thereby advertising that the area was occupied.

Because of low level aboriginal hunting pressure (Chapters 4 and 7) and the pristine state of the forest in central Siberut Island, Tenaza (1975, 1976), Tilson (1981, 1982; Tenaza and Tilson, 1977; Tilson and Tenaza, 1982) and Whitten (1982b,e) were able to document processes of group formation, fragmentation, and territorial behavior by *Hylobates klossii* in a habitat that was free of logging and roads.

The basic social unit of Kloss gibbons is the monogamous adult pair with 0-4 youngsters (Tenaza and Hamilton, 1971; Tenaza, 1975 et seq.; Tilson, 1981 et seq.; Whitten 1982a et seq.). Means for family size in central Siberut vary between 3.5 individuals ($n = 16$) and 4.2 ± 1.4 individuals ($n = 12$) (Tilson and Tenaza, 1982; Tilson, 1981).

In addition to monogamous families, Tenaza (1975, 1976) observed the following fragmentary social units at Simimuri: (a) 2 unmated (perhaps widowed) females, one accompanied by a subadult female and the other by a juvenile male; (b) 2 unmated males that each held a territory; (c) a courting pair; and (d) 4 single "floating" males that lived in a riverine forest that was peripheral to the hill forest which was dissected into gibbon territories.

Trespassing females are chased by the resident female but are tolerated by the male. He chases trespassing males while she is more tolerant of them (Tenaza, 1975). Females lead group travel through their range while males follow along and guard them against predators (Tenaza, 1975).

Although Tenaza's (1975) initial study indicated that Kloss gibbons occupy very small areas ($r = 5$-8 ha), later research revealed that their home ranges are comparable with those of other hylobatid species. Tilson (1981) documented ranges between 15 and 20 ha at Tenaza's Sirimuri locality and

Whitten (1981b) found that 3 Kloss groups in the Paitan area had home ranges between 31 and 35 ha. A greater density of gibbons in the Sirimuri area may have resulted in their having smaller ranges. This is evidenced by the higher levels of territorial behavior at Sirimuri than at Paitan (Whitten, 1981b).

Whitten (1981b) suggested that because the boundaries of gibbon territories commonly lie along ridge tops and rarely, if ever, along valley bottoms, the shapes of their ranges may be determined partly by topography. Further, like MacKinnon and MacKinnon (1977) he speculated that gibbon territories and home ranges may be more permanent than the groups that inhabit them. After the male of his focal study group was shot by a local hunter, the female and juvenile dispersed and another group of 3 gibbons entered the home range from beyond his study area (Whitten, 1981b).

Tilson (1981) provided the fullest account of group formation in Kloss gibbons, based on a 21-month study in which he focussed on social change, especially the transition from unmated to mated status. During a 30-month period, 5 Hylobates klossii disappeared from the Sirimuri area. Although one senile female may have died undramatically, the other 4 were probably killed by hunters during Tilson's (1981) leaves. Despite their wariness of humans he partially habituated 4 out of 15 groups to his presence.

Tilson (1981) confirmed Tenaza's (1975) observation that in Kloss families adults are increasingly aggressive toward the like-sexed youngsters from 4 years of age until they are excluded from the group, generally by their eighth year. Over the span of his study, 3 subadult males and one subadult female were excluded from family groups and 4 others (3 males and 1 female) were peripheralized. Females were not as markedly peripheralized as the males were, probably because the latter must establish territories into which they can seduce mates with their songs (Tilson, 1981).

Although they aggressively repel maturing family members, adult Kloss gibbons are also disposed to help them form partnerships of their own. In Tilson's (1981) study, 2 out of 3 successful pairings were effected with assistance by adults which presumably were related to one partner.

After the adult residents of a territory disappeared, 2 peripheral males from adjacent groups entered the area and competed for it with calls and chases. Then a third subadult male entered with his family and joined the contest. There followed a period wherein the family returned to their own territory, leaving him behind; he rejoined them; the entire group returned to the area and left him behind, etc. Eventually, a solitary female joined him. After a period of breaks and reunions, they were bonded and produced an infant 13 months later (Tilson, 1981).

In the second case, a lone male established himself in a space adjacent to a group that contained a peripheralized subadult female. The group expanded their range toward him. He began to follow them and even entered their home territory without incurring the wrath of the resident male. After he embraced the squealing subadult female, they began to spend more time together. Eventually she joined him and directed her song displays toward her home group. They returned to their original territory; the new pair mated and henceforth successfully defended the boundary between the two territories (Tilson, 1981).

At Sirimuri the third successful new group formed when a solitary male established a territory and attracted a subadult female to it from an adjacent group. They produced an infant 16 months after the courtship began. The female returned to her home group several times before settling in the new territory (Tenaza, 1975; Tilson, 1981).

Another mechanism for a new group formation and territorial establishment is suggested by a foiled adventure of Kloss gibbons. A family accompanied their peripheralized subadult male member into a neighboring territory where the males fought with the resident male and displaced his group. But their triumph was temporary. When the subadult was left to defend the conquest alone, the usurped group returned and, after several encounters between the groups, took it back. The subadult male then disappeared from the area (Tenaza, 1975; Tilson, 1981).

A subadult male replaced the male in his family after the latter had disappeared, probably into a hunter's pot. Three solitary males visited the area and commenced countersinging. When one of them moved on the female, she chose not to be chosen and fought vigorously in the manner of male-male confrontations. The invader then chased the subadult male but failed to usurp him. Five days later, after 6 song bouts, the couple copulated twice. They continued to sing and mate and produced an infant 15 months after the patriarch had disappeared (Tilson, 1981).

A subadult male that had been excluded from his group for 13 months rejoined them after the male disappeared. He embraced the youngsters and was tolerated by the female. They did not mate before the study was terminated (Tilson, 1981).

Widows lacking subadult family males with which to mate can hold their territories for some time against invading males. But whether they can do so indefinitely and whether they eventually accept extra-familial males is unknown (Tilson, 1981).

Tenaza and Tilson (1977) concluded that there is high probability that Kloss gibbons in neighboring territories are close relatives. A major sociobiological advantage of this arrangement is that the far-carrying sirening and alarm trills of males (Chapter 7) warn neighboring, as well as dependent, relatives about the presence of aboriginal hunters. Their speculation that this may hold true also for other hylobatid species is not supported with data.

Tilson's assumption that the subadult males which replaced the males in their families were the sons of the females is reasonable but unsupported. Recall that the younger siblings of a male usurper joined the new family of

Thai *Hylobates lar* (Treesucon and Raemaekers, 1984). One of them might later replace him, giving short-term observers the erroneous impression that the couple exemplified mother-son inbreeding.

Tenaza (1975) might have overgeneralized the role of intrasexual aggression in the peripheralization of subadults in hylobatid species. For example, it appears that among siamang the male is responsible for peripheralizing subadults of both sexes (Chivers, 1974; Fox, 1972). Further long-term studies on group formation in all species are needed before we can generalize about hylobatid monogamy and other social relations. Hopefully, this can be accomplished without the unintentional assistance of hunters.

Why Monogamy?

Unlike the Aves, in which monogamy is common, in the Mammalia it is a relatively infrequent mating pattern (Crook and Goss-Custard, 1971; Kleiman, 1977; Emlen and Oring, 1977; Wrangham, 1979a; Wittenberger, 1979; Wittenberger and Tilson, 1980). Further, within the Primates, monogamy is uncommon (Crook and Gartlan, 1966; Eisenberg et al., 1972; Clutton-Brock and Harvey, 1977; Rutberg, 1983). The monogamy of hylobatid apes is even more outstanding among primates now that Sussman and Kinzey (1984) have effectively excluded the marmosets and tamarins (Callithricidae) from the ranks of monogamous species.

Brockelman and Srikosamatara (1984; Brockelman, 1984) attempted to explain the adaptive significance of hylobatid monogamy vis-à-vis the wealth of new field data, albeit in several currencies, that is available. They notably complemented the work of other theoretical biologists who puzzled over obligate monogamy in the Hylobatidae. And, perhaps most importantly, they acknowledged the limitations of their model, proffered several caveats, and thereby laid the groundwork for future field work, including further experimentation.

Brockelman and Srikosamatara (1984) and Mitani (1984) concluded that in most, if not all, species, the behavior of females is the key to understanding hylobatid territoriality and obligate monogamy. Because gibbons are selective feeders on high quality foods that are not generally abundant (Chapter 3) females stake claims to sections of the forest that are large enough to support themselves and their progeny, yet ones that are small enough to defend against other females. Males, which are similarly limited by available resources, cannot command larger territories with more than one female/young unit, so they become obligately mated monogamously, repel other males, and often assist to eject other intruders.

Female acceptance of a male for the long haul is most easily understood in siamang, a species in which the male carries and otherwise cares for youngsters. But siamang are singular among lesser apes in this character. Predator defense is another possibility, particularly in Kloss gibbons

(Chapters 4 and 7). The role of lodge trees (Chapter 4), as an ecological factor selecting for gibbon social behavior, was not explicitly addressed by Brockelman and Srikosamatara (1984). However, they proposed that a low degree of interspecific competition and niche overlap, which certainly could include choice sleeping sites, are probably major factors selecting for their territoriality.

While surely serving to announce their presence to others, the morning duets of most species (Chapter 7), seem to be too elaborate for intergroup communication alone. They could serve also to bond the pair, but precisely how this might work is a mystery. Brockelman (1984) sensibly called for greater concentration on the behavior between males and females within mated pairs instead of continued emphasis on intergroup dynamics in the field. Indeed it is sobering to realize that we still have few clues to how obligate monogamy is enforced in one of the most thoroughly studied and most socially cohesive of arboreal mammals, the siamang (Brockelman and Srikosamatara, 1984).

ORANGUTANS

In marked contrast with the hylobatid apes, adult orangutan males and females are rarely seen together in the forest. Instead the most common social units are individual females with their dependent youngsters, lone adult and subadult males, and various small groupings of adolescents or singletons. Mother/young units and older youngsters sometimes join temporarily and sizeable aggregations are seen on rare occasions at large fruit sources (Carpenter, 1938; Schaller, 1961; Harrisson, 1962; Yoshiba, 1964; Davenport, 1967; de Silva, 1971; MacKinnon, 1971, 1974a,b, 1979; Cohen, 1975; Horr, 1972, 1975, 1977; Rodman, 1973, 1977, 1979, 1984; Mitani, 1985c,d; Galdikas, 1978, 1979, 1981, 1983, 1984, 1985a-c; Rijksen, 1975, 1978; Schürmann, 1982).

In captivity, family groups are readily assembled and live together as compatibly as other primates for which cohesive bisexual families are the norm. Captive orangutans can be exceptionally affectionate, playful and tolerant of one another (Edwards and Snowdon, 1980; Edwards, 1982; Maple, 1980; Becker and Hicks, 1984). Hence a major question for field behavioralists is, how do orangutans become so solitary.

On the basis of cross-sectional data, Horr (1977, pp. 305-316) constructed a chronicle on the ontogenetic distancing of young Bornean orangutans from their natal units. During its first year, the baby orangutan is totally dependent upon its mother for nourishment, transport, and protection in the forest. It clings to hair on her side (Figure 5) and nests with her at night. She grooms and plays with it briefly and erratically and introduces it to solid foods by allowing it to sample her meals, including premasticated tidbits from her mouth.

During its second and third years, the infant learns to build nests (Chapter 4) and becomes increasingly adventurous in climbing about trees while its mother feeds and rests. If an older sib is present, they may play together briefly, often with little body contact. But a sibling's attempt to touch a young infant are normally rebuffed by the mother. Often by the time the youngster is independent enough to play vigorously (i.e. 2.5-3 years old), the older sibling is no longer with them. If the mother/young unit encounters other orangutans, the infant usually is not handled by them (Horr, 1977).

Juveniles are totally, and apparently unwillingly, weaned from the mother's breast (Chapter 7) and move on their own except for an occasional assist between trees by a motherly body bridge (Chapter 2). Until it is about 5 or 6 years old the youngster will remain in the vicinity of its mother. Though initially very attentive to the mother's movements, as it matures, the juvenile begins to lag further behind and to engage in distinct activities at its own pace. It is now seldom groomed. Horr (1977) saw a mother offer food to her female juvenile offspring and to share food sources with her. Juveniles begin to sleep in their own nests, especially after repeated rebuffs from the nesting mother. This may induce tantrums. Older juvenile males seem to strike out on their own more readily than females, which tend to remain near their mothers (Horr, 1975). It is not known whether this is due to stronger negative behavior from the mother toward male offspring.

As older juveniles begin to roam on their own, they increase the chance of meeting other young orangutans. Horr (1977) noted that they did not engage in long social play sequences. One roving juvenile male even seemed to be upset by the attentions of a juvenile female and her mother. Horr (1977) concluded that given the general absence of male role models, the development of male behavior may be under more direct genetic control than that of females is.

Rodman (1973) found that the 4 mother/young units (n = 9 individuals) in the Kutai study area had extensively overlapping home ranges and that a sometimes independently roaming juvenile female remained within the boundaries of her natal unit range. The ranges of the 2 adult males did not overlap one another but extended widely to include parts of 2 or 3 female ranges. Because 2 of the mother/young units used the same paths and fruit trees where their ranges overlapped and they met, fed rested and moved together for brief periods, Rodman (1973) concluded that the adult female of one unit might be the offspring of the other. A male and another mother/young unit also shared the area but used different resources within it.

Rodman (1973) and his assistants observed secondary groupings of the 6 primary units on 13 occasions during a span of 15 months. They accounted for a mere 1.65% of the total observation time. Eight of them were temporary associations of a male with one or two mother/young units; one of these was clearly sexually motivated while the other 7 were chance encounters at food sources. The remaining 5 secondary groupings were between the 2 mother/young units that were mentioned hereabove as possibly being related matrilineally.

MacKinnon (1971, 1974a) provided a somewhat different account of Bornean orangutan ranging and social structure. But, like Rodman (1973), he noted that the average size of the[social units that he encountered is 1.8. When units met, the adult females ignored one another while the more independent youngsters might play.]Juveniles sometimes joined other groups for a few hours in order to play with peers. Grooming was rare and mostly confined to mothers grooming their infants.]

In 1971, MacKinnon concluded that the 160 orangutans of the Ulu Segama region were[nomadic]and wandered over "enormous" areas. In 1974a (p. 16) he modified this observation as follows:

> ...although some animals may have been nomadic, many of the Bornean orang-utans occupied definite home ranges. The ranges of adult males were large, certainly several square kilometers, but two of the resident adult females may have had rather smaller ranges. Ranges of both sexes showed considerable overlap but there was some evidence that males defended range boundaries against others of their sex.

MacKinnon's study site was near an area of intensive logging and recently had suffered tectonic disturbances which may have disrupted the subjects. Further, during 16 months in Borneo he was unable to habituate individuals to his presence so that he could follow them continuously enough to plot actual ranges. nevertheless, his observation that at some times the units were more clumped than at other times stands as a caveat against overgeneralizing about orangutan ranging on the basis of other studies. It appeared to MacKinnon (1974a, p. 17; 1974b, p. 210) that, although loosely banded, Segama orangutan social units maintained contact and coordinated their movements through the area with reference to calls of male leaders.

The information gathered by Galdikas (1978) and her assistants during 6 years (10,000 observation hours) in the Tanjung Puting forest basically supports the inferences of Horr and Rodman concerning Bornean orangutan ranging patterns, especially those of females. They recognized 58 individuals, including 12 solitary adult males, 9 subadult males, 14 adult females and their dependents (i.e. 7 infants and 7 juveniles), and 9 adolescents (Galdikas, 1981b, 1984; 1985a-c).

[Whereas Tanjung Puting females ranged over 5-6 km², adult males ranged over at least 12 km². Female ranges were stable over the years and overlapped extensively (Galdikas, 1979). Independently roaming juveniles stayed within their natal unit ranges](Galdikas, 1984). A male disappeared from the study area for a few years and then returned to follow his old pathways (Galdikas, 1978, 1979).

Social units at Tanjung Puting were in contact much more (19% of 6,804 observation hours) than the Kutai units were (Galdikas, 1984). Galdikas (1979, 1985b) stated that demographic and ecological differences between the regions could be responsible but she did not elaborate. She considered that Tanjung Puting orangutans were more similar socially to Sumatran orangutans, as described by Rijksen (1975), than to populations studied in Sabah and Kutai. And she ultimately concluded that orangutans are only semisolitary (Galdikas, 1985b). The largest temporary groups (n = 2) at Tanjung Puting included 9 individuals, none of which were adult males (Galdikas, 1984, 1985b).

Galdikas (1978, 1985b,c) observed dyadic contacts between adult males only 6 times; half of them culminated in combat and a fourth fomented a chase. Subadult males avoided adult males.

Relations between adult females were more variable. Upon meeting, some females ignored one another, others reacted aggressively, and still others were associative (Galdikas, 1984). Galdikas (1978) believed that adult females remembered one another even though they had not met for many months. The mutual reactions of specific pairs were consistent during successive encounters. Some mother/young units seemed to travel together for more than a week after meeting (Galdikas, 1982b). But the longest associations between 2 mature females spanned only 3 days and 2 nights (Galdikas, 1984). The peer contacts of youngsters in the groups appeared to be "as important, if not more so, than direct geneological ties in determining female associations in adulthood" (Galdikas, 1978, p. 294).

[Bornean orangutans are probably most social during immaturity.] Galdikas (1978; 1985a,b) noted that 76% of the time adolescent females and 41% of the time subadult males were with non-natal units. Only adolescent females were observed grooming individuals outside their natal units (1.5 out of 10,000 hours). They groomed each other and, more frequently, adult males (Galdikas, 1978, p. 294).

Although his study of Sumatran orangutans was less than half the duration of his Bornean project (9 mos. vs. 20 mos.), MacKinnon (1974a, p. 18) commented that the picture of their social organization is clearer. In particular, the small social units were more clumped in Sumatra so that encounters between them were seen more often. Again, he believed that they were coordinated by male long calls.

Because of nearly complete overlap of subunit ranges MacKinnon (1974a) concluded that there was a group range instead of numerous discrete subunit ranges. At Ranun he saw 14 individuals (4 adult males, one sub-adult male, 4 females and 5 youngsters) feeding in the same tree. The high frequency of males accompanying females with infants led him to infer that they were complete family units instead of temporary consortships or casual groupings.

MacKinnon (1974a, p. 66; 1979, p. 264) explained the greater social cohesiveness of Sumatran orangutans as a consequence of large predators (especially tigers and panthers*), which are absent from Borneo, and the tendency for siamang to attack young orangutans (Chapter 3). He speculated that Sumatran orangutans represent a stage between hypothetical group living ancestral orangutans and the more highly "desocialized" Bornean apes.

During a 3-year study at Ketambe in which he recognized 22 individuals (3 adult males, 4 subadult males, 6 adult females and 9 youngsters), Rijksen (1975, 1978) confirmed that the ranges of Sumatran orangutan male and female social units partly or entirely overlap each other. Males generally roamed more widely than females did (Rijksen, 1978).

Average group size at Ketambe is 1.5 individuals (Rijksen, 1978, p. 166). Forty-six percent of observed orangutans were loners or mother/young units. The majority (54%) were in larger social groups (17%) or temporary arboreal feeding associations (37%) (Rijksen, 1975, 1978). Most social groups, i.e. subjects moving together in a coordinated manner, were composed of subadults (8-13 or 15 year olds) and especially adolescents (5-8 year olds). Adults minimized direct social contact. Rijksen (1975) found that whereas adolescents of both sexes were in social groups 33% of the observation time, adult males were encountered thus only 5% and females 6% of the time.

Even though the population density of Ketambe orangutans was high, Rijksen (1975, 1978) noted no tendency for adults to band together into larger social groups or to coordinate their movements through the area with reference to calling by male leaders. He stressed their individualism and dispersal and proposed that the wider interindividual social contacts that they had during immaturity probably provided the base for stable relationships during chance encounters and temporary feeding associations in adulthood. During immaturity, dominance relations might be established which order adult social intercourse (Rijksen, 1975, 1978). The presence of a provisioning station might have affected the population density and frequency of encounters between social units at Ketambe.

Rijksen's (1978, pp. 284-292) descriptions of dyadic reactions between members of different age/sex classes of Sumatran orangutans are similar to those for Bornean subjects. He rarely observed grooming by wild and rehabilitant Ketambe orangutans. Play was more common than grooming, with immature individuals participating more than adults; young males played together more frequently and roughly than young females did. Females were more inclined to invite males to play with them than vice versa (Rijksen, 1978, p. 253).

---

*The presence of *Panthera pardus* on Sumatra is not well documented (Rijksen, 1978, p. 101).

Although inventive captives engage in a good deal of it (MacKinnon, 1974a; Maples, 1980; Nadler, 1977; Rijksen, 1978), solitary sex is perhaps no more satisfying to orangutans than it is to most adult humans. Indeed mounting evidence indicates that libidinous female and male orangutans will devote considerable effort to pursue mating partners (Chapter 7).

Horr, Rodman and MacKinnon had gathered a wealth of data on the socioecology of Bornean orangutans immediately before the centennial of Darwin's (1871) *The Descent of Man and Selection in Relation to Sex*. Horr and Rodman studied at Harvard University where Trivers (1972), Wilson (1972) and others synthesized sociobiological theory. Horr (1972) and Rodman (1973) were particularly astute in viewing the social organization and extreme sexual dimorphism of orangutans as a consequence of intrasexual competition for resources. Later students (especially Galdikas, 1979; 1981b) filled in important details and Wrangham (1979a,b) extended the approach to other ape societies.

Horr (1972, 1975) suggested that because adult orangutans are large animals that can consume most of the fruit in a particular tree, they forage in small units. They have no "serious natural predators" to induce them to seek group protection (Horr, 1972, p. 49). Since males are unencumbered by youngsters, they can roam more widely for food and have a better chance to breed with estrous ones. The main limitation to the movements of adult females, even ones with clinging infants, would seem to be juvenile tagalongs, which have trouble negotiating sections of the arboreal highway (Horr, 1975). If bulk alone were the drag, adult males would be expected to have smaller ranges than the females do.

According to sociobiological theory, the sex with the greater parental investment in offspring will be a limiting resource for the opposite sex (Tivers, 1972; Williams, 1975). As Rodman (1973, p. 202) slickly put it: "Male orang-utans manifestly invest nothing in their offspring but the energy of sperm production. ... " Hence dispersed females are a limiting resource for males while the patchy distribution of fruit is the main limiting resource for females. Because the birth interval is rather long and estrus lasts only 5-6 days per month, a male must travel widely and advertise his location in order to contact a ready, willing and able female. The large size and elaborate epigamic features of adult males (Figure 23) are used primarily for intrasexual competition (Rodman, 1973, 1977; Horr, 1975). According to this model, orangutans are a prime example of Darwinian sexual selection.

Rodman (1977, pp. 410-411; 1979, p. 249) observed that a resident male orangutan seemed to subsist on less preferred food than the females did and that he moved away from them when fruit was scarce in their common area. He inclined toward the idea that this represents altruism by the male.

Alternatively, males may seek the best source of food during times of shortage. Their knowledge of a larger chunk of the habitat may enable them to move off until the common area is more fruitful for all. Further, they tend to feed at particular trees longer than the females do (Rodman, 1979). In lean fruiting periods this would predispose them to seek richer sources of less preferred foods which their powerful jaws are able to process (Chapter 3). It would be interesting to know what happens when one or more females come into estrus during a low fruiting period; this, of course, assuming that estrus and food supply are not coupled.

Rodman (1979) elaborated why an incompatibility of lifeways separates the sexes most of the time. If a male were to try to keep up with a female, he would have to travel arboreally more often between feeding bouts because females change feeding locations more often than males do. And if she were to follow him faithfully, she would sacrifice a degree of selectivity among food trees and at so-so sources because of his formidable bulk and greater appetite. Rodman (1979, p. 251) observed that while a male settled down for a big breakfast his temporary consort moved off, presumably to sample a more varied menu. Like a talented starlet, after insemination, she had little to gain and much to lose by remaining with the boob.

MacKinnon (1971) introduced the subject of rape (Estep and Bruce, 1981) anent the sex lives of orangutans. Seven of the 8 matings that he witnessed at Segama entailed aggressive males chasing and assaulting screaming, resistant females. Plucky youngsters of the victims also screamed in addition to biting, striking and pulling the attacker's hair. The males hunched the females irrespective of their positions and did not seem to achieve vaginal intromission. The most dramatic battle between the sexes began in a treetop and bottomed on the ground. The eighth copulation was between youngsters (MacKinnon, 1979).

MacKinnon (1974a, p. 56; 1979, pp. 262-264) also stressed that Bornean and Sumatran orangutans form "apparently stable consortships." At Segama half the sexually mature males consorted with females in the more disturbed area and 23% were engaged thus in the less disturbed area. At Ranun half the adult and subadult males were involved in consortships. In both localities more subadult than fully adult males were observed with females (subadults: 75% at Ranun, 41% at Segama; adults: 54% at Ranun[*], 22% at Segama).

One can interpret MacKinnon's (1974a, pp. 56-57) figures on and few descriptions of intersexual interactions of orangutans in other ways.The subadult males may have been tagalongs that were intent upon contacting the females regardless of their sexual receptivity. Indeed one 2-day association was punctuated by 3 "rapes." In a second example, the subadult

---

[*]This is based on MacKinnon, 1974a, p. 56:" ... 6 out of 11.... " In MacKinnon, 1979 (p. 264) the figures "7 (41%) of 17" Sumatran adult males is probably an erroneous repetition of the figures for Bornean subadults thereinabove.

male tried to inspect an adolescent female's genitalia but did not attempt copulation.

MacKinnon (1974a, p. 57) reported that "consorting orang-utans usually nested in the same tree at night" and that Sumatran males were seen twice grooming their companions during the day. Both sexes initiated foreplay. This is more persuasive evidence for consortship. However, MacKinnon's (1974b, p. 209) further claims for long-term relationships between adult males and females, to the extent that bisexual family units exist, are not supported by longitudinal studies on known individuals.

MacKinnon (1979, p. 264) confirmed Rodman's (1973, 1979) observation that full-bodied males exert a dragging effect on the foraging females that keep them company. This led him to hypothesize that males must adopt different reproductive tactics as they grow from spermatic lightweights to full-blown adults. His model of male sexual strategy differs in some features from that of Horr and Rodman.

Because he saw more subadult than adult males with adult females and never saw "old adult males" copulating, even though they called and performed aggressive displays, MacKinnon (1979, p. 268) concluded that "sexual vigor and potency decline earlier in the male's lifetime than does agonistic vigor." According to MacKinnon (1979) a subadult male either establishes long consortships with cooperative females or rapes uncooperative ones. He provided only circumstantial evidence for the former, which would have the best chance for reproductive success, especially if the female remained faithful to him when she came into estrus. MacKinnon's (1979, p. 270) claim that "adult males also sometimes indulge in rape" is unsupported by narrative. He agreed that prime males mainly attract females and defend their priorities with calls. As they decline from prime, males more often adopt the role of guardian for the products of their youthful escapades. Their calls and aggressive behavior repel intruders that would compete with their progeny for food (MacKinnon, 1979).

Much of MacKinnon's model is not supported by Galdikas's (1979, 1981b, 1983, 1985c) long-term studies of known individuals. She concluded that despite their more frequent associates with females and fumbling copulatory forays subadult males are virtually excluded from reproduction because the females prefer large, padded adults that fight for their sexual prerogatives. Although, unlike MacKinnon, Galdikas (1979) noted vaginal intromission during some "rapes," none of the victims became pregnant therefrom. If they are with females, subadult males beat a hasty retreat, often on the ground, at the approach of a mature male. Solitary subadults simply stay or move out of the way of closely passing adult males (Galdikas, 1979, 1981b, 1984, 1985a-c).

Most male-female encounters at Tanjung Puting did not culminate in copulation. Adults of opposite sexes generally seemed to ignore and to avoid one another. Whereas subadult males sometimes orally or digitally explored the perineums of passing females, adult males were not seen to do so. Estrous females did not overtly avoid the males that they encountered. Pregnant females refrained from copulating with subadult interlopers which followed them. One new mother went out of her way to avoid a subadult male. Galdikas (1979) calculated that the interval between births is usually 4.5-5 years.

Galdikas (1979, p. 216) inferred that females as a class tend to prefer fully mature males as sexual partners over the subadult males that more often were associated with them. Adolescent females were particularly attracted to the big fellows (Galdikas, 1981b). They sometimes eagerly approached, groomed and orally and manually contacted the genitalia of would-be mates. One even urinated on her stoic sex object (Galdikas, 1981b). Adult females rarely engaged in such brazen behavior, but they willingly submitted to the advances of mature consorts. The adult males appeared to be more interested in receptive adult than adolescent females (Galdikas, 1979, 1981b).

Galdikas (1979, 1981b) confirmed that orangutans form consortships wherein they travel together for 3-8 ($\bar{x} = 5.4$) days, during which they mate a number of times ($\bar{x} = 2.9$; $r = 1$-6). Their copulations are generally conducted ventroventrally while suspended from overhead branches (Figure 24). They also mate in nests. The male usually mouths or handles the female's genitalia before the main event and variably utters a long call or part thereof as part of the breeding bout (Chapter 7). Sexually successful consortships were characteristic of adult pairs; but persistent adolescent females sometimes also succeeded in establishing them with adult males.

Subadult males were responsible for 95% of non-consort copulations, 86% (19/22) of which were forced (Galdikas, 1981b; 1984a). And they are most persistent; 83% of subadult male social time was spent with adult females (Galdikas, 1985a). Galdikas (1979, 1981b, 1985a) reasonably suggested that quite apart from the tactic which might be optimal vis-à-vis the overall reproductive strategy of an individual, he or she might be expected also to exhibit behavior based on personal idiosyncrasies or past social experiences.

Unlike Galdikas, Rijksen (1978) rejected the Horr-Rodman model in favor of MacKinnon's (1974a) suggestion that the extreme arboreality and solitariness of recent *Pongo* is the result of interactions with *Homo*. Thus, when faced with food competition and depredation by hunting *Homo sapiens* and perhaps *Homo erectus*, previously more terrestrial and banded *Pongo* took to the trees and dispersed in the canopy where they would be less conspicuous and more distant targets.

Still Rijksen's (1978) data on the sexual behavior and interpretations on the reproductive tactics of female and male orangutans at Ketambe are in close agreement with those of Horr, Rodman and Galdikas. He too stressed female choice and intermale competition for reproductive access to them.

Adult females that were most likely to be receptive, viz. childless ones and others with juveniles, were more often with adult than with subadult males. The latter more frequently accompanied female-infant dyads, the adults of which would not be expected to cycle menstrually due to lactational amenorrhea (Rijksen, 1978).

Whereas the 4 consortships that engaged adult males were initiated by the females, the 4 consortships that involved subadult males were initiated by them. None of the 8 couples were seen *in copulo*. Rijksen (1978, pp. 274-275) reported that, guided by long calls, estrous females actively seek the highest ranking male in the area and induce consortships with him. He appears to be interested in a female only after she has presented herself for inspection. She terminates the relationship, presumably after insemination. Subadult males are chiefly reduced to rape, which Rijksen (1978, p. 257) speculated was related as much to dominance as to sexual behavior. Some bouts included vaginal penetration and left evidence of insemination (viz. ejaculate on the vulva). Rijksen (1978) noted that some females were more prone to be raped than others were and ascribed this to their social status; it certainly cannot be attributed to their attire. He observed only 5 rapes (out of 35 encounters) between wild subadult males and wild females. Adult males were never caught at it.

Although Rijksen (1978) clearly signalled when he was referring to wild versus rehabilitant subjects, his inferences appear to be influenced by both classes. For instance, he stated that consorting Ketambe orangutans travelled together for days, sometimes even months, but provided no quantification that can be compared with other studies on wild subjects. His narrative reveals that the longer pairings (3 and 7 months) were between a wild subadult male and rehabilitant female.

Subadult male rehabilitants were involved in more rapes (n = 22) than wild males and their behavior was more bizarre. They raped every new female that they encountered at the feeding station. When 2 couples met in the forest each subadult male promptly raped the female that was with the other male. Such observations on rehabilitants, which had had unorthodox, human-dominated childhoods, are an improper base from which to conclude that in orangutans rape is largely an expression of dominance behavior.

Schürmann (1981, 1982) confirmed that Ketambe females prefer the most dominant adult male as a mating partner. He chronicled the sexual history of a young female, which at first accepted subadult males but later climbed to ever higher ranking mates. As an adolescent she was keenly interested in the most dominant male but it took nearly 5 years for her to cement a consortship with him. He gradually began to return her interest, e.g. by allowing her to take bits of fruit from his hand and mouth. She placed her genitalia close to his face; sometimes he sniffed. She also masturbated before him and handled his penis. For several months he ignored or gently rebuffed her advances but finally began to consummate their union. She manually assisted penile intromission and performed the pelvic thrusts while he remained inert.

Several months after their first copulation they established consortships (spanning about 16 and 21 days) during which the male chased away other males that approached them. Copulations occurred midway (over 9 days and 10 days) during consortships, never at the beginnings or near their terminations (Schürmann, 1982). Even during the most heated moments of their consortships, the female usually initiated mating. She sometimes changed position from the common ventroventral one to lateroventral and dorsoventral ones and broke off to manipulate and mouth his penis. Ultimately, her efforts were rewarded with pregnancy and a firstling (Schürmann, 1982).

The adult male could claim reproductive victory over the 4 subadult males that had pursued, raped and peacefully poked her before the consortship periods. Schürmann (1982) noted that although the 4 subadults appeared to be more active sexually than the adult male, in fact, during one consortship he mated 25 times with his eager mate. This was more often than any of the subadult males had scored during one year.

Individual mating bouts lasted longer when the female initiated them, which she never did with subadults. If the dominant male uttered a long call, his consort presented to him, he sniffed the gift, mounted her and quickly climaxed. Once the female crashed a snag to which he responded with a long call and quick mating bout. It seemed to Schürmann (1982) that she had used this tactically as a turn-on. Males respond to loud noises nearby with long calls (Chapter 7) and are sexually excited when they do so. Schürmann (1982) saw penile erection, pre-copulatory ejaculation, and solitary nonfrictional ejaculation by the calling male. However, the fact that the female had pushed the snag after a rebuff recommends an alternative interpretation. If she were venting her frustration and the snag fell by chance, we need not ascribe to her as much intention as Schürmann (1982) did. One crash cannot confirm a Cleopatra.

Mitani's (1985c) observations on the mating behavior of Kutai male orangutans led him to question the certainty of current ideas about the relative reproductive success of the dominant, subordinate, and subadult males. He forthrightly laid out the limitations of his own substantial data base and concluded that further research is essential to clarify the issue. The absence of a paternity test is especially frustrating. Although the dominant male regularly supplanted other males that were with females, and sometimes copulated with his conquests, he probably was not the sire of the only infant that was born during Mitani's (1985c) tenure at Kutai.

Mitani and his assistants observed copulation during all daylight hours. Most matings (177 of 179 bouts) were arboreal though 2 were terrestrial. Only one copulation was instigated by a female, which poised her perineum in front of the male's face. The rest were initiated by the male partner.

Ventroventral postures were most popular but dorsoventral and latero-ventral ones also occurred occasionally. Males sometimes inspected the females' genitalia orally or manually before positioning them for intromission. Mitani (1985c) did not report intermissions for ancillary sex play.

At Kutai, adult, as well as subadult, males practiced forced copulations, as indicated by lengthy struggles, including slapping, biting and grabbing by the males and kiss-squeaking, grumphing, whimpering, crying, squealing and grunting by the females (Chapter 7). Forced copulations ($\bar{x} = 17.77 \pm 10.25$ min) lasted significantly longer than pacific couplings did ($\bar{x} = 9.09 \pm 4.67$ min. Humping males had to restrain struggling females by grasping their limbs and/or trunk. Resistant females were almost equally successful in escaping the clutches of subadults (8% of forced copulations) and adult males (7%) (Mitani, 1985c).

Whether an individual engaged in forced or complaisant copulations seemed to be related to male rank and size. Most (95% of 151) copulations between subadult males and adult females were forced. The dominant male perpetrated only one forced copulation and engaged in 8 that were unforced. Four subordinate adult males performed 12 forced matings and 7 that were amiable. A poorly pouched and padded lightweight among them accounted for three-fourths of their forced copulations (Mitani, 1985c).

Sometimes during male-female associations the initial copulation was forced while subsequent ones were peaceful. But in other instances of serial copulations all were forced. During one association, the dominant male copulated pacifically 6 times with his partner.

Like Galdikas (1985a), Mitani (1985a) found that subadult males are most persistent in their associations with fertile females. However, regardless of the fact that they performed the majority of copulations, none of their efforts resulted in a conception at Kutai. Mitani (1985c) agreed that subadult males might be missing the females' ovulations or might be shooting blanks from overindulgence.

COMMON CHIMPANZEES

Difficult though it was to decipher the basic social patterns of Asian apes, answers emerged at a respectable rate proportionately to field effort. This stands in marked contrast with the egnima of chimpanzee social organization. Yerkes's most social of apes defied comprehension for a number of years during which they were studied intensively.

Nissen (1931, p. 17) rejected as gratuitous the designation "family" for Guinean chimpanzee groups; it merely reflected the anthropocentric beliefs of the local people and gullible Europeans. He observed discrete groups of between 4 and 14 subjects ($\bar{x} = 8.5$, n = 25), which sometimes contained several adult males and females, but confessed that he had gained no knowledge of their social cohesiveness. In short, his 2.5-month study carried science but little beyond the Yerkeses's (1929) summary on the topic.

Beginning in 1960, field studies in eastern Zaire (Kortlandt, 1962), Uganda (Reynolds, 1965; Reynolds and Reynolds, 1965), and western Tanzania (Goodall, 1962 et seq.; Lawick-Goodall, 1965 et seq.; Azuma and Toyoshima, 1961-2; Izawa and Itani, 1966; Itani and Suzuki, 1967; Suzuki, 1969; Nishida, 1968 et seq.; Izawa, 1970; Nishida and Kawanaka, 1972; Kawanaka and Nishida, 1974; Uehara, 1981; Uehara and Nyundo, 1983; Kawanaka, 1982 et seq.; Hasegawa and Hiraiwa-Hasegawa, 1983; Hiraiwa-Hasegawa et al., 1984; Nishida and Hiraiwa-Hasegawa, 1985; Nishida et al., 1985) rekindled expectance that the secrets of chimpanzee society would be revealed.

Kortlandt's (1962, p. 6) bird's-eye view of chimpanzees emerging from the forest into a banana and papaw plantation led him to challenge the idea that they lived in closed "harem" groups. Instead he noted 2 sorts of aggregation by chimpanzees when they were travelling: (1) sexual groups, which consisted chiefly of adult males and childless females and a minority of females with youngsters; and (2) nursery groups, which were composed mostly of females and youngsters and sometimes one or two adult males. The sexual groups generally had more than 20 members; the nursery groups usually were less than 15. The former were the more wide-ranging and demonstrative. Individuals seemed to join other groups freely and were not confined to one type of group. Fission and fusion were at the forefront of their foraging rounds on the plantation. There seemed to be no particular order for indivudals during group progression (Kortlandt, 1962).

Tracking spooky chimpanzees in the intimidating Budongo Forest, Reynolds and Reynolds (1965; Reynolds, 1965) confronted even worse problems than Nissen had in Guinea. These undermined their primary goal to understand the social organization of the subjects (Reynolds and Reynolds, 1965, p. 370). Nevertheless, they also concluded that chimpanzees do not live in closed social groups. Typically there were fusions and fissions of chimpanzee aggregations and scattered individual arrivals and departures, particularly at food sources. They termed these loosely organized, unstable congeries "bands" (Reynolds and Reynolds, 1965, p. 396). Larger bands (n > 15) were noted during major fruit seasons than when fruit was scarce (n = 3-4; singletons also).

Reynolds and Reynolds (1965) recognized (1) bisexual adult bands that occasionally included adolescents but no younger subjects; (2) exclusively male bands; (3) mother bands, which occasionally included also females unaccompanied by youngsters; and (4) mixed bands which were amalgams of (2) and (3). Like Kortlandt, they detected that mothers generally travelled less widely than adults without dependents did. Although they observed occasional instances of aggression, Reynolds and Reynolds

(1965) inferred that Budongo chimpanzees lacked male and female linear hierarchies and permanent group leaders. However, they prophetically entertained the idea that they possess a highly developed social organization that can be sustained without continuous visual contact among its members. The chimpanzees literally drummed this into their heads (Chapter 7). It took a while longer for the message to be broadcast.

Goodall (1965; Lawick-Goodall, 1967, 1968a,b) perpetuated the new image of chimpanzee society as loose and fluid. She believed that the only stable unit consisted of a mother with her dependent youngsters and that there was no identifiable larger community.

During 2 years of study before intensive provisioning, Goodall (1965) observed that 91% of 498 Gombe "groups" contained 9 or fewer subjects, including 64 singletons. The largest group was 23; there were 3 groups of 20 individuals. Thirty percent of them were mixed bands; 28% were male bands and lone males*; 24% were mother bands; and 18% were bisexual adult bands.

Provisioning allowed Goodall to observe and film details on the interrelationships of known individuals and the ontogeny of social behaviors that would have been acquired much more slowly, if at all, from shy peripatetics. Once subjects were habituated, they could be followed from camp to see how the behaviors were expressed in the forest. Whether provisioning perturbed their social relationships in degree or in kind could not be discriminated readily. We probably never will know how typical Goodall's subjects were of pristine Gombe chimpanzees (Reynolds, 1975; Lawick-Goodall, 1975, p. 83; Goodall, 1983; Wrangham, 1974a; Wrangham and Smuts, 1980, p. 16; Riss and Busse, 1977; Ghiglieri, 1984a,b).

Dominant individuals generally regulated movements of small groups with which they travelled. This could be accomplished from anywhere in the group unless they moved rapidly, in which case the leader would be at the forefront. Leaders were not apparent in large mixed groups and bachelor parties (Lawick-Goodall, 1968a).

Rank was evidenced by freedom from attack by other chimpanzees, right of way on the paths and priority of access to favored foods. Lawick-Goodall (1968a) concluded that healthy adult Gombe males were dominant over all adult females and both classes were dominant over youngsters. Nishida (1970, p. 73) observed that some Kasoje females were dominant over 2 adult males. The Gombe males did not have a strictly linear hierarchy. There was one top male, followed by 6 others whose ranks depended upon what other males were present. Four more males were subordinate to the first 7; relative rank among them also depended upon which other males were present. Some individuals tended to assist one another if one of them

---

*Lawick-Goodall (1968a, p. 211) revised this to "unisexual groups" (10%) and "lone individuals" (18%), which were sometimes female but more commonly were male.

was threatened by conspecifics or baboons (Lawick-Goodall, 1968a; Hamburg, 1971; Riss and Goodall, 1979; Bygott, 1979). Similar intermale alliance relationships have been noted elsewhere (Nishida, 1983a; Nishida and Hiraiwa-Hasegawa, 1985; de Waal, 1978, 1982, 1984; de Waal and van Roosmalen, 1979).

Although interfemale alliances were rare, Gombe females also showed a relatively clear dominance ordering. Their boldness seemed to depend upon the presence of an adolescent son and perhaps by their estrous cycle; perineal tumescence seemed to be coupled with assertiveness. The status of youngsters depended upon the presence and status of their mothers (Lawick-Goodall, 1968a, 1975).

Lawick-Goodall (1968a; p. 251) stressed the long-term ties that exist between chimpanzee females and their offspring. These are probably developed during the relatively long period of infant and juvenile dependency. Weaning is not completed until youngsters are between 4.5 and 7.0 years old (Lawick-Goodall, 1973a) and individuals may continue to travel with their mothers and siblings until adolescence or early adulthood (Lawick-Goodall, 1975). Whereas infant Gombe males were observed to mount and hunch with their mothers, older sons left such overtly oedipal behavior in the realm of Sophocles (Lawick-Goodall, 1968a). Sugiyama and Koman (1979, p. 336) also observed male youngsters hunching their mothers at Bossou, Guinea.

The bonds between mothers and their offspring are reinforced not only by tender loving care, including fiddling with their infant's genitalia, but also gentle play, which decreases in frequency, and mutual grooming, which intensifies, as they mature. Grooming behavior occupied a good deal of the leisure time in Gombe chimpanzees and was common among presumably unrelated adults, as well as siblings and parents. Although this was partly an artifact of provisioning (Lawick-Goodall, 1968a, p. 263), it is probable that in general wild chimpanzees are more inclined to allogroom than Asian apes are. Because both play and grooming entail intimate physical contact and mutualism they probably contribute to the somewhat greater cohesiveness of chimpanzee versus orangutan society.

Lawick-Goodall (1968a) observed grooming clusters of up to 10 individuals and allogrooming sessions which lasted up to 2.5 hours. However, most allogrooming occurred in pairs during less than one hour spans. Adult males got more grooming than they gave while adolescent males generally groomed more often than they were groomed. Adolescent females nearly broke even and youngsters received more grooming than they performed, largely through maternal attentions. Females with juvenile or older offspring spent more time grooming than being groomed. Indeed they were the secondmost common class of groomers. Estrous females were groomed more than anestrous ones (Lawick-Goodall, 1968a).

One of the more remarkable findings, especially in comparison with orangutans, is that high-ranking Gombe males spent a lot of time grooming one another. They were the most common groomers and groomees on the provisioning ground. The upmost ape (Mike) and recently deposed alpha male (Goliath; Chapter 5) seemed to prefer each other as grooming partners (Lawick-Goodall, 1968a).

Simpson (1973) conducted an intensive yearlong study on the grooming behavior of 11 adult males that visited the Gombe banana ground. The average grooming session lasted one hour; 3 hours was the longest duration. The 7 elders groomed one another more than the 4 younger males did. High status males, as measured by their higher frequencies of displaying, eliciting pant-grunts, and especially supplanting others, were most often engaged in grooming bouts. Although he was groomed a lot, the top male (Mike) reciprocated for long periods and often was first to groom the males that presented to him. Other high-ranking males usually groomed subordinates less than they were groomed but most bouts were characterized by some degree of mutualism. Bouts were initiated by both lower and higher ranking partners.

Simpson (1973, p. 498) insightfully surmised that grooming facilitates social cohesion among the excitable males. No specific dyads were seen for more than 60% of the time that the individual males were at the feeding ground. Nevertheless, individuals clearly showed consistent preferences for some males over others during the year. In short, although Gombe male society appeared to be relatively open and fluid it was not haphazard.

Following a preliminary survey by Itani in 1960, a research team from Kyoto University inaugurated studies on the socioecology of chimpanzees at several localities south of Gombe National Park. In early reports, they portrayed chimpanzee society much as Kortlandt, Reynolds and Goodall did, viz. as rather wide-ranging small to medium size groups of variable composition that mix and subdivide peacefully according to the abundance of preferred food at clumped versus diffuse arboreal sources (Azuma and Toyoshima, 1961-2; Izawa and Itani, 1966).

Itani and Suzuki (1967; Suzuki, 1969) provided the first solid evidence for a more complex social organization among chimpanzees. They observed a group of 43 subjects progressing together over sparsely treed territory between two patches of riverine forest at Filabanga. The central section of the procession consisted of 7 adult males. Congeries of females and youngsters preceded and followed them. They dispersed in smaller groups in the forest. Three additional sizeable (n = 31, 21, 32) mixed groups at other localities also progressed with a clump of males in the middle. Small all male groups were encountered routinely by Itani and Suzuki (1967).

Surveys of the surrounding area suggested that the Filabanga 43 were the only chimpanzees in the region. Itani and Suzuki (1967) concluded that they constituted a discrete group and inferred that, universally, chimpanzees are divided into groups of 20-50 individuals, of which 4-6 were adult males, about 10 were adult females, and the remainder were youngsters. They indicated that young adult males were absent or rare in the groups and speculated that they may leave the natal group and then eventually join a neighboring group. The only unit lower than the large group was the mother and infant (Itani and Suzuki, 1967).

Kano (1971), who succeeded Itani and Suzuki at Filabanga, noted 3 large groups but he could not count them accurately. He doubted that young males are driven from their natal groups.

Nishida (1968) coined the name unit-group for the basic social unit of chimpanzees. He conducted a long-term study of unit-groups at Kasoje in the Mahale Mountains, south of Filabanga. Subjects were provisioned with sugar cane and bananas, but not in the quantities that characterized the Gombe banana bonanza. At least 5 unit-groups* resided in the area. Nishida and his co-workers focussed on 2 of them (Kawanaka and Nishida, 1974).

At Kasoje, unit-groups contained between 20 and 106 members. As at other localities, the unit-group is generally divided into temporary sub-groups of variable size ($\bar{x} = 8$ members, $r = 1$-28), composition and stability. Subgroups were larger at the end of the dry season when food was most abundant. The adult males tended to stay together more consistently than females did. Nishida (1979) estimated that male cohesiveness was twice that of females. Forty-six percent of grooming occurred between males, 39% between males and females, and only 10% between females. Estrous females were most commonly seen in sizeable, mixed subgroups of more than 11 subjects (Nishida, 1968).

Nishida (1968, 1979) concluded that adult males are the stable core of the unit-groups, whereas females, especially estrous nulliparous ones, commonly emigrate to neighboring unit-groups. Indeed most young females ultimately reside outside their natal unit-groups. For example, over an 8-year span, 23 out of 29 original members of K(ajabara)-group remained therewith. Two males probably died, 4 subadult females disappeared from K-group, presumably to join another unit-group, and 6 new ones immigrated into it. Of the 9 adult females in K-group, 3 appeared to be attached strongly to the core males while the other 6 seemed to be less so, but remained in the area (Nishida, 1979).

Kasoje unit-groups seemed to have well delineated traditional boundaries which they would not cross, though large sections of their home ranges overlapped. The larger M(imikere)-group, which ranged south of K-group, was dominant to the latter, as evidenced by its displacement northward when M-group moved seasonally into the southern part of their range (Nishida and Kawanaka, 1972). M-group used about half of the home range of K-group, which in turn used about one-third or one-fourth of the home

---

*The P-group is now thought to have been erroneously distinguished from the B-group (McGrew and Collins, 1985, p. 59).

range of M-group (Nishida, 1979). Transfers of estrous adult females between the unit-groups occurred most often during the 2 short annual migratory periods which constitute 20-30% of the yearly round. Estrous adolescent and subadult females more commonly transferred to another unit-group during sedentary phases when unit-groups were farther apart (Nishida and Kawanaka, 1972).

Whereas female transferees are gradually accommodated in a new social milieu, males of different unit-groups react violently toward one another. For instance, when a lone old male of M-group appeared at the feeding ground, he was attacked by 3 of the 4 males from K-group that were already there. He fled with a bitten thigh instead of a bite to eat (Kawanaka and Nishida, 1974). Generally, however, when in the common section of the 2 ranges, chimpanzees of K-group retreated rapidly and silently upon hearing calls from M-group. Retreats seemed to be premised on which subgroup included the fewer adult males.

Lawick-Goodall (1973c, 1975; Goodall, 1983) accepted the concept of unit-group but used the term community in her discussions of Gombe society. She and her co-workers (Goodall et al., 1979; Bygott, 1979; Halperin, 1979; Pusey, 1979, 1980; Teleki et al., 1976) documented instances of female transfer and territorial behavior in chimpanzees. Over a a 7-year span adult Gombe males did not change communities (Pusey, 1979). Instead they went on group boundary patrols during which they were unusually quiet, cohesive, and systematic in investigations of their habitat. If they encountered aliens in their range, they often called, displayed, and chased them. When they caught vulnerable transgressors, they set upon them most murderously. Several individuals have been mortally wounded during these encounters. Like Caligula Caesar, the alpha male seemed to be less inclined to patrol and attack aliens than his subordinates were (Goodall et al., 1979).

Goodall et al. (1979) speculated that the patrols may serve as a mechanism whereby young females are recruited into the resident community. Long-term observations at Kasoje have documented the near extinction of K-group due to expansions of the range of M-group and loss of males from K-group, some of which may have been killed by males of M-group (Itani, 1980). By 1979, K-group had been reduced to a one-male enterprise from which cycling females moved to M-group while the male attempted to "herd" them within his shrunken range (Nishida and Hiraiwa-Hasegawa, 1984). In 1982, the last adult male and a subadult male disappeared, leaving only 2 adult females, an infant and an adolescent male in "K-group" (Nishida et al., 1985). Then the M-group included 11 adult males, 39 adult females, 19 adolescents, 14 juveniles and 22 infants, and received visits from 12 adult females with 3 adolescents and 5 juveniles. Its range extended over 14 km² (Hiraiwa-Hasegawa et al., 1984).

With one possible exception (Sugiyama and Koman, 1979a; Sugiyama,

1981, 1984), studies outside Tanzania (Sugiyama, 1968, 1969, 1973a,b; Suzuki, 1975; Ghiglieri, 1984a,b; McGrew et al., 1982; Tutin et al., 1983; Baldwin et al., 1982) have either confirmed or at least not countered the hypothesis that androcentric unit-groups are universally characteristic of wild chimpanzees. The question of widespread territoriality is unresolved from available data. For example, Sugiyama (1968, 1973a,b) reported that Budongo chimpanzees from neighboring regional populations mixed freely and maintained friendly relationships.

At Mt. Assirik in Senegal there was only one unit-group of 25-30 *Pan troglodytes verus* so there was no opportunity to observe inter-unit-group responses and female transfers. Interestingly, all male subgroups were rarely seen [4 (2%) of 247 parties] and they consisted only of a pair of adults or an adult and an adolescent (Tutin et al., 1983). Border patrols would be unnecessary in an isolated population.

As at other localities, mixed bands were most common; and, on average, those containing estrous females were significantly larger ($\bar{x} = 10$; $r = 3$-22) than those without estrous females ($\bar{x} = 7$; $r = 2$-20). Whether their large unit-group range (228 km²) is a consequence of food availability, predator avoidance, or the absence of competing territorial unit-groups is unclear. However, Tutin et al. (1983) reasonably relate the greater tendency for chimpanzees at Mt. Assirik to forage and travel in large, mixed parties to reducing the risk of predation by 4 formidable species of Carnivora (*Panthera leo, Panthera pardus, Lycaon pictus* and *Crocuta crocuta*) in the park.

The nonprovisioned chimpanzees at Bossou, Guinea, may constitute an exception to the androcentric model that was formulated on the basis of Tanzanian chimpanzees. Over a decade the Bossou group remained about the same size (n = 19-21) but more than half the adult males disappeared from the core area of 6 km² (Sugiyama, 1984). During a 6-month study, Sugiyama and Koman (1979a) noted that 2 males "from other groups" joined the Bossou group for short periods. Because no females emigrated from the group, they concluded that, like macaques, among Bossou chimpanzees, males are the transferring sex. They suggested that provisioning may be the cause of female transfers in Tanzania.

Three years later, Sugiyama (1981) restudied the Bossou group for 3 months. Its size had decreased to 19 due to the loss of 5 individuals, which was only partly offset by 3 births. An adolescent male and 2 adult males, including the previous alpha one, had disappeared during the recess in observations. No adolescent or adult female had joined or left the Bossou group. This induced Sugiyama to reiterate his charge that provisioning had altered the natural societal pattern of the Gombe and Kasoje (Mahale) unit-groups.

But Nishida et al. (1985) found that when the number of adult males in the Mahale K-group fell to 3, only one adolescent female immigrated into it.

And when there were only 2 males, the resident females began to leave K-group. The Bossou group had no more than 2 adult males during the study periods (Sugiyama, 1984).

Apart from the relative brevity of their study, there are still other reasons to doubt that the Bossou chimpanzees require dismantlement of the androcentric model. Sugiyama and Koman (1979a, pp. 327-328) observed no actual emigrations. The 2 males, an adolescent and an oldster, that visited the Bossou group were transients, not immigrants. The adolescent remained with the group less than 24 hours. Behavior between the old male and the resident alpha male indicate that they knew one another (cf. Bauer, 1975, 1979). The oldster departed after 20 days. While Sugiyama and Koman inferred that the 2 visitors were solitary males, it seems more likely that they had been travelling together. Recall that a similar casual pairing occurred at Mt. Assirik. Further, they stated that although many of the Bossou chimpanzees stayed together, a few of them "kept apart from the major part of the group and moved independently" (p. 332). Both focal and transient subjects at Bossou may be part of a single isolated community, like that at Mt. Assirik (Kawanaka, 1984).

Ghiglieri (1984a,b) reported that unprovisioned chimpanzees in the Kibale Forest, Uganda, evidenced stronger affiliations among females than one might expect according to an exclusive androcentric model. Indeed he observed that Ngogo females "frequently ranged together, almost as if they were members of a female community" (1984a, p. 182). However, because of his limited observational spans Ghiglieri's study, like Sugiyama's, can only stand as a caveat against universal application of Nishida's model for common chimpanzee society.

Informed critics (Ghiglieri, 1984a; Itani, 1980; Kawanaka, 1982b, 1984; Pusey, 1980; Uehara, 1981; Goodall, 1983; Nishida et al., 1985) promptly rallied and came down soundly on Wrangham's (1979b) model, which virtually excluded female chimpanzees from the broader social framework. He proposed that, like orangutans, female chimpanzees are distributed in core areas in order to maximize access to preferred food for themselves and their youngsters while the males range more widely to gain access to a number of sexually receptive females (Wrangham and Smuts, 1980). A major difference from the orangutan pattern is that male chimpanzees travel in groups and act together to prevent incursions by extracommunity males. This also conserves food resources in their area for the females, their youngsters, and themselves. Rank ordering among the males reduces the frequency of intragroup competition over the limited resources (Wrangham, 1979b).

Wrangham (1979b) interpreted the reported unit-group transfers of females at Kasoje as core males of a neighboring unit-group having moved into the ranges of more or less resident females instead of the females having immigrated to a new unit-group. Accordingly, only male chimpanzees are bonded into cohesive social units (i.e. communities) while females generally exist in the same asocial state that is ascribed to orangutans by most observers.

Uehara (1981) acknowledged that the males of the Kasoje K-group and M-group were discrete in their associations and ranging patterns. Some females resided within the boundary of K- or M-group and associated exclusively with its males. Others ranged in the area or areas of one or both unit-groups and interacted alternately with males of both groups in the region where the ranges of the unit-group males overlapped. But no female was forever fickle. Diachronic observations revealed that with maturity they settled in one community or the other. Consequently, Uehara (1981) dismissed Wrangham's male model in favor of a bisexual community for common chimpanzees.

Initially Kawanaka (1982b) provided circumstantial, yea anecdotal, evidence to counter Wrangham's males-only model. He observed an encounter between 4 adult females and a juvenile of K-group and 2 adult females and 2 young adult males of M-group, which was characterized by tension, manifested by loose bowels, but no outright agonism. Indeed some individuals greeted and groomed one another despite apparent nervousness. Kawanaka (1982b) concluded that a donnybrook was avoided because fully adult males were absent from both groups. And behavior of the females during the meeting suggested that they identified with their own unit-groups, counterpart members of which would not evoke so much colonic activity.

A more thoroughgoing proximity matrix analysis of data from a yearlong study of the two unit-groups revealed that except for 3 cycling females, one of which was accompanied by her juvenile male son, all females remained exclusively with either K-group or M-group. The female/juvenile dyad ranged further than any other subjects, including the adult males. The others did not confine themselves to core areas within the ranges of unit-group males but instead roamed widely within the entire range of at least one unit-group (Kawanaka, 1984).

Ghiglieri (1984, p. 183) found that although Kibale chimpanzee males exhibited the greatest affinity among themselves (based on travel companionship and allogrooming), the females were more social toward one another than toward the males and they associated with males in ways that are not directly relatable to mating. Thus, he too favored the concept of a bisexual community versus the Wrangham model.

Although he did not contest Wrangham's model, Halperin (1979) noted that 5 anestrous Gombe females with dependents were homophilically social. Mothers with single dependents preferred to be with other females in nursery groups while females with 2 dependents seemed more inclined to go it alone. Perhaps the social needs of the latter females were satisfied by their offspring, whereas females with singletons are induced to seek the company of other adult females (Halperin, 1979, p. 497).

Pusey (1979, p. 479) initially agreed with Wrangham but soon abandoned his model in favor of the bisexual unit-group. She noted that Gombe females sometimes participate with unit-group males on border patrols and they do not always accept immigrant females peacefully into their community. Further, the males of a unit-group will attack females of another unit-group unless they are in estrus (Pusey, 1980).

Itani (1980) pointed out that Wrangham's model fails to account for the cohesive trek of the Filabanga 43 and a later, presumably seasonal, migration from the area for a span of at least 7 months.

Studies at Gombe National Park have led to intriguing hypotheses about the reproductive strategies and sexual tactics of female and male chimpanzees. The work of Tutin (1975, 1979a,b, 1980) and McGinnis (1979; Tutin and McGinnis, 1981) is outstanding in this complex problem area. Early reports (Lawick-Goodall, 1968a, 1973a, 1975) led to an image of Gombe females as highly promiscuous and voracious. Indeed, if placed on a human scale, they would rival Valeria Messalina.

Lawick-Goodall (1968a, p. 220) reported that estrous females in large groups were normally mounted in rapid succession by up to 8 males, especially during periods of general social excitement. Young females were less popular sex objects than older females were. For instance, during her perineal swells one oldster attracted retinues of between 6 and 10 mature and 4 or 5 adolescent males. Out of 54 observed copulations, young males only made it with her twice (Lawick-Goodall, 1968a).

Adult males generally accepted the gang approach of other mature males but randy adolescents were subjected to frequent interruptions by male elders and thus had to be sexually discreet in their presence. They seldom joined the mating frenzies that were triggered by group reunions and banana booms. Instead most of their mating occurred peripherally during pacific moments. Male mating behavior is almost fully developed during the first two years of infancy. But they are not socially mature for an additional 9 years at least (Lawick-Goodall, 1975). Hence males have a long wait for the opportunity to fulfill their breeding potentials.

In addition to gang sex, Lawick-Goodall (1968a, pp. 219-220) and McGinnis (1979) noted occasions when males seemed to force females, including anestrous ones, to accompany them for spans up to 4 days. But in other instances, individual estrous females seemed to travel voluntarily with a male, perhaps to avoid serial gang sex. And Lawick-Goodall (1968a) showed even more explicitly that Gombe females are not always passive objects of male attentions; 17% (n = 37) of 213 copulations and attempts thereof were initiated by the female partner.

Tutin (1975, 1979a,b, 1980; Tutin and McGinnis, 1981) documented that over a 16-month span the majority of Gombe females (9 out of 14) were impregnated during restrictive associations with single males versus gang sex with several opportunists. She recognized two forms of restrictive mating patterns: short-term possessiveness by high-ranking males, which prevented estrous females from mating with other males, and temporary consortships, in which breeding pairs actively avoided the company of other chimpanzees. The former was solidly a male dominated relationship while the latter required notable female cooperation wherein they could exercise choice. Further the brevity of possessive mating periods suggests that female choice was operant therein also. Pairs usually remained together in possessive associations for a day or less while the consortships had a mean duration of 9.5 days (r = 3 hrs - 28 days; median = 7 days). The majority (85% of 209) of unsuccessful courtship sequences occurred because the females avoided or did not respond to the males. Only 4% failed because of male unresponsiveness (Tutin and McGinnis, 1981, p. 250).

Female chimpanzees must choose mates strategically because they have a limited reproductive potential and a high level of parental investment (Tutin and McGinnis, 1981). Based on an examination of a 19-year accumulation of data from Gombe, Tutin (1980) concluded that the theoretical lifetime reproductive potential of females is 5 or 6 offspring; in fact, the median is 3 births per female; and, only 2 of them are expected to survive to reproductive age. Teleki et al. (1976) cautioned that the banana bonanza may have reduced the birth rate at Gombe National Park.

Consortships accounted for 7 of the 9 conceptions that occurred during restrictive matings. This is remarkable considering that only 2% of the 1,137 observed copulations involved consorting pairs. The vast majority (73%) of copulations occurred opportunistically and the remaining 25% were effected with possessive males (Tutin, 1979a).

A male establishes a consortship by remaining close to an estrous female until there is an opportunity to lead her silently in a different direction when the foraging group beings to travel. The female may choose not to be chosen simply by not following the suitor or by responding to the calls of other males (Tutin and McGinnis, 1981, pp. 257-258). Rebuffed males make no attempt to prevent the female from joining calling fellow travelers.

Consorting estrous females are rewarded with higher frequencies of grooming ($\bar{x} = 14.8\%$ of waking hours) with the male than the average (5.3%) in larger mixed groups (Tutin and McGinnis, 1981). They probably also profit from decreased competition for food while staying with only one other adult (Wrangham, 1975, 1977).

The factors that determine female choice are unknown. However, the rather long periods of adolescent sterility during which females copulate promiscuously may provide clues about which of the males might be the best partners when they are physiologically equipped to conceive (Tutin and McGinnis, 1981). Equally puzzling is what induces young females generally to conduct their sexual experimentation outside the natal community where initially they would know much less about the personalities of the males (Pusey, 1980).

Estrous females prefer males that will groom, share food, and spend time with them (Tutin, 1975, 1979a). Only the alpha male appears to gain reproductively from the position that he earned through energetic displays, coalitions, and occasional strong-arm tactics. Other males are reduced to gang banging, which apparently is not the most reliable means for impregnation; breeding with nulliparous young females, which commonly are less fertile than mature females are; and wooing prime breeders away from the maddening crowd where copulation can really count.

Tutin (1979a) concluded that except for the alpha male, consortships are the best reproductive arrangement for adult males and females. The major disadvantage to consorting couples is that in order to gain privacy they sometimes venture to the periphery of the community range and beyond, where they could be set upon by border patrols from neighboring communities. Their quietness, especially regarding pant-hooting, may serve not only to isolate them from their own community but also to avoid detection by a meaner lot.

Another disadvantage for males that engage in prolonged consortships is that, among chimpanzees, absence makes the heart grow harder. Returnees may be challenged aggressively and can lose their dominance ranking when they attempt reunion with the group.

The alpha male would seem to have the best of both worlds. He can enjoy reproductive prerogatives and sustain his status while subordinates deflect potential extracommunity rivals. It remains to be discovered whether possessive or consorting males produce the most offspring during their lifetimes and whether some combination of the two tactics is the fittest long-term strategy for a given male.

The universality of the Gombe chimpanzee mating pattern is unknown. Studies in the Mahale Mountains indicate that it is at least broadly comparable in other communities of Tanzanian chimpanzees. Nishida (1968, p. 199) noted female choice. Indeed he reported that almost all of the sexual interactions that he observed were initiated by females and that frequently the males were unresponsive. Though rare, instances of overt intermale sexual competition were dramatic. But during the initial 22 months of his study, Nishida (1968) witnessed only 14 copulations between members of K-group and 3 between other group members.

In a subsequent yearlong study of K-group, Nishida (1983a) reported that a male in the alpha position exhibited possessive behavior wherein he maintained proprietary copulatory rights with the estrous females. The 2 subordinate adult males of K-group engaged in consortships but apparently most of the time their partners were anestrous. After he had been bloodied and usurped by a coalition between the beta- and gamma-males his sexual privilege was forfeited. His successor became the sexual powerhouse of K-group. The contests that led to his fall always occurred when at least one estrous female was present (Nishida, 1983a).

A 28-month study in the Mahale Mountains by Hasegawa and Hiraiwa-

Hasegawa (1983) explicitly confirmed Tutin's conclusion that common chimpanzees engage severally in opportunistic, possessive and consort matings. They concentrated on the populous M-group, which included 12 adult males, 40 adult females and 54 youngsters. They made supplementary observations on K-group, whose potential spermatic contingent had dwindled to include only a postprime male and a young adult male.

The vast majority (92%) of 660 copulations were opportunistic and a mere 7% occurred during possessive relationships. Hasegawa and Hiraiwa-Hasegawa (1983) observed only one consortship. But their data is biased because of difficulty in contacting pairs in the bush and their absence from the feeding station. Opportunistic matings occurred more frequently in large groups than in small ones. The alpha male could not monopolize all of the estrous females in M-group. However, 75% of 20 possessive mating bouts involved him. All of his possessive matings were with older females. Two other high ranking males and a middle ranking male effected the remaining 4 possessive matings. Whereas none of the 3 subordinate males engaged in possessive couplings for more than a day, most of the alpha-male's possessive bouts lasted 2 days or more ($r = 1$-7 days). Further, the subordinates managed possessiveness only when the alpha male was absent or engaged himself with another estrous female.

Immature males usually copulated with young nulliparous females and sometimes with adult females that were unlikely to be ovulating at the time. Indeed mature females were more restrictive in their mating as their perineal swells ripened. Whereas 35 (7%) of 501 matings involving mature females were restrictive, only 1 out of 115 copulations by younger females occurred in a possessive relationship. Young females copulated with young males more than older females did.

The disintegration of K-group, which occurred during their study, allowed Hasegawa and Hiraiwa-Hasegawa (1983) to compare the mating patterns of female transferees with those of resident females. They found that resident females generally mated more restrictively than newcomers did. Further the alpha-male restricted his possessive matings to resident females of M-group. Herein he differed from an alpha-male of K-group which most frequently mated with transferees (Nishida, 1979a). Whether this variation reflects personal preferences by the two alpha-males or was due to behavior by residents of M-group and K-group toward transferees cannot be discerned from available reports. Other high-ranking males of M-group also seemed to prefer resident females as mating partners (Hasegawa and Hiraiwa-Hasegawa, 1983).

BONOBOS

Latest to be studied and extra elusive with humans, bonobos are the supreme bafflers anent the nature and adaptive meaning of ape social organization and sexual behavior. Our knowledge is still sketchy and preliminary despite more than a decade of research at several localities in Zaire.

Early studies near Lake Tumba were busts because the investigators had little or no direct contact with bonobos. Nishida (1972a) conveyed hunters' reports that they are often seen in sizeable groups (n = 15-40) and sometimes travel in pairs or alone. Per contra, Horn (1980), whose sightings during a 2-year period were ephemeral, reported that bonobos of the Lake Tumba region live in small groups. The largest consisted of 4 members (an adult bisexual pair with 2 youngsters). Horn (1980) also saw a lone male and a group of 3, comprised of an adult bisexual pair with a youngster. He suspected that they occasionally aggregated into larger congeries.

In 1973, Nishida and Kano surveyed the Salonga National Park in central Zaire. Based on large clusters of nests and 14 encounters with bonobos, Kano (1979) speculated that they live in groups that are larger than those of common chimpanzees. Although he counted no more than 22 bonobos in a group, he estimated that total sizes ranged between 10 and 90 individuals.

The first major advance in studies of bonobo social and sexual behavior came with the establishment of a base for long-term studies near Wamba village in the Salonga National Park (Kuroda, 1979, 1980, 1984; Kano, 1980, 1982a; Kano and Mulavwa, 1984; Kitamura, 1983; Mori, 1983, 1984).

Like *Pan troglodytes*, Wamba *Pan paniscus* are organized into unit-groups of between 50 and 120 individuals (Kano, 1982a), which usually forage in smaller subunits ($\bar{x}$ = 17 individuals; r = 2-54; n = 147) (Kuroda, 1979). Lone individuals were seen infrequently (n = 10). Small foraging subunits (i.e. <10 individuals) were usually quiet, non-vocal, unaggressive toward one another and sexually abstemious. They became excited and vocalized in response to larger parties. The latter were clamorous, demonstrative and given to copulatory frenzies and other sex acts that outdo the orgiastic reunions of common chimpanzees. However, excited bonobos tend to be less aggressive than their Tanzanian congeners during reunions (Kuroda, 1979).

Kuroda (1979) observed that nearly three-quarters of 163 subunits were mixed bands (sensu Reynolds and Reynolds, 1965). And he suspected that a good portion of an additional 10% of groups with uncertain composition were probably mixed bands. Bisexual mixes characterized both small and large groups. The smallest mixed band consisted of a male, an anestrous female and an infant (Kuroda, 1979).

Other sorts of subunits were rare at Wamba. Kuroda (1979) saw only 4 adult bands (n = 3-13 individuals), 3 of which included 2 or 3 estrous females and between 4 and 6 adult males. He witnessed a single copulation in an adult band, which happened to be the largest one.

Kuroda (1979) noted 4 male bands, all of which were near larger groups and consisted of fully adult individuals. He doubted that they were truly separate from other bonobos.

The compositions of 8 mother bands ranged from an adult female/young juvenile pair to one with 7 adult females with infants, 4 juveniles and a young female. Only one estrous female was seen in a mother band. Half the mother bands were within 100 m of mixed bands. In one instance they exchanged calls and probably fused (Kuroda, 1979). Two adolescents that Kuroda (1979) scored as loners were near groups. All lone females were anestrous. A stray female with an ophthamological disorder was sighted thrice. The lone males (n = 4 observations) were all young adults or adolescents.

Kuroda (1979) concluded that bonobos have stronger affiliation between males and females than common chimpanzees do; the core of the group is bisexual; and there is little difference between the ranging patterns of males and females. Males appeared to initiate travel between food sources more often than females did (67% of 39 progressions). Within traveling parties there was notable clumping of adult males with females. Young males tended to rest and nest at the periphery of mixed bands (Kuroda, 1979).

Unlike common chimpanzees, female bonobos associate more frequently and intimately with one another than males do. Nevertheless, interfemale affiliations are second to bisexual ones. Forty-nine percent of grooming pairs were bisexual, 25% were interfemale and 8% were intermale. Bisexual grooming bouts lasted longer ($\bar{x} = 21.3$ min) than interfemale ($\bar{x} = 15.5$ min) and intermale ($\bar{x} = 14.5$ min) bouts. Males groomed females for longer spans than they were groomed by them. Although males groomed most often with estrous females, they also groomed anestrous females a bit more than the females groomed each other (Kuroda, 1980).

The sexes are quite tolerant of one another; interfemale interactions are particularly pacific. Kuroda (1980) observed 56 agonistic episodes, all of which occurred arboreally. The aggressors were adult males or could not be identified. The targets were mostly males, ranging in age from juvenile to adult. Only a quarter of the victims were adult females. Commonly the mum attacker leaped or otherwise dashed through the trees, chased and kicked at his target. The attacks were free of the pummeling, stomping, hair-pulling and biting that characterize chimpanzee muggings. The victims were seldom bloodied.

If the screaming victim presented, it was most often mounted. Occasionally the attacker merely touched, held, groomed or touched his rump against that of the crouching offender (Figure 27b). Two such chases culminated in copulation.

Kuroda (1980) concluded that adult males were dominant over all females and he suspected that there was a dominance order among Wamba bonobo males. But his limited ability to identify individuals, especially during arboreal attacks, precluded its delineation.

Kano (1980, 1982a), who succeeded Kuroda at Wamba, provisioned some members of a unit-group with sugar cane. This allowed greater individual identification of the subjects and observations of their behavior on the ground. He basically confirmed Kuroda's conclusions about bonobo sociality. Out of 156 encounters with bonobos, 85% were in mixed bands. Adult bands (n = 2), mother bands (n = 1) and loners were quite rare. Estrous females were absent from the adult and mother bands. The singletons were males, one of which was seen also with mixed groups (Kano, 1980). In a later study on provisioned bonobos, Kano (1982a) reported that out of 180 sightings, 96% were of mixed bands; the others were male bands (n = 2) or lone males (n = 6).

Social grooming underscored the special affinities between adult males and females and among females. Fifty-four percent of 84 adult allogrooming bouts were between males and females; 34% involved females only; and 12% were intermale. During an average adult grooming session, the partners switched roles 3 or 4 times; some alternated as many as 10 times during a session. In addition to grooming reciprocally with adult males and other females, mothers unidirectionally groomed their infants and young juveniles. In fact the highest number of grooming session (n = 102) was of this sort (Kano, 1980).

In addition to grooming data, Kano (1980) cited propinquity while feeding, food-sharing (Chapter 3), and bizarre homosexual hunching (Figure 27a) as evidence for strong affiliations among bonobo females. Wamba females tended to aggregate peacefully in large food trees while the males bickered and dispersed themselves more widely so that ultimately a choice resource would contain only one or a few males with many females (Kano, 1980).

Nearly 61% of interadult food-sharing was between females. Kano (1980) regarded their behavior during begging and sharing as predominantly friendly, though not totally free of tension. Kuroda (1980) and Kano (1980) reasoned that the tension is alleviated by ventroventral rubbing of the genitalia between pairs of females. Kuroda (1980, p. 189) termed the behavior "genito-genital (GG for short) rubbing" and Kano (1980) simplified it to "genitals rubbing." Because the former calls to mind the elusive spot that Ladas et al. (1982) set us in search of, and the latter is ungrammatical, another term—homosexual hunching—is used here.

Homosexual hunching occurred most frequently during the first 5 minutes after females had arrived at a feeding locality. Then it decreased, except when begging clusters formed. As a female joined a group of cadgers she serially solicited or was solicited by others to embrace ventroventrally whereupon they rapidly and rhythmically rubbed their pudenda together sidewise for spans between a few seconds and more than 60 seconds (Kuroda, 1980; Kano, 1980; Mori, 1984).

The solicitor stares at the intended partner's face and perhaps pedally

Figure 27: (a) How sweet it is. Homosexual hunching by female bonobos on a bed of sugar cane. Note that the topmost female has a prominent perineal swell. (b) Mutual rump touching by bonobos. (Photos courtesy of Takayoshi Kano.)

touches her foot or knee. Then one of them reclines or hangs from a branch, the other closes the embrace, and they hunch. Usually both partners are silent but occasionally a copulatory scream is uttered. As attested by the absence of agonistic signals, they rarely rub each other the wrong way.

Most homosexual hunchers are estrous (i.e. sporting perineal swells) but anestrous females also participate. Despite engagement of the genitalia Kano (1980) concluded that homosexual hunching is not sexual behavior per se but a form of female greeting behavior that eases tension during crowding. It was infrequently observed during rest periods and travel.

After identifying 130 mature bonobos individually, Kano (1982a) established that there were 4 unit-groups in the Wamba region. Kano and Mulavwa (1984, p. 234) revised this to 5. Kano (1982a) successfully provisioned one of them (E[langa]-group) and was able to identify all 63 of its mature and immature members. Except for 2 adolescent females that moved between unit-groups, the bonobos associated peacefully with members of a single unit-group. The E-group included 15 adult males (>13 yrs.), 16 adult females (>13 yrs.), 5 adolescent males (8-13 yrs.), 8 adolescent females (8-13 yrs.), 9 juveniles (2-7 yrs.), and 10 infants (<2 yrs.). It was the third largest unit-group in the region.

The ranges of neighboring unit-groups overlapped, especially where food was plentiful (Kano, 1982a). When bands from different unit-groups came into audible range, they avoided one another. This contrasts with parties of the same unit-group, which generally merged noisily within 1 to 10 minutes of mutual awareness. Kano (1982a) inferred that vocal exchanges allowed band members to determine whether another band belonged to their unit-group and, if not, whether they were rivals that could be supplanted or should be avoided. Thereby direct agonistic engagements were usually averted.

Kano (1982a) reported that the average size of mixed bands at the sugar cane field was 19 (r = 5-37; n = 172). This is not appreciably larger than the bands that Kuroda (1979) reported for unprovisioned Wamba bonobos. Kano confirmed that both sexes are nearly equally represented in mixed bands, except in ones that have fewer than 10 members. Virtually all (98%) of them included at least one estrous female. Indeed approximately a fourth of the members in most mixed bands were estrous females.

Kano (1982a) decided that there were 4 bisexual subunits of E-group, whose members tended to forage and travel together in particular sections of the E range. Only young nulliparous females did not associate regularly with a single subunit. Kitamura (1983), whose study overlapped that of Kano, confirmed that there are cohesive bisexual subgroups within the unit-group. But he identified only 2 (a northern and a southern) primary subunits in E-group. Further, Kitamura (1983) noted that female members of the unit-group generally range less widely than some of the males do. He proposed that female clusters are the focal points of the bisexual groups.

Kano (1982a; Kano and Mulavwa, 1984, p. 234) claimed that in addition to coteries of females, cliques of males and a mother and her male offspring may serve as cores of the various subunits in bonobo unit-groups. He inferred kinship from facial likeness (by which criterion I could be Bob Newhart's younger brother).

As at Gombe, provisioning increased the frequency of intragroup agonistic behavior among the subjects at Wamba (Kano and Mulavwa, 1984; Mori, 1984; Kuroda, 1984). It was most frequent between males and some individuals were excluded from the cane field. However, multiparous females boldly approached the feeding spots of high-ranking adult males and supplanted them with no overt aggression. It would be interesting to know whether they were closely related to the supplanted males or had powerful sons in the vicinity.

Kano's (1983) observations at Yalosidi, near the southeastern limit of the range of *Pan paniscus*, basically corroborate the portrait of their society derived from Wamba. Most of the 68 hours of direct observations occurred in areas of secondary forest. Kano (1983, 1984b) recognized several individuals by physical characteristics, e.g. mutilations from wire traps. Mixed bands accounted for 70% of the foraging groups that he encountered. On average they contained 8.5 members (r = 3-21). The ratio of females to males was 1.6.

In addition to mixed bands, Kano (1983) sighted the following units: lone males (n = 5); a pair of males, which he followed for 2 days; 2 bands of subadult males; and 2 mothers travelling only with their youngsters. As at Wamba, Yalosidi males tended to feed in separate trees while females did not avoid one another and aggregated near a few individual males.Thus, Kano (1983) commonly saw one and occasionally 2 males in a fruit tree with the females and youngsters. Other males were scattered nearby.

Kano (1983) noted 107 allogrooming sessions. Bisexual dyads were most common, followed by female pairs. Adult males were not seen to groom one another. During the average 6-7 minute bout, the partners took turns as groomer and groomee several times. The 3 longest grooming bouts (45-77 min) were between males and females. Once Kano (1983) witnessed 3 spells of homosexual hunching by females.

Intragroup relations were generally peaceful at Yalosidi. Idyllically, juveniles often played while the adults fed. One juvenile entered the day nest of a male and stayed with him for an hour, during which they groomed one another reciprocally (Kano, 1983).

The Lomako Forest is the northernmost locality where bonobo sociality has been studied (Badrian and Badrian, 1977, 1980, 1984a,b; Badrian and Malenky, 1984; MacKinnon, 1976; Thompson-Handler et al., 1984). The subjects there were not provisioned.

Initially, Badrian and Badrian (1977) sighted mostly (80%) small groups (n = 2-4) and single bonobos. MacKinnon (1976) also noted small groups, which, like those of the Lake Tumba area, were bisexual. Subsequently, during an 18-month intensive study, the Badrians (1984b) and co-workers found that socially and sexually the Lomako bonobos are remarkably like those at Wamba, except that they characteristically forage in smaller groups.

Two unit-groups, with overlapping ranges, live in the Lomako study area. The B[akumba]-group consists of approximately 50 members; the E[yengo]-group was not censused totally. Vocal "contests" facilitate their mutual avoidance (Badrian and Badrian, 1984b).

The modal group of Lomako bonobos consisted of 2-5 individuals ($\bar{x}$ = 7.6; r = 1-50; n = 268). Bands with 6-10 members were secondmost commonly observed. Indeed three-quarters of the encounters were with loners and bands with 10 or fewer subjects. As in other areas, small bands were relatively quiet in the forest while larger groups were given to vocalizations and displays both toward observers and among themselves (Badrian and Badrian, 1984b).

As at Wamba, bisexual groups were the rule in the Lomako Forest. They tended to have approximately equal numbers of adult females and males. However, as group size increased, females tended to outnumber males. Out of 191 completely censused bands, 68% were mixed and 8% were bisexual adult bands. Most (60%) of the adult bands consisted of adult pairs, in 7 of which there was an estrous female. Five bisexual adult bands had 2 adult males with between one and 3 females, many of which were in estrus (Badrian and Badrian, 1984b).

Male bands were seen on 9 occasions; 7 were in pairs and 2 had 3 members. Female and mother bands were also infrequent (5.2% of encounters) and small (2-5 members). Lone males were encountered 24 times and 12 adult females, 5 of which were with dependent youngsters, were seen apart from other adult bonobos. Badrian and Badrian (1984b, p. 334) acknowledged that some of the "lone" bonobos were probably laggards from silently retreating bands. Vigilant males were especially inclined to lag.

Although the composition of Lomako bonobo bands and congeries vary according to the fruiting of certain trees (Badrian and Malenky, 1984; Badrian and Badrian, 1984b), there is evidence for notable cohesiveness among some adults. For instance, Badrian and Badrian (1980; Badrian et al., 1981) observed a band of 2 males, a female, an infant and a juvenile continuously for more than 2 weeks before a recess in their observations. As at other localities, reunions of large bands entailed clamor, display and sex more often than encounters between small bands at food trees did (Badrian and Badrian, 1984b).

Allogrooming, which usually occurred during rest periods, was also observed more frequently in large groups (n = 10-20 individuals) than in smaller ones (Badrian and Badrian, 1984b). Half of 72 sessions involved mother/youngster dyads. Males rarely (1% of sessions) groomed youngsters.

The next most common partnerships (25%) were between adult males and females. The former served as groomers more often than the latter did. Copulations were a common component of their sessions together. Females (17% of sessions) groomed one another more often than males did (7% of sessions). Yet the longest session (125 min) involved a pair of males (Badrian and Badrian, 1984b).

Badrian and Badrian (1984b) provide no data on vegetal food-sharing by Lomako bonobos. They noted that females and youngsters acquired bits of young bushbuck from the males that had captured them (Chapter 3). But Badrian and Malenky (1984, p. 296) stated that the meat was not shared voluntarily. Female cadgers copulated with the meaty males and indulged in homosexual hunching during the consumatory period.

Like Kitamura (1983), Badrian and Badrian (1984b) concluded that the core of bonobo society consists of strongly bonded females and the males that are associated with them. The survey of sites hereabove supports the further conclusion that bonobo society is unique among ape societies and should not be considered a mere variant of common chimpanzee society or a compromise between their social pattern and that of gorillas (Badrian and Badrian, 1984b).

The reproductive tactics and strategies of bonobos are still veiled in mystery. Basically we only have descriptions of sex acts that would wilt leaves in the perfumed garden and observations from 2 sites indicating that female bonobos are probably the most estrous of apes.

Kano (1982a) concluded that estrus is the usual state of postpubertal female bonobos. Adolescent females are receptive and attractive to males during the greater part of the menstrual cycle. And parous females resume cycling within a year after delivery even though they are nursing their infants. Approximately a fourth of observed copulations involved females with their youngsters clinging ventrally or dorsally (Kano, 1980).

Individual mating histories were not available for Wamba bonobos. Kano (1982a) leaves the impression that promiscuity prevails. For example, he noted that when 2 parties met, the males copulated with newly arrived females and with their current companions.

Kano (1980) witnessed 121 copulations. Forty percent occurred during feeding periods, 35% during bouts of travel, and 26% during rest periods. He obtained some evidence for restrictive mating, i.e. females that copulated at least twice a day with a single male. But he also saw females copulate with more than one male per day.

Nearly two-thirds of adult copulations were performed dorsoventrally; 35% were ventroventral; and 3% commenced canine and climaxed missionary. Seventeen copulations paired adults with youngsters. Immature females (n = 2) were not penetrated. Immature males (n = 15) achieved intromission with their adult partners. The relationships of the couplers were unknown so the state of incest cannot be assessed.

Studies by Thompson-Handler et al. (1984) in the Lomako Forest provide important complementary information on bonobo sexual behavior. Unfortunately, the figures in their text (pp. 351-352 and 362) are a bit muddled. Consequently new figures, based on computations from their Table I (Thompson-Handler et al., 1984, p. 352) are employed in the discussion herebelow. Because their subjects were unhabituated and the span of the project was 18 months, they did not obtain individual mating histories. Most matings were witnessed during the morning feeding peak (0700-0900 h).

Adult males performed 84% of heterosexual copulations; adolescent, juvenile and infant males were the penile partners in the remaining 7%, 1% and 8% of observed copulations. The youngsters preferentially mated ventroventrally (83% of 12 cases), usually with their mothers. Per contra, adult males rarely (14% of 58 copulations) mated ventroventrally, preferring instead the dorsoventral position. Seven of the 8 adult missionary matings were with the mothers of dependent youngsters (Thompson-Handler et al., 1984).

Females with somewhat less than complete perineal swells copulated most frequently (60% of 52 cases); those without swells were not seen to copulate. The secondmost common female copulators were ones with early swells (27%). Fully tumescent females were least often seen to copulate (13%), apart from the anestrous ones. Mothers engaged in 44% (n = 32) of 73 copulations (Thompson-Handler et al., 1984).

We should not be surprised if future studies on free-ranging bonobos reveal that, like common chimpanzees, they engage in temporary consortships. Indeed Badrian and Badrian (1984b, p. 332) entertained this possibility. The following observations recommend this hypothesis.

In all study areas, bisexual adult pairs have been seen regularly. The fact that the female partner commonly has dependent youngsters does not argue against consortships because we know that often bonobo mothers are estrous and willing to copulate while their youngsters cling to their bodies. The apparent relatively low frequency of copulation by Lomako females in the peak of estrus might be due to their absence from observed groups in order to engage in consortships. Although copulation is common at feeding sites, that which counts most could be occurring away from the promiscuous crowd, which includes not only libidinous males but also hunching females. Recall that it was only after observers could follow known individuals that the importance of chimpanzee consortship was discovered at Gombe.

Thompson-Handler et al. (1984) not only confirmed the universality of homosexual hunching among bonobo females but also observed instances of "genital contact" between males. The former (n = 25) was fivefold more frequent than the latter (n = 5). Although most (64%) female homosexual hunching occurred during feeding bouts, it was also seen during travel (16%), reunions (8%), and play between infants (12%).

The latter indicates that female homosexual hunching, like the heterosexual behavior of male bonobos, begins during infancy. Mothers do it with their daughters. Clinging infants were commonly seen sandwiched between hunching pairs. Immatures were one or both partners in 36% of female hunching sessions; infants accounted for somewhat more than half of these.

Mothers and other adult females with penultimately puffed perinei were the most frequent homosexual hunchers. They constituted one or both partners in 60% of bouts. Full-blown females were engaged in 20% of bouts and those that had early swells were involved in 24% of bouts (based on data in Table II, Thompson-Handler et al., 1984, p. 356).

Thompson-Handler et al. (1984) concluded that female homosexual hunching probably serves multiple functions in bonobo society. They noted that it seemed to stimulate heterosexual copulation in a group, viz. by piquing the interest of males. On several occasions it was interspersed with proper copulations. They agreed with Kano (1980) and Kuroda (1980) that it might also serve to alleviate intragroup tension. For instance, homosexual hunching and heterosexual copulation occurred during one of the meat-eating episodes at Lomako.

Thompson-Handler et al. (1984) briefly described 4 of 5 cases of "male-male genital contact," only half of which properly belong under this label. Pace certain sodomites, the anus is not a genital organ (International Anatomical Nomenclature Committee, 1977, p. A49). Hence dual rump rubbing (1 instance) and dorsoventral mounting with 30 seconds of pseudocopulatory thrusting (1 instance) do not qualify as "genital-genital" contact. In the other 2 instances, males left off feeding to engage in a sort of ventroventral mounting. Both sets of partners, viz. a pair of juveniles and an adult with an adolescent, had erect penes. Thompson-Handler et al. (1984) provided no further details. But they speculated that dominance seemed to play a greater role in these interactions than it did with the homosexual hunching of bonobo females.

GORILLAS

Donisthorpe (1958), Kawai and Mizuhara (1959-1960), Osborn (1963), Emlen (1960), and Schaller (1963, 1965; Emlen and Schaller, 1960a,b, 1963) provided the first scientific accounts on the social grouping and behavior of free-ranging gorillas in natural habitats. They primarily concentrated on *Pan gorilla beringei*. The most thoroughgoing follow-up studies are also focussed on the mountain gorilla (Fossey, 1970, 1971, 1972, 1974, 1979, 1981, 1982, 1983; Fossey and Harcourt, 1977; Elliott, 1976; Caro, 1976; Harcourt, 1977, 1978a,b, 1979a,b,c,d; Harcourt and Curry-Lindahl, 1978; Harcourt and Fossey, 1981; Harcourt, Fossey and Sabater Pi, 1981; Harcourt, Fossey et al., 1980; Harcourt and Groom, 1972; Harcourt and Stewart, 1977, 1981, 1983, 1984; Harcourt, Stewart and

Fossey, 1976; Stewart, 1977; Vedder, 1984; Watts, 1984, 1985a,b; Weber and Vedder, 1983).

We know much less about western and Grauer's gorillas. But available studies indicate that their sociality is at least grossly similar to that of the mountain gorillas.

During an 8-month study in the Kigezi Gorilla Sanctuary of southwestern Uganda, Donisthorpe (1958) sighted several groups, which contained between 3 and 12 members. Osborn (1963), who had participated in a failed provisioning scheme before Donisthorpe's arrival, also observed small groups in the region: 2 with at least 3 members, one with 4 members, and one with 4 or 5 members. Their successors, Kawai and Mizuhara (1959-60), encountered 39 gorillas in 4 groups (n = 5, 7, 9 and 18). Further, Kawai and Mizuhara (1959-60) censused 2 sizeable groups (n = 17 and 13) of mountain gorillas in the Bwindi Forest Reserve of Uganda. The largest group contained a silverback, a blackbacked male, 4 adult females, 6 youngsters, and 5 members that could not be identified accurately anent age and sex.

Twenty years later, Harcourt (1980-81) found that the mean and median size of 4 Bwindi gorilla groups was 9 individuals (r = 6-11). His census was premised on nests and the sizes of fecal boluses therein. An average group included a silverback, 3-4 adult females and 4-5 youngsters. Harcourt (1980-81) also found the spoor of 3 lone males.

The brevity of studies on Ugandan mountain gorillas and the ephemeral nature of encounters between them and the observers greatly limited data on intragroup and intergroup social dynamics. Donisthorpe (1958, p. 216) concluded that large groups sometimes divide into subgroups, with irregular compositions, and reunite later. Per contra, Kawai and Mizuhara (1959-60) stressed the cohesiveness of mountain gorilla "polygamous families," which serve as the cores of larger groups. They noted that each group had a silverbacked male leader and protector. They further inferred that some silverbacks had particularly strong bonds with individual adult females in their groups. During their study, a secondary young silverback became peripheral to the group with which he had been associated. Although the ranges of adjacent groups overlapped, they tended to avoid one another.

Schaller (1963, 1965) was the first scientist to habituate mountain gorillas to the presence of a human observer. He focussed his study on 11 groups in the Parc des Virungas (formerly Albert National Park) of eastern Zaire. Six of them tolerated his presence for spans between 1 and 7 hours. He was rarely able to follow them on complete daily rounds. And the dense foliage greatly limited his ability to monitor their individual and social behaviors.

Schaller (1963, 1965) found that the average group size (17; r = 5-27; n = 10) of Zairean mountain gorillas at Kabara was more than twice the average size (7-8; r = 2-15; n = 6) of Ugandan mountain gorilla groups that he had surveyed. Seven of the 10 groups that he completely censused at Kabara had one silverback; one group included 2 silverbacks; and 2 groups contained 4 silverbacks when he first encountered them. In addition, 8 of the groups had one or more blackbacks. Nevertheless, in all except one Kabara group, adult females outnumbered the spermatic males.

Schaller (1965, p. 339) concluded that groups tended to have stable compositions over many months, except for births and deaths of infants and the arrivals and departures of males (Figure 28). Fission into subgroups was rare, the units generally remained near one another, and reunion occurred shortly thereafter.

In addition to the 10 relatively cohesive, stable groups at Kabara, Schaller (1963, 1965) also encountered at least 4 silverbacks, only one of which appeared to be past prime, and 3 blackbacks, which were solitary most or all of the time. He did not see female counterparts of these "lone males." In the group (IV) that he knew best, only the dominant silverback remained in the group yearlong. Another 3 silverbacks visited and departed from group IV. Of the 6 groups that Schaller (1963, 1965) knew well only 2 were popular with the loners. He concluded that the visitors were quite welcome because there was no overt aggression between them and other males in the group. However, he did describe an instance in which the dominant male "stared threateningly" at an approaching silverback, which subsequently avoided his group (Schaller, 1965, p. 339).

Despite observations of charging and glaring by the dominant males during some group encounters and uprooted tufts of hair where groups had met, Schaller (1963, 1965) concluded that mountain gorillas are not territorial. Usually the members of different groups ignored one another and occasionally they mingled peacefully for several minutes before going their separate ways. Nevertheless, focal groups were variably tolerant of different neighbors (Schaller 1963, pp. 111-120).

Schaller's (1963, 1965) portrait of the silverback gorilla stresses attractiveness as much as peacefulness. He is the hub of the cohesive group. Females and youngsters cluster about him while other spermatic males are peripheral and peripatetic. All group members appeared to be attuned to the silverback's activities; he leads them through their daily rounds of feeding, travel, resting and nesting. When extra silverbacks are with a group, its members usually react only to the dominant silverback as their leader (Schaller, 1963, p. 238).

During rapid progressions, the silverback headed the group. If danger threatened, he would drop back to protect the rear of his fleeing charges. Once a startled silverback grabbed a juvenile and carried it 10-13 yards away from Schaller (1965, p. 347).

Because dominance is largely correlated with body size, the silverback supplanted all other group members at choice sitting and feeding sites and

Figure 28: Visoke mountain gorillas (group 5). (a) The silverback (on right) plays with his 8-year-old son. An adult female sits to the left of the silverback and an infant female is to his right. (b) Silverback (reclining on right) with 2 adult females and 2 infants. (Photos by David P. Watts.)

on narrow trails. When more than one silverback was in a group, they formed a linear hierarchy. Blackbacks and females were dominant over youngsters. The females did not seem to have a stable hierarchy among themselves and their relations with blackbacks were variably dominant and subordinate (Schaller, 1963, 1965).

The cohesiveness of gorilla groups seems not to be maintained by allogrooming or regular, acrobatic sex. Schaller (1963, 1965) reported that Kabara gorillas rarely groomed and that grooming was never reciprocal. He concluded that adult allogrooming was chiefly utilitarian, that is for the removal of irritants from otherwise inaccessible regions of the groomee. Two-thirds (n = 89) of 134 instances entailed females grooming youngsters. A silverback was groomed once (by a juvenile) and he groomed only infants (3 times). The blackback never groomed and he was groomed only twice (once each by a female and a juvenile). Females rarely groomed each other (3.7% of bouts). Accounting for a fourth of the bouts, juveniles were the secondmost frequent groomers. They focussed their attentions mainly upon infants and other juveniles. They also groomed females 9 times. Schaller (1965, p. 348) noted that juveniles used grooming to initiate contact with females and infants.

Youngsters are more inclined to play than to allogroom. All except 5 of 156 subjects that engaged in 96 play bouts were youngsters. Often (43% of observations) they played solitarily. Social play bouts were relatively brief ($\leq$ 15 min) and never led to quarreling or injury even though the partners were sometimes grossly different in size (Schaller, 1965).

Based on cross-sectional data, Schaller (1963, 1965) concluded that infant mountain gorillas remain physically close to their mothers for about 3 years. They nest with and are carried by them, especially during escapes from nuisances. Nutritional weaning begins during the infant's first year.

Because Fossey and co-workers established more intimate relationships with their subjects than Schaller did and conducted long-term studies, a wealth of detailed information on gorilla sociality has been collected in the Parc des Volcans, particularly on the slopes of Mt. Visoke. Unfortunately, political turmoil prevented Fossey (1970, 1983) from continuing her work on Mt. Mikeno, where Schaller was based. Further, human encroachments and depredations have periodically disrupted the gorilla studies in Rwanda and have drastically reduced the populations of mountain gorillas since 1960 (Harcourt and Groom, 1972; Harcourt, 1977; Harcourt and Curry-Lindahl, 1978; Harcourt et al., 1981; Fossey, 1981, 1983).

These tragic events have accelerated new group formation among the Visoke gorillas. We cannot always discern which aspects of the process are characteristic of gorilla social dynamics in the absence of destructive agencies. We can only hope that some groups will be allowed to stabilize and develop naturally in large undisturbed tracts of Rwanda, Zaire and Uganda.

In an early report based on censuses of 9 groups, Fossey (1971) noted that the average group of mountain gorillas had 13 members (r = 5-19 individuals). She confirmed that each group is ruled by one silverback. Additional silverbacks, blackbacks, females and youngsters comprised the remainder of most cohesive groups (Fossey, 1970, 1971). However, she also located an unusual unit, designated group 8, that initially consisted of 5 spermatic males, headed by a dominant silverback, and an aged female that died shortly after Fossey encountered them.

Later one silverback became solitary. Then the others took females from a bisexual group and group 8 experienced further additions, including a birth, and departures as it developed into a small bisexual group more like other gorilla groups (Elliott, 1976; Caro, 1976; Fossey and Harcourt, 1977, p. 421; Fossey, 1981, 1983, xxi).

Like Schaller, Fossey (1970, p. 65; 1971, p. 577) stressed the tranquility of gorilla life as attested by the near absence of aggressive behavior, though she noted that the tertiary silverback left group 4 when the dominant one died and was replaced by the secondary silverback in the group. Further, the new leader led his group away from group 8, which then began to chivy group 9. Aggression increased among the silverbacks of group 8 and between groups 8 and 9 (Fossey, 1971, p. 582).

Fossey (1974) documented notable overlap among the ranges of 4 study groups (4, 5, 8 and 9). When aware of one another's presence, groups exchanged vocalizations and sometimes engaged in displays. However, they commonly also approached one another and occasionally intermingled.

Lone silverbacks ranged over large sections of the ranges of their natal groups (Fossey, 1974; Caro, 1976; Elliott, 1976). They tended to overuse the core sections of their ranges, which Fossey (1974, p. 578) inferred was a form of defence against encroachments by other gorillas. This would seem to be counterproductive in the attraction of females for mating and new group formation.

During a 13-week period, 2 lone silverbacks roamed no more widely than their natal groups did and there was nothing peculiar about their daily rounds. Interactions were predominantly agonistic between a lone silverback from group 8 and another lone silverback and neighboring groups, including group 8 (Caro, 1976). This is not surprising since Fossey (1974) discovered that lone silverbacks establish new groups by taking young females from other groups.

Harcourt (1978a,b, 1979a-d) published the most thorough studies of social interaction and relationships among Visoke gorillas. He and Fossey agreed with Schaller that mountain gorillas are not territorial in the sense that they defend specific tracts of habitat (Fossey and Harcourt, 1977). Indeed although dominant silverbacks defend the integrity of their groups, overlap between group ranges and those of lone males is too extensive for them to fit the classic definition of territoriality, viz. to defend a section of habitat against conspecific social units for one's own exclusive use.

Harcourt's 20-month study was focused on Visoke groups 4 and 5, each of which included a silverback and a blackback. The numbers of adult females (n = 2-4 in group 4 and 3-6 in group 5) and youngsters (n = 4-5 in group 4 and 4-8 in group 5) varied due to female transfers, births, deaths, and maturation (Harcourt, 1979a, p. 188; 1978a, p. 124).

Harcourt (1978a, 1979a-d) solidly confirmed that the silverback is hands down the focal member of the mountain gorilla group (Figure 28). The proximity of individuals, particularly adult females, to one another is commonly attributable to their attraction to the silverback instead of mutuality among themselves. Females with dependent offspring are especially inclined to stay near him (Harcourt, 1979c).

Some, but not all females, in each group groomed with their silverbacks. In group 4, all subadult females and the youngest adult female solicited and were groomed occasionally by the silverback. They did not reciprocate. In group 5, the 2 adult females that spent the least time near the silverback groomed him most often. Only one of them was groomed reciprocally by him. Unlike Schaller, Harcourt (1979c) concluded that adult grooming facilitates affinitive interactions between the silverback and the females, which initiate most of the interactions. Often grooming females are ones having the least stable relationship with a silverback.

Although most females were in proximity with one another because they were all attracted to the silverback, related females and those that had been together since they were youngsters exhibited more mutuality than unrelated, immigrant females did. Young females also spent greater than average spans with their counterparts in the 2 study groups (Harcourt, 1979b). On rare occasions, estrous, and even less commonly anestrous (and pregnant), female mountain gorillas engaged in homosexual behavior (Harcourt, 1979b; Stewart, 1977).

Nulliparous females are quite interested in the neonates of other females and tend to stay near them. Females with infants rarely touched one another while related females and those that had been together since childhood had more frequent bodily contact. Usually the younger of a pair was the initiator. A daughter groomed her mother much more frequently than she reciprocated.

Harcourt (1979b) could not document a dominance hierarchy among the females of either group even though some females tended to avoid the aggressive approaches of others, which they, in turn, never displaced. Watts (1985b) documented that there were linear dominance hierarchies among the females of Visoke groups 4 and 5. Aggression usually erupted mildly in the form of pig-grunting (Chapter 7). Louder, more demonstrative displays and attacks were rare. Group females seemed to be as tolerant of newcomers as they were to resident females, perhaps because of the silverback's presence and his acceptance of female transferees. He stifled female bickering as well as the rowdiness of youngsters (Harcourt, 1979b; Fossey, 1979).

Relationships between the silverbacks and blackbacks in the two study groups were dramatically different. Whereas in group 4 they rarely interacted, in group 5 they spent notable time together and the blackback even groomed the silverback. The silverbacks freely encroached on the blackbacks' feeding spaces but the reverse was rare (Harcourt, 1978a, 1979a,d).

Intimate interactions between the spermatic males of group 5 were mostly initiated by the blackback. Whereas they had a history of close association since the birth of the younger male and might even have been father and son, no such affinities existed between the males of group 4 (Harcourt, 1979d).

The silverback was highly attractive to youngsters, which also enjoyed the company of peers. They freely used him as a play mound during adult rest periods. Older youngsters preferred the silverback over their mothers when the group moved and fed. Even more remarkably, 4 youngsters, ranging from 2.5 to 3.8 years survived the deaths or abandonments by their mothers because the silverback continued to protect them and even bedded them in his night nest. It is doubtful that they could have survived the nocturnal chill on Mt. Visoke without the silverbacked furnace (Harcourt, 1978a, 1979a; Fossey, 1979, 1983).

Harcourt (1978b) and Fossey (1983, 1984) shattered Schaller's image of the silverback as an overly gentle giant that is highly tolerant with his neighbors. Apart from the brevity of his contacts with the Kabara subjects, it is possible that Schaller observed only stable groups that had well established dominance relationships with one another during his visit. Per contra, on Mt. Visoke all hell broke loose during the formation of new groups (Harcourt, 1978b; Fossey, 1981, 1983, 1984).

Harcourt (1978b, p. 409) concluded that aggression is the main tactic used by a resident male against others to prevent their association with his females. Fossey (1984, p. 222) noted that female transfers between established groups or from a group to a lone silverback almost always entails overt aggression between the dominant male in the female's current group and the seductive silverback. The presence of estrous females and copulations within groups seemed to attract visitors and to set off displays and fights between the males of the different social units (Fossey, 1982).

After inter-unit encounters, males sometimes sported wounds on their heads and shoulders, presumably from their antagonists' canines (Harcourt, 1978b). But the brunt of several contests was borne even more heavily by infants in the resident groups, at least 7 of which were bitten to death by impetuous young silverbacks in pursuit of mates (Fossey, 1981, 1983, 1984; Harcourt et al., 1981). During a 15-span of research on Mt. Visoke, 38% (n = 13) of infant deaths are attributable to infanticide and another 23% of infant deaths also may have been infanticidal (Fossey, 1984), though not necessarily all perpetrated by silverbacks (Chapter 3).

Based on censuses at 2 localities in Equatorial Guinea, average groups of *Pan gorilla gorilla* contain 6 or 7 individuals (r = 2-12; n = 13). The populations there were harassed by human hunting and deforestation. Jones and Sabater Pi (1971) noted that although the gorilla groups were more stable than those of sympatric chimpanzees, occasionally they intermingled temporarily with other gorilla groups and loners. One singleton was a silverback. Jones and Sabater Pi (1971) did not specify the age and sex classes of the other lone gorillas that they encountered. From nests, Bütlzer (1980) estimated that 4 groups of Cameroonian *Pan gorilla gorilla* contained 4, 7, 10 and 11 members.

Goodall's (1977) main study group of 20 *Pan gorilla graueri* in the Kahuzi-Biega National Park, Zaire, contained one silverback, one silvering blackback, 3 blackbacked males, 4 adult males, 10 immature individuals and a semiperipheral silverback. Casimir's (1975, 1979) study group in the Kahuzi-Beiga National Park also contained about 20 individuals. Murnyak (1981) located 14 groups and 5 lone males in the Kahuzi-Biega National Park. Based on nest counts he determined that the mean group size is 16 individuals (r = 6-37). But most groups contained between 6 and 15 individuals (median = 12.5). The mean is elevated by 2 large groups containing 34 and 37 members.

Yamagiwa (1983), who overlapped the second half of Murnyak's tenure in the Kahuzi-Biega National Park, observed that Goodall's study group (G) had grown to 42 members and Casimir's study group (C) had declined to 9 members in 1978. He also censused 10 other groups by counting nests. The mean size of 12 Kahuzi groups is 14 members (r = 3-42). The G-group contained one silverback, 4 blackbacks, 17 adult females, and 9 independent and 11 dependent youngsters. The neighboring C-group had one silverback, 2 blackbacks, 3 adult females and 3 independent youngsters (Yamagiwa, 1983).

Three years before Yamagiwa's 5-month study, the leader of C-group was challenged by a young lone silverback and died from an infected wound. The challenger took over the group. The highest ranking blackback in C-group left with several members but lost them to G-group (MacKinnon, 1978, p. 60). There were no recorded births in C-group after the demise of their leader in 1975 and at least through March, 1979. Yamagiwa (1983) concluded from these events that male gorillas must be fully mature in order to retain females and to reproduce.

Yamagiwa's brief study evidences that intragroup relationships of the Kahuzi Grauer's gorillas are closely similar to those of the Visoke mountain gorillas. Kahuzi blackbacks kept their distance from the silverback and tended to associate mostly with females that were free of offspring. During travel they kept away from the middle of the procession and were often associated with independent youngsters (Yamagiwa, 1983).

Kahuzi females with dependent youngsters stayed near the silverback and were thereby closer to one another than nulliparous females were. Yamagiwa (1983) conjectured that because youngsters are drawn to the tolerant silverback their mothers are also near him. This was especially true during rest periods. Mothers often left their youngsters to play around the silverback while they foraged. The youngsters also approached the silverback when he initiated night nesting. During travel, females with dependent infants clustered in the middle of the procession, while the silverback either led or served as a rear guard for the group.

Yamagiwa (1983) suggested that the cohesiveness of gorilla groups may be premised on the strong affinities between the dominant silverback and youngsters. Females are bound to him through their children. Young females without youngsters have looser ties and are thus disposed to emigrate. Adventurous young silverbacks can break male-female bonds by killing the infants of parous females. Yamagiwa's (1983) scenario leaves unexplained how the father-daughter bond loosens, if, in fact, she was bonded to him as a youngster.

Among mountain gorillas, and presumably also western and Grauer's gorillas, members of both sexes usually emigrate from their natal groups during early adulthood. Males live alone or temporarily with a few other males, for several years until they can acquire fertile females. Only one male, a blackback, is reported to have transferred to another group. Females transfer to neighboring groups and beyond, but commonly visit their natal groups, before settling down to produce infants with a silverback. Sometimes a son remains in his natal group and inherits the harem upon his father's demise. However, he may fail to keep the recently beheaded group together in the face of forays by silverback competitors and the wanderlust of its females (Harcourt et al., 1976, 1980, 1981; Harcourt and Stewart, 1977, 1978b, 1981; Harcourt, 1978b, 1979a-d, 1981; Fossey, 1982, 1983, 1984; Yamagiwa, 1983).

The dynamics of male departure are clearer than those of spontaneous female emigration. Indeed Harcourt (1978b, p. 408) commented the emigration of female gorillas was always slightly surprising because it occurred so abruptly.

The dominant silverback prevents subordinate males in his group from breeding with estrous primiparous and multiparous females. He is more tolerant of their copulations with immature, nulliparous and pregnant females, none of which are likely to be impregnated therefrom (Harcourt et al., 1980, 1981; Harcourt and Stewart, 1978b, 1981). Consequently, if the leader remains robust, young spermatic males must leave the group and somehow obtain their own females in order to become full-fledged members of the breeding population. They may have to fight silverbacks in order to secure females for themselves. If a young silverback kills a female's infant, she will likely join his budding breeding unit (Fossey, 1984).

In brief, 3 tactics are available to a male gorilla that would head his own group: (1) wait to inherit the group of a senescent or deceased silverback that is probably his father (Harcourt and Stewart, 1981), (2) depose the silverback of an established group, or (3) emigrate and attract transferring females and perhaps accelerate the process by raiding established groups. The third tactic is the one most commonly employed by young silverbacks on Mt. Visoke. Whatever tactic is adopted, the males are about 15 years old before they begin to breed productively (Harcourt et al., 1976).

Although initially silverbacked leaders may try avoidance or display to keep would-be raiders at bay, they commonly fight ferociously to hold their mates. At Visoke, 79% (15/19) of observed encounters between strange males led to violent threat displays. During half (8 of 16) the encounters in which Harcourt et al. (1981) could determine whether or not a fight had occurred, battles were manifest.

Silverbacks tend not to herd their females or otherwise to direct overt efforts toward them that would keep them from transferring once they have chosen divorce. Despite these vicissitudes the dominant males of 2 Visoke study groups produced 4-6 and 8-10 surviving offspring between 1967 and 1977 (Harcourt et al., 1981). And Fossey (1982) reported that one dominant silverback had sired at least 19 offspring.

On average, female Visoke gorillas were menarcheal in their seventh years (r = 6.4-8.6 years; n = 13) (Fossey, 1982). Menstruation and estrus are very difficult to detail in wild gorillas. Nulliparous females exhibit small labial swells but parous females sport no visible perineal clues to their estrous conditions. Adolescent females are receptive for 3-5 days and their monthly cycles may be irregular. On the basis of observed copulations, regularly cycling females were inferred to be receptive for only 1-3 days per month. Several pregnant females also were estrous, though infrequently (Fossey, 1982; Harcourt and Stewart, 1978b).

Visoke females experience variable spans between menarche and first pregnancy. Indeed emigration, which generally occurs shortly after menarche, may delay conception. The average age at which 9 transferees first conceived was 10 years. The ages at which females became primiparous varied between 8.7 and 12 years (Fossey, 1982).

Like males, for young females emigration is the rule, to which there are some exceptions. Over a 13-year span 26 Visoke females transferred between groups 43 times (Fossey, 1982). Most of them were of low status vis-à-vis the silverbacked leader. The dominance of Visoke females largely depends upon the order of their acquisition by the dominant silverback. Over a 13-year span, only 2 of 15 adult females transferred out of the groups in which they had conceived infants with its leaders. Both had occupied relatively low statuses. Their solicitations to copulate had been rebuffed, perhaps because they had nearly 4-year old youngsters that continued to nurse occasionally. Their statuses improved notably after emigration since they were the first and secondly acquired females of their respective new silverbacks. One of them promptly produced an infant with her new mate (Fossey, 1982).

A minority of young females, which remain in their natal groups, may be bred incestuously by males that have matured into silverbacks and have not emigrated. Over a 15-year span, Fossey (1984) documented the birth of one infant from a brother-sister coupling. And at least 3 infants were produced by half-sibs.

Unlike males, females transfer directly to potential mates and do not experience prolonged periods of singleness. They generally passed from their natal or latest groups to new units very quickly; one female remained alone for 1.5 days (Harcourt, 1978b). Peaceful as well as violent transfers usually occurred when social units were close together and interacting. Transferring females spent between 3 days and 41 months (median, 3.5 mos) in alien units before settling down (Harcourt, 1978b).

Transferring females prefer new groups and lone silverbacks over established groups (Harcourt, 1978b, p. 405). As mentioned earlier, the intragroup status of females depends directly upon the order in which they become mates of the dominant silverback. Thus the first mate has highest status, the second is beta, and so forth (Fossey, 1982, 1984).

There are several possible reasons for female choice among available silverbacks. Harcourt (1978b) stressed the quality of the male's range and especially the success of the female in rearing offspring with the male as reasons for her staying in a new unit. If the male cannot protect the female against infanticidal males, she will leave him for one with recently demonstrated prowess, yea even the murderer of her child (Harcourt, 1978b; Fossey, 1984). Primiparous Visoke females were 3 times more likely to lose their infants than multiparous females were. Their infants were more often killed by males (Fossey, 1982, 1984).

Fossey (1984) noted that females commonly (10 of 18 individuals) transfer in pairs from their natal units to the same new groups either simultaneously or within a few months of one another. She speculated that such pairs are more likely to be integrated into new groups and they will be more successful reproductively because the reduced tendency to roam would limit their exposure to infanticidal males. Further, a single bisexual pair of gorillas seems not to be a viable breeding unit (Harcourt et al., 1976).

Females are attracted to the silverbacks that not only can copulate upon command and protect their infants but probably also because they take an active part in parenting. Harcourt (1979c) and Fossey (1979) noted that youngsters stayed near the silverback while their mothers moved off to feed. Indeed during travel-feed periods some youngsters spent as much time near the silverback as near their mothers (Harcourt, 1978a, 1979c). Even more than tolerating the presence and capers of 3- and 4-year-olds whose mothers had new infants, the silverback sometimes groomed and cuddled them (Fossey, 1979, p. 183).

Like the silverback's penis, gorilla copulatory bouts are remarkably brief (Short, 1979, 1980, 1981; Harcourt et al., 1980, 1981). The estrous female generally (79% of 439 copulations) initiates couplings by slowly and hesitantly approaching the dominant silverback, which commonly seems oblivious to her over-the-shoulder glances and brief manual contacts with his body (Fossey, 1982). But once he agrees to copulate, the seated silverback acts decisively. As the female abruptly turns her rump towards him and backs up, he draws the squatter onto his lap, often while holding her about the waist. Hence their posture is a dorsoventral one. They adjust their perineums to achieve intromission and thrust, on average, for 1.5 min; the copulatory position is maintained for spans between 15 sec and 19.75 min (median, 96 sec) (Harcourt et al., 1980).

The precopulatory behavior of mature gorillas lacks faciogenital and manual-genital contacts. Usually both partners thrust and vocalize (Chapter 7) but the silverback normally sustains these activities longer than his mate does (Harcourt et al., 1980). The female usually (89% of 514 copulations) breaks the embrace and moves away first (Fossey, 1982).

Unlike chimpanzees, the copulations of silverbacks with choice mates very rarely receive interference from other group members (Harcourt and Stewart, 1978b; Fossey, 1982). Most (80% of 67) copulations took place during travel-feed periods when group members were well separated.

Usually only one female of a group is estrous on a single day (Harcourt et al., 1980). Only 5% of 580 copulations occurred on days when more than one female in the group were estrous. One day, a silverback copulated with an elder female 5 times and with a young female once during a 4-hour span. In this and other instances of multiple copulations, the silverback's vigor flagged with each successive copulatory bout (Fossey, 1982, p. 103).

Sexual experimentation begins early in mountain gorillas. Fossey (1982) tabulated 240 mountings that involved youngsters. The youngest mounter and mountee were 2.5 and 0.5 years old, respectively. Both ventroventral and dorsoventral positions were employed by the playful pairs. Unlike chimpanzees, young gorillas did not mount their mothers. And unlike silverbacks, immature male gorillas did attend to the pudenda of estrous females, probably in response to olfactory cues (Fossey, 1982).

Harcourt et al. (1981) observed 39 mountings that involved youngsters. The youngest mounter was 3.1 years old and the youngest mountee was 1.8 years old. In about a third of the mountings both partners were immatures, with the elder virtually always topside. The remainder of the episodes entailed adults, especially a blackback, mounting youngsters. Exceptionally, a subadult male hunched the shoulder and side of a parous female.

Harcourt et al. (1981; Harcourt and Stewart, 1979b) observed 10 instances of homosexual hunching between female mountain gorillas. Unlike bonobos, dorsoventral hunching (n = 6) was more common than ventroventral hunching (n = 3). Usually at least one partner uttered copulatory vocalizations. One individual would approach another; they would embrace, hunch, and part. Far from being turned on by their behavior, a silverback attacked a pair of hunchers, the estrous one of which he had bred an hour before.

Harcourt et al. (1981) concluded that female homosexual hunching is sexual behavior among mountain gorillas because, typically, estrous females were the active hunchers and they had copulated with a silverback not long before switching to a fellow female. They also saw 3 cases of males mounting males but inferred that these were agonistic instead of sexual acts.

SEXUAL SELECTION IN THE PONGIDAE

The mass of detailed information on the social structures and mating behaviors of great apes has greatly facilitated informed interpretations on certain puzzling somatic and genital features of the Pongidae.

Although male gorillas are twice the bulk of female gorillas (Chapter 1), they have relatively tiny penes and testes. Further, whether tumescent or flaccid, their penes are pathetically inconspicuous. Females too sport no readily visible clues to their sexual condition.

Short (1979) reasoned that the male gorilla's somatic massiveness is essential for him to establish and maintain a harem of fertile females vis-à-vis the competition of other silverbacks. But once a group is established, neither he nor his females need to flash gaudy genitalia in order to realize their breeding potentials. Females are estrous for very brief periods and are always close to the silverback. They need only approach him seductively and he needs only close the embrace in order for them to reproduce themselves. Small testicles and concomitant modest sperm counts are sufficient to impregnate the females, particularly considering the infrequency of estrus and absence of intragroup male competition when it does occur in gorilla groups (Short, 1979).

Females may be cued to try more fertile fields if the male repeatedly fails to respond sexually or if he actively rebuffs them. Homosexual hunching may suffice for the short haul, but eventually estrous females would be expected to emigrate, especially if the silverback ignored their first invitations to copulate.

Like silverbacks, adult male orangutans are twice the size of conspecific adult females (Chapter 1), have relatively diminutive penes and testes (Short, 1979), and will fight fiercely with rivals. However, unlike silverbacks, at least in some areas, males compete by territorially commanding the range in which resident females forage instead of bonding with them to form stable cohesive breeding units. Further, though their pink penes are secret when flaccid, they are readily visible upon erection (Short, 1979). Still a male's cheek pads are probably a better clue to his sexual capacity (cf. testosterone titre) than his penis is.

Like gorillas, female orangutans sport no visible signs of estrus. The male may attract receptive females by vocal advertisements and grotesque epigamic features. If he is the only show in town or has somehow demonstrated himself to be the best, he will breed with all, or at least most, of the females in his territory or range. The postural versatility and patience of both sexes facilitate accommodation of a diminutive penis in an unswollen pudenda.

The longer period of estrus in orangutans (by comparison with that of gorillas) allows time for the prospective mates to find one another and to consort productively. Further, the sexual vigor of prime male orangutans seems to be greater than that of mountain gorillas, at least as attested by Schürmann's Sumatran dynamo that swung with a persistent sexpot.

Chimpanzees stand in marked contrast with gorillas and orangutans anent many features that are the focus of sexual selection. The males are not inordinately larger than the females (Chapter 1) and the sexes can be confused facially (Short, 1979). Although styloid, the erect pink chimpanzee penis is long and conspicuous, especially when flipped (Chapter 7). Adult male chimpanzees have relatively titanic testicles and astronomical sperm counts (Short, 1979).

All of this is coupled with a greater intragroup competition for mates in chimpanzee society. Swollen females compete to attract the most robust males, which in turn compete with one another to inseminate the females, particularly those that have shown their fertility. In an arena where gang sex is common, a male must quickly flood the field with sperm in order to have a chance to impregnate a female that will probably copulate straightway with other males (Short, 1979; Harcourt, 1981).

Even though he can keep other males in his purview from mating, the alpha male also needs to be sexually vigorous and speedy because he may have a number of estrous females to service over a short period of time and cannot dally with one of them lest the others cheat by going into consortship or behind a bush with another male. The rather long estrous period of the female chimpanzee must further challenge the sexual energies of the alpha male. Of course, it may give her time to attract and hold a choice mate in consortship.

9

Synoptic Comparison of Apes

"Our haste and anxiety to have solutions to many critical human problems may have led to many premature generalizations in the behavioral sciences, thus actually retarding rather than advancing the development and formulation of proved principles of sociobiology. Until basic principles have been established by evidence, uncertainties and indecision will characterize the applied professional arts of psychiatry, psychoanalysis, anthropology, and applied psychology." (Carpenter, 1962, p. 287)

The Great Apes (Yerkes and Yerkes, 1929) ends with interape comparisons, contrasts between their behavior and scant knowledge about other primates, and comments on the status and future of research with apes. The centerpiece of the Yerkeses' synopsis is an extensive table of characters, including not only behavioral traits but also numerous anatomical features, which they had not discussed previously.

It seems to be appropriate also to conclude *Apes of the World* with a synoptic comparative table. Like the Yerkeses, I have included information that was not specifically discussed in the preceding text. Many new characters are added and many of the Yerkeses' choices are not repeated in Table 3. They did not regularly compare apes in the body of their book; I have done so.

Historians might be interested to compare the two documents for a quick overview regarding what we have and have not learned since 1929 and the nature of questions that behavioral scientists ask. One should note immediately that I have not ranked the apes anent most features. The Yerkeses were constrained by the prevailing concept of a Scale of Nature, leading from lower creatures through the apes to us, the highest form of earthling.

Table 3: Synoptic Comparison of Gibbons (Including Siamang) and Great Apes

Character		Gibbons	Orangutans	Common Chimpanzees	Bonobos	Gorillas
Geographic distribution		Southeastern Asia (Fig. 1)	Northern Sumatra and coastal Borneo	equatorial Africa from Gambia to W. Uganda and NW. Tanzania (Fig. 2a)	central Zaire (Fig. 2a)	equatorial Africa from Cameroon-Uganda & Rwanda (Fig. 2b)
Conservation status		some species abundant; others in jeopardy	endangered but can be saved	endangered in some areas; protected in other areas	declining but can be saved	some western populations safe now; eastern sspp. very precarious
Habitats		Tropical rain-forest	Tropical rain-forest	Primary, secondary, and riverine forests; woodlands; forest-savanna mosaics	Tropical rain-forests	Tropical rain-forest; montane forest, and secondary forest
*Adult body weights (lb)	M	9 - 28	75 - 200	70 - 154	81 - 134	290 - 480
	F	9 - 28	72 - 100	58 - 110	59 - 84	150 - 215
Diploid chromosome number		44 (most spp); 38 (hoolocks); 50 (siamang); 52 (concolors);	48	48	48	48
Age classes (yrs)						
Adult	M	>6	>13	>15	>13	>12
	F	>6	>8	>13	>13	>8
Subadult	M	6 - mated	8 - 13+	12 - 15	–	8 - 12
	F	6 - mated	–	10 - 13	–	–
Adolescent	M	4 - 6	5 - 8	7 - 12	7 - 13	6 - 8
	F	4 - 6	5 - 8	7 - 10	7 - 13	6 - 8
Juvenile		2 - 4	2.5 - 5	5 - 7	2 - 7	3 - 6
Infant		<2	<2.5	<5	<2	<3

*Values for African apes are based on Jungers and Susman (1984, p. 143) and references in Chapter 1.

(continued)

Table 3: (continued)

Character	Gibbons	Orangutans	Common Chimpanzees	Bonobos	Gorillas
Gestation (days)	210	245 - 275	228	231 - 244	255 - 265
Birth interval (yrs)	2 - 3	4.5 - 5	3 - 7.6	—	3 - 5
Longevity (yrs)	>23	>55	>50	—	>50
Locomotion between food sources	brachiation; climbing and bipedalism	versatile climbing; some terrestrial quadrupedalism	terrestrial knuckle-walking; climbing	terrestrial knuckle-walking; versatile climbing	terrestrial knuckle-walking
Feeding postures	suspensory; seated; squatting	seated; squatting; suspensory	seated; squatting; bipedal; suspensory	seated; squatting; suspensory; bipedal	seated; squatting; bipedal
Sleeping postures	seated (roosting)	reclining in nest	reclining in nest	reclining in nest	reclining in nest
Foods and other ingesta	figs and other small or soft fruits; new leaves; arthropods; soil and feces (rarely); honey	figs; durian and other hard-rinded fruits; leaves; bark; vines; arthropods; soil; one gibbon	fruit; leaves; arthropods; small vertebrates; bark; dried fruits; honey; soil; feces (rarely); eggs; resin	fruit; leaves; fibrous ground plants; arthropods; honey; small vertebrates; soil	leaves; stalks; vines; thistles; nettles and other herbage; bark; wood; fruit; roots; soil; occasional arthropods and feces
Drinking	hand-dipping in tree bowls; lick coat and vegetation	hand-dipping in tree bowls; lick coat and vegetation	hand-dipping; sops; sucking directly from streams	sucking directly from standing water	sucking free water; lick vegetation

(continued)

Table 3: (continued)

Character	Gibbons	Orangutans	Common Chimpanzees	Bonobos	Gorillas
Cannibalism	none reported	none reported	≤3 year olds	none reported	<3 year olds
Subsistence and toilet tools (in wild)	none reported	perineal scratcher	probes; levers; hammers; sops; wipes	none reported	none reported
Food-sharing	none reported	mother/young; insects between males and females	notable, especially meat but also including vegetal foods	notable, especially vegetal foods	none reported
Night lodging	roost high arboreally and cohesively	nest arboreally and deployed widely	nest arboreally with some clumping	nest arboreally with clumping	nest together terrestrially and arboreally
Range size	11 - 57 ha	M ≤12 km^2 F 5 - 6 km^2	6 - > 278 km^2	22 - 58 km^2	6 - 34 km^2
Insightful performance of instrumental tasks	some experimental evidence	experimental evidence	experimental evidence	experimental evidence	experimental evidence
Agonistic branch dropping, throwing and/or brandishing	some evidence	manifest	manifest	rare	manifest
Handedness	poorly documented	poorly documented	poorly documented	poorly documented	sparsely documented

(continued)

Table 3: (continued)

Character	Gibbons	Orangutans	Common Chimpanzees	Bonobos	Gorillas
Correct perception of photographs and veridical drawings	not reported	experimental evidence	experimental evidence	not reported	not reported
Transfer index	monkey-level performance	high-level passes	high-level passes	untested	high-level passes
Abstraction	not demonstrated	manifest	manifest	untested	manifest
Long-term memory	untested	well developed	well developed	untested	well developed
Intentionality	untested	untested	manifest	untested	untested
Concept of conservation	untested	not demonstrated	manifest	untested	untested
Numeration	untested	untested	probable (n = <7)	untested	untested
Cross-modal skills	untested	manifest	manifest	untested	untested
Self-recognition in reflecting surfaces	not demonstrated	manifest	manifest	untested	not demonstrated
ASL command	untested	>56 signs	>180 signs	untested	>246 signs
Natural vocalizations	9 - 11	$\geqslant17$	13	14	16
Facial expressions	12	not enumerated	6 categories	} 47	not enumerated
Body signals	19	not enumerated	not enumerated		not enumerated

(continued)

Table 3: (continued)

Character	Gibbons	Orangutans	Common Chimpanzees	Bonobos	Gorillas
Percussions	—	snag crashing	drumming, ground-slapping, and thumping	ground-slapping	chest-beating, ground-slapping
Social units	cohesive monogamous families	males solitary; females and adolescents semisolitary	androcentric multimale bisexual unit-groups with fluid unisexual and bisexual subgroups (bands)	bisexual multimale unit-groups with fluid bisexual subgroups	cohesive bisexual polygamous groups centered about silver-backed male; lone males
Group size (adults and youngsters)	2 - 6	1 - 3	19 - 106	50 - 120	2 - 42
Mating pattern	bonded bisexual pair	quasipromiscuous; cooperative copulation by dominant male; forced copulation by others; consortships	commonly promiscuous, including gang sex, but also consortships, and possessive matings by high-ranking males	promiscuous; perhaps some consortships	restricted largely to dominant silverback, though others also score occasionally
Perineal estrous swellings	no	no	yes	yes	no
Copulatory posture	dorsoventral and ventroventral while suspended	ventroventral, dorsoventral and versatile while suspended and in nest	dorsoventral, usually on ground	dorsoventral and ventroventral in trees and on ground	dorsoventral and ventroventral on ground
Female homosexual behavior	not observed	not observed	not observed	common	manifest but episodic
Territoriality	manifest	evidenced by males in some areas, but not foolproof	evidenced in Tanzania	not documented	probably not

(continued)

Table 3: (continued)

Character	Gibbons	Orangutans	Common Chimpanzees	Bonobos	Gorillas
Dominance	male and female generally co-dominant	M: hierarchy among males in region F: not well documented	M: hierarchy and coalitions F: less stable hierarchies	males thought to be dominant over females; further details unclear	M: dominant silverback F: hierarchy
Leadership	variable; notable female leadership in several groups	inconsequential in absence of permanent groups	core males of unit-group	bisexual core of adults, perhaps with females most central	dominant silverback of group
Social grooming	yes, but not prolonged	rare	common, especially between males, bisexual pairs, and mother/young groups	common between females, in bisexual dyads, and mother/young units; some intermale	uncommon; focused on silverback; mother/young
Male parental care	siamang carries, grooms and roosts with youngster	none	tolerance but little overt caretaking	share nest and some grooming	notable, including active care of motherless youngsters and baby-sitting
New group formation	mutual attraction; usurpation of male; succession upon death of resident	—	female emigration from natal group; female recruitment by core males of group	not observed; some evidence for female emigration	female emigration to join young silverbacks; raids on established groups; perhaps succession upon demise of dominant silverback
Infanticide	not reported	not reported	committed by males and females	not reported	committed by males and perhaps females

Today we attempt to understand the adaptive complexes of each species in its own right. Subsequent comparisons are made in order to establish guidelines about how organisms and populations adapt in general and to particular kinds of environments. This may allow us to speculate in a highly informed manner about how humans acquired certain characteristics, assuming, of course, that the paleoenvironments and skeletal biology of our ancestors are discovered by the paleoanthropologists.

References

Abordo, Enrique J. 1976. The learning skills of gibbons, in: *Gibbon and Siamang* (D.M. Rumbaugh, ed.), Vol. 4, pp. 106-134, Basel: Karger.

Akeley, Carl E. 1922. Hunting gorillas in Central Africa. *World's Work*, 44:169-183, 307-308, 393-399, 525-533.

Akeley, Carl E. 1923a. *In Brightest Africa*, Garden City, New York: Doubleday, Page & Co.

Akeley, Carl E. 1923b. Gorillas—real and mythical. *Natural History*, 23:428-447.

Albrecht, Helmut 1976. Chimpanzees in Uganda. *Oryx*, 13:357-361.

Albrecht, Helmut, and Dunnett, Sinclair Coghill 1971. *Chimpanzees in Western Africa*, Munchen: R. Piper and Co.

Aldrich-Blake, F. Pelham G., and Chivers, David John 1973. On the genesis of a group of siamang. *American Journal of Physical Anthropology*, 38:631-636.

Almquist, Alan J. 1974. Sexual differences in the anterior dentition in African primates. *American Journal of Physical Anthropology*, 40:359-368.

Anderson, James R. 1983a. Responses to mirror image stimulation and assessment of self-recognition in mirror- and peer-reared stumptail macaques. *Quarterly Journal of Experimental Psychology*, 35B:201-212.

Anderson, James R. 1983b. Mirror-image stimulation and short separations in stumptail monkeys. *Animal Learning & Behavior*, 11:139-143.

Anderson, James R. 1984a. Ethology and ecology of sleep in monkeys and apes. *Advances in the Study of Behavior*, 14:165-229.

Anderson, James R. 1984b. Monkeys with mirrors: some questions for primate psychology. *International Journal of Primatology*, 5:81-98.

Anderson, James R. 1984c. The development of self-recognition: a review. *Developmental Psychobiology*, 17:35-49.

Anderson, James R.; Williamson, Elizabeth A.; and Carter, Janis 1983. Chimpanzees of Sapo Forest, Liberia: density, nests, tools and meat-eating. *Primates*, 24:594-601.

Andrew, Richard J. 1962. The situations that evoke vocalization in primates. *Annals of the New York Academy of Science*, 102:296-315.

Andrew, Richard J. 1963a. Evolution of facial expression. *Science*, 142:1034-1041.

Andrew, Richard J. 1963b. The origin and evolution of the calls and facial expressions of the Primates. *Behaviour*, 20:1-109.

Andrew, Richard J. 1963c. Trends apparent in the evolution of vocalization in the Old World monkeys and apes. *Symposia of the Zoological Society of London*, no. 10, pp. 89-101.

Andrew, Richard J. 1964. The displays of the Primates, in: *Evolutionary and Genetic Biology of Primates*. (John Buettner-Janusch, ed.), Vol. 2, pp. 227-309, New York: Academic Press.

Andrew, Richard J. 1965. The origins of facial expressions. *Scientific American*, 213:88-94.

Andrews, Peter J. 1978a. A revision of the Miocene Hominoidea of East Africa. *Bulletin of the British Museum of Natural History (Geology)*, 30:85-224.

Andrews, Peter J. 1978b. Taxonomy and relationships of fossil apes, in: *Recent Advances in Primatology*, Vol. 3, *Evolution* (D.J. Chivers and K.A. Joysey, eds.), pp. 43-56, London: Academic Press.

Andrews, Peter 1982. Hominoid evolution. *Nature*, 295:185-186.

Andrews, Peter, and Groves, Colin P. 1976. Gibbons and brachiation, in: *Gibbon and Siamang* (D.M. Rumbaugh, ed.), Vol. 4, pp. 167-218, Basel: S. Karger.

Andrews, Peter; Hamilton, William Roger, and Whybrow, P.J. 1978. Dryopithecines from the Miocene of Saudi Arabia. *Nature*, 274:249-250.

Andrews, Peter, and Simons, Elwyn 1977. A new African Miocene gibbon-like genus, *Dendropithecus* (Hominoidea, Primates) with distinctive postcranial adaptations: its significance to origin of Hylobatidae. *Folia Primatologica*, 28:161-169.

Andrews, Peter, and Tekkaya, Ibrahim 1976. *Ramapithecus* in Kenya and Turkey, in: *Les Plus Anciens Hominidés* (P.V. Tobias and Y. Coppens, eds.), pp. 7-25, Paris: Centre National de la Recherche Scientifique.

Andrews, Peter, and Tekkaya, Ibrahim 1980. A revision of the Turkish Miocene hominoid *Sivapithecus meteai*. *Palaeontology*, 23:85-95.

Andrews, Peter, and Tobien, Heinz 1977. New Miocene locality in Turkey with evidence on the origin of *Ramapithecus* and *Sivapithecus*. *Nature*, 268:699-701.

Asano, Toshio; Kojima, Tetsuya; Matsuzawa, Tetsuro; Kubota, Kisou; and Murofushi, Kiyoko 1982. Object and color naming in chimpanzees (*Pan troglodytes*). *Proceedings of the Japan Academy*, 58:118-122.

van den Audenaerde, Dirk F.E. Thys 1984. The Tervuren Museum and the pygmy chimpanzee, in: *The Pygmy Chimpanzee: Evolutionary Biology and Behavior* (R.L. Susman, ed.), pp. 3-11, New York: Plenum Press.

Avis, Virginia 1962. Brachiation: the crucial issue for man's ancestry. *Southwest Journal of Anthropology*, 18:119-148.

Azuma, Shigeru, and Toyoshima, Akisato 1961-2. Progress report of the survey of chimpanzees in their natural habitat, Kabogo Point Area, Tanganyika. *Primates*, 3:61-70.

Badrian, Alison, and Badrian, Noel 1977. Pygmy chimpanzees. *Oryx*, 13:463-468.

Badrian, Alison, and Badrian, Noel 1980. The other chimpanzee. *Animal Kingdom*, 83:8-14.

Badrian, Alison, and Badrian, Noel 1984a. The bonobo branch of the family tree. *Animal Kingdom*, 87:39-45.

Badrian, Alison, and Badrian, Noel 1984b. Social organization of *Pan paniscus* in the Lomako Forest, Zaire, in: *The Pygmy Chimpanzee. Evolutionary Biology and Behavior* (Randall L. Susman, ed.), pp. 325-346, New York: Plenum Press.

Badrian, Noel; Badrian, Alison; and Susman, Randall L. 1981. Preliminary observations on the feeding behavior of *Pan paniscus* in the Lomako Forest of Central Zaire. *Primates*, 22:173-181.

Badrian, Noel, and Malenky, Richard K. 1984. Feeding ecology of *Pan paniscus* in the Lomako Forest, Zaire, in: *The Pygmy Chimpanzee: Evolutionary Biology and Behavior* (Randall L. Susman, ed.), pp. 275-299, New York: Plenum Press.

Baldwin, Lori A., and Teleki, Geza 1982. Patterns of gibbon behavior on Hall's Island, Bermuda, in: *Gibbon and Siamang* (D.M. Rumbaugh, ed.), Vol. 4, pp. 21-105, Basel: Karger.

Baldwin, Pamela J.; McGrew, William C.; and Tutin, Caroline E.G. 1982. Wide-ranging chimpanzees at Mt. Assirik, Senegal. *International Journal of Primatology*, 3:367-385.

Baldwin, Pamela J.; Sabater Pi, Jorge; McGrew, William C.; and Tutin, Caroline E.G. 1981. Comparisons of nests made by different populations of chimpanzees (*Pan troglodytes*). *Primates*, 22:474-486.

Bassett, Beth Dawkins 1983. The language continuum. *Emory Magazine*, 59:6-13.

Bauer, Harold R. 1975. Behavioral changes about the time of reunion in parties of chimpanzees in the Gombe Stream National Park, in: *Contemporary Primatology* (S. Kondo, M. Kawai, and A. Ehara, eds.), pp. 295-303, Basel: Karger.

Bauer, Harold R. 1979. Agonistic and grooming behavior in the reunion context of Gombe Stream chimpanzees, in: *The Great Apes* (D.A. Hamburg and E.R. McCown, eds.), pp. 394-403, Menlo Park, California: The Benjamin/Cummings Co.

Bauer, Harold R., and Philip, M. Michelle 1983. Facial and vocal individual recognition in the common chimpanzee. *The Psychological Record*, 33:161-170.

Beatty, Harry 1951. A note on the behavior of the chimpanzee. *Journal of Mammalogy*, 32:118.

Beck, Benjamin B. 1967. A study of problem solving by gibbons. *Behaviour*, 28:95-109.

Beck, Benjamin B. 1980. *Animal Tool Behavior: The use and manufacture of tools by animals*, New York: Garland STPM Press.

Beck, Benjamin B. 1982. Chimpocentrism: bias in cognitive ethology. *Journal of Human Evolution*, 11:3-17.

Becker, Clemens, and Hick, Uta. 1984. "Familienzusammenfürung" als soziale Beschäftigungstherapie und Aktivitätssteigerung bei sieber Orang-Utans (*Pongo p. pygmaeus*) im Kölner Zoo. *Zeitschrift des Kölner Zoo*, 27:43-57.

Belleville, Richard E.; Rohles, Frederick H.; Grunzke, Marvin E.; Clark, Fogle C. 1963. Development of a complex multiple schedule in the chimpanzee. *Journal of the Experimental Analysis of Behavior*, 6:549-556.

van Bemmel, A.C.V. 1969. Contributions to the knowledge of the geographic races of *Pongo pygmaeus* (Hoppius). *Bijdragen tot de Dierkunde Amsterdam*, 1968:13-15.

Berdecio, Susana, and Nash, Leanne T. 1981. Facial, gestural and postural expressive movement in young, captive chimpanzees (*Pan troglodytes*). *Arizona State University Anthropological Research Papers*, No. 26.

Berkson, Gershon 1962. Food motivation and delayed response in gibbons. *Journal of Comparative and Physiological Psychology*, 55:1040-1043.

Bernstein, Irwin S. 1961a. The utilization of visual cues in dimension-abstracted oddity by primates. *Journal of Comparative and Physiological Psychology*, 54:243-247.

Bernstein, Irwin S. 1961b. Response variability and rigidity in the adult chimpanzee. *Journal of Gerontology*, 16:381-386.

Bernstein, Irwin S. 1962. Response to nesting materials of wild born and captive born chimpanzees. *Animal Behaviour*, 10:1-6.

Bernstein, Irwin S. 1967. Age and experience in chimpanzee nest building. *Psychological Reports*, 20:1106.

Bernstein, Irwin S. 1969. A comparison of nesting patterns among the three great apes, in: *The Chimpanzee* (G.H. Bourne, ed.), Vol. 1, pp. 393-402, Basel: Karger.

Bertrand, Mireille 1969. The behavioral repertoire of the stumptail macaque. *Bibliotheca Primatologica*, No. 11, Basel: Karger.

Biegert, Josef 1973. Dermatoglyphics in gibbons and siamangs, in: *Gibbon and Siamang* (D.M. Rumbaugh, ed.), Vol. 2, pp. 163-184, Basel: Karger.

Bingham, Harold C. 1932. Gorillas in a native habitat. *Carnegie Institute of Washington Publications*, 426:1-66.

Birch, Herbert G. 1945a. The role of motivational factors in insightful problem-solving. *Journal of Comparative Psychology*, 38:295-317.

Birch, Herbert G. 1945b. The relation of previous experience to insightful problem-solving. *Journal of Comparative Psychology*, 38:367-383.

Bishop, Alison 1964. Use of the hand in lower primates, in: *Evolutionary and Genetic Biology of Primates* (J. Buettner-Janusch, ed.), Vol. II, pp. 133-225, New York: Academic Press.

Blakemore, Collin B., and Ettlinger, George 1968. The independence of visual and tactile learning in the monkey. *Journal of Physiology*, 196:127P.

Blank, Marion; Altman, L. Dorel; and Bridger, Wagner H. 1968. Crossmodal transfer of form discrimination in preschool children. *Psychonomic Science*, 10:51-52.

Blanshard, Roberta Yerkes 1977. Home life with chimpanzees, in: *Progress in Ape Research* (G.H. Bourne, ed.), pp. 7-13, New York: Academic Press, Inc.

Boesch, Christophe 1978. Nouvelles observations sur les chimpanzes de la Foret de Tai (Côte d'Ivoire). *La Terre et la Vie*, 32:195-201.

Boesch, Christophe, and Boesch, Hedwige 1981. Sex differences in the use of natural hammers by wild chimpanzees: a preliminary report. *Journal of Human Evolution*, 10:585-593.

Boesch, Christophe, and Boesch, Hedwige 1982. Optimisation of nut-cracking with natural hammers by wild chimpanzees. *Behaviour*, 83:265-286.

Boesch, Christophe, and Boesch, Hedwige 1984a. The nut-cracking behavior and its nutritional importance in wild chimpanzees in the Tai National Park, Ivory Coast. *International Journal of Primatology*, 5:323.

Boesch, Christophe, and Boesch, Hedwige 1984b. Possible causes of sex differences in the use of natural hammers by wild chimpanzees. *Journal of Human Evolution*, 13:415-440.

Boesch, Christophe, and Boesch, Hedwige 1984c. Mental map in wild chimpanzees: an analysis of hammer transports for nut cracking. *Primates*, 25:160-170.

Bolster, R. Bruce 1978. Cross-model matching in the monkey (*Macaca fascicularis*). *Neuropsychologia*, 16:407-416.

Bolster, R. Bruce 1981. The function of inferoparietal cortex during cross-model matching in the monkey: analysis of event-related potentials. *Neuropsychologia*, 19:385-394.

Bolwig, Niels 1959. A study of the nests built by mountain gorilla and chimpanzee. *South African Journal of Science*, 55:286-291.

von Bonin, Gerhardt 1962. Anatomical asymmetries of the cerebral hemispheres, in: *Interhemispheric Relations and Cerebral Dominance* (V.B. Mountcastle, ed.), pp. 1-6, Baltimore: Johns Hopkins University Press.

de Bonis, Louis; Bouvrain, Genevieve; Geraads, D.,; and Melentis, J. 1974. Première découverte d'un Primate hominoïde dans le Miocéne supérieur de Macédoine (Grèce). *Comptes Rendus de l'Academie des Sciences, Paris* 278, D:3063-3066.

de Bonis, Louis; Bouvrain, Genevieve; and Melentis, J. 1975. Nouveaux restes de Primates hominoïde dans le Vallésien de Macédoine (Grèce). *Comptes Rendus de l'Academie des Sciences, Paris* 281, D:379-382.

de Bonis, Louis, and Melentis, J. 1976. Les Dryopithécinés de Macédoine (Grèce): leur place dans l'evolution des primates hominoides du Miocéne, in: *Les Plus Anciens Homininés* (P.V. Tobias and Y. Coppens, eds.), pp. 26-38, Paris: Centre National de la Recherche Scientifique.

de Bonis, Louis, and Melentis, J. 1977a. Un nouveau genre de Primate hominoïde dans le Vallésien (Miocéne supérieur) de Macédoine. *Comptes Rendus de l'Academie des Sciences, Paris* 284, D:1393-1396.

de Bonis, Louis, and Melentis, J. 1977b. Les Primates hominoïdes du Vallésien de Macédoine (Grèce). Étude de la Mâchoire inférieure. *Geobios*, 10:849-885.

de Bonis, Louis, and Melentis, J. 1978. Le Primates hominoïdes du Miocène supérieur de Macédoine. Étude de la mâchoire supérieure. *Annales de Paleontologie (Vertebres)*, 64:185-202.

Bourlière, François; Minner, E.; and Vuattoux, R. 1974. Les grand mammifères de la région de Lamto, Côte d'Ivoire. *Mammalia*, 38:433-447.

Bourne, Geoffrey H. 1971. *The Ape People*, New York: G.P. Putnam's Sons.

de Bournonville, D. 1967. Contribution à l'étude du chimpanzé in République de Guinée. *Bulletin de l'Institut Fondamental d'Afrique Noire*, 29:1188-1269.

Boutan, Louis 1914. Les deux méthodes de l'enfant. *Actes Société linnéenne de Bordeaux*, 68:217-360.

Bradley, Mary Hastings 1922. *On the Gorilla Trail*, New York: D. Appleton & Co.

Bradshaw, John L., and Nettleton, Norman C. 1983. *Human Cerebral Asymmetry*, Englewood Cliffs, New Jersey: Prentice-Hall, Inc.

Braggio, John T.; Hall, Anthony D.; Buchanan, James P.; and Nadler, Ronald D. 1982. Logical and illogical errors made by apes and children on a cognitive task. *Journal of Human Evolution*, 11:159-169.

Brésard, B., and Bresson, F. 1983. Handedness in *Pongo pygmaeus* and *Pan troglodytes*. *Journal of Human Evolution*, 12:659-666.

Brockelman, Warren Y. 1975. Gibbon populations and their conservation in Thailand. *Natural History Bulletin of the Siam Society*, 26:133-157.

Brockelman, Warren Y. 1978. Preliminary report on relations between the gibbons *Hylobates lar* and *H. pileatus* in Thailand, in: *Recent Advances in Primatology* Vol. 3, *Evolution* (D.J. Chivers and K.A. Joysey, eds.), pp. 315-318, London: Academic Press.

Brockelman, Warren Y. 1981. Field research on primates in Thailand. *Journal of the Science Society of Thailand*, 7:9-17.

Brockelman, Warren Y. 1984. Social behaviour of gibbons: introduction, in: *The Lesser Apes: Evolutionary and Behavioural Biology* (H. Preuschoft, D.J. Chivers, W.Y. Brockelman, and N. Creel, eds.), pp. 285-290, Edinburgh: Edinburgh University Press.

Brockelman, Warren Y., and Gittins, S. Paul 1984. Natural hybridization in the *Hylobates lar* species group: implications for speciation in gibbons, in: *The Lesser Apes: Evolutionary and Behavioural Biology* (H. Preuschoft, D.J. Chivers, W.Y. Brockelman, and N. Creel, eds.), pp. 498-532, Edinburgh: Edinburgh University Press.

Brockelman, Warren Y., and Srikosamatara, S. 1984. Maintenance and evolution of social structure in gibbons, in: *The Lesser Apes: Evolutionary and Behavioural Biology* (H. Preuschoft, D.J. Chivers, W.Y. Brockelman, and N. Creel, eds.), pp. 298-323, Edinburgh: Edinburgh University Press.

Brown, Diana P.F.; Lenneberg, Eric H.; and Ettlinger, George 1978. Ability of chimpanzees to respond to symbols of quantity in comparison with that of children and of monkeys. *Journal of Comparative and Physiological Psychology*, 92:815-820.

Brown, Roger W. 1981. Symbolic and syntactic capacities. *Philosophical Transactions of the Royal Society, London*, B 292:197-204.

Bruce, Elizabeth J., and Ayala, Francisco J. 1978. Humans and apes are genetically very similar. *Nature*, 276:264-265.

Bryant, Peter E.; Jones, P.; Claxton, V.; and Perkins, G.M. 1972. Recognition of shapes across modalities in infants. *Nature*, 240:303-304.

Buchanan, James P.; Gill, Timothy V.; and Braggio, John T. 1981. Serial position and clustering effects in a chimpanzee's "free recall." *Memory and Cognition*, 9:651-660.

Busse, Curt D. 1977. Chimpanzee predation as a possible factor in the evolution of red colobus social organization. *Evolution*, 31:907-911.

Busse, Curt D. 1978. Do chimpanzees hunt cooperatively? *American Naturalist*, 112:767-770.

Butler, Harry 1966. Some notes on the distribution of primates in the Sudan. *Folia Primatologica*, 4:416-423.

Butynski, Thomas M. 1982. Vertebrate predation by primates: a review of hunting patterns and prey. *Journal of Human Evolution*, 11:421-430.

Bützler, Wilfried 1980. Présence et répartition des gorilles, *Gorilla gorilla gorilla* (Savage & Wyman, 1847), au Cameroun. *Säugetierkundliche Mitteilungen*, no. 1, pp. 69-79.

Bygott, J. David 1972. Cannibalism among wild chimpanzees. *Nature*, 238:410-411.

Bygott, J. David 1979. Agonistic behavior, dominance, and social structure in wild chimpanzees of the Gombe National Park, in: *The Great Apes* (D.A. Hamburg and E.R. McCown, eds.), pp. 404-427, Menlo Park, California: The Benjamin/Cummings Publishing Co.

Caldecott, Julian Oliver 1980. Habitat quality and populations of two sympatric gibbons (Hylobatidae) on a mountain in Malaya. *Folia Primatologica*, 33:291-309.

Caldecott, Julian Oliver, and Haimoff, Elliott H. 1983. Female solo singing by a wild lar gibbon in Peninsular Malaysia. *Malayan Nature Journal*, 36:167-173.

Calvert, Julie J. 1985. Food selection by western gorillas (*G. g. gorilla*) in relation to food chemistry. *Oecologia (Berlin)*, 65:236-246.

Campbell, Scott S., and Tobler, Irene 1984. Animal sleep: a review of sleep duration across phylogeny. *Neuroscience and Biobehavioral Reviews*, 8:269-300.

Cantfort, Thomas E. Van, and Rimpau, James B. 1982. Sign language studies with children and chimpanzees. *Sign Language Studies*, 34:15-72.

Caro, Timothy M. 1976. Observations on the ranging behaviour and daily activity of lone silverback mountain gorillas (*Gorilla gorilla beringei*). *Animal Behaviour*, 24:889-897.

Carpenter, Clarence Ray 1934. A field study of the behavior and social relations of howling monkeys. *Comparative Psychology Monographs*, 10:1-168.

Carpenter, Clarence Ray 1935. Behavior of red spider monkeys in Panama. *Journal of Mammalogy*, 16:171-180.

Carpenter, Clarence Ray 1937. An observational study of two captive mountain gorillas (*Gorilla beringei*). *Human Biology*, 9:175-196.

Carpenter, Clarence Ray 1938. A survey of wild life conditions in Atjeh, North Sumatra, with special reference to the orang-utan. *Netherlands Committee for International Nature Protection*, Amsterdam, Communications no. 12, pp. 1-34.

Carpenter, Clarence Ray 1940. A field study in Siam of the behavior and social relations of the gibbon (*Hylobates lar*). *Comparative Psychology Monographs*, 16:1-212.

Carpenter, Clarence Ray 1962. Field studies of a primate population, in: *Roots of Behavior: Genetics, Instinct, and Socialization in Animal Behavior* (Eugene L. Bliss, ed.), pp. 286-294, New York: Harper & Bros., Publishers.

Carpenter, Clarence Ray 1976. Suspensory behavior of gibbons *Hylobates lar*, in: *Gibbon and Siamang* (D.M. Rumbaugh, ed.), Vol. 4, pp. 1-20, Basel: Karger.

Carpenter, Clarence Ray, and Durham, Norris M. 1969. A preliminary description of suspensory behavior in nonhuman primates. *Proceedings of the 2nd International Congress of Primatology*, Vol. 2, pp. 147-154, Basel: Karger.

Carpenter, Clarence Ray, and Nissen, Henry W. 1934. An experimental analysis of some spatial variables in delayed reactions of chimpanzees. *Psychological Bulletin*, 31:689.

Casimir, Michael J. 1975. Feeding ecology and nutrition of an eastern gorilla group in the Mt. Kahuzi region (République du Zaïre). *Folia Primatologica*, 24:81-136.

Casimir, Michael J. 1979. An analysis of gorilla nesting sites of the Mt. Kahuzi Region (Zaïre). *Folia Primatologica*, 32:290-308.

Casimir, Michael J., and Butenandt, Eckardt 1973. Migration and core area shifting in relation to some ecological factors in a mountain gorilla group (*Gorilla gorilla beringei*) in the Mt. Kahuzi region (République du Zaïre). *Zeitschrift für Tierpsychologie*, 33:514-522.

Chasen, F.N., and Kloss, C.B. 1927. Spolia Mentaweiensia: Mammals. *Proceedings of the Zoological Society, London*, 1927:797-840.

Chevalier-Skolnikoff, Suzanne 1973. Facial expression of emotion in nonhuman primates, in: *Darwin and Facial Expression, a Century of Research in Review* (Paul Ekman, ed.), pp. 11-89, New York: Academic Press.

Chiarelli, Brunetto 1968. Caryological and hybridological data for the taxonomy and phylogeny of the Old World primates, in: *Taxonomy and Phylogeny of Old World Primates with References to the Origin of Man* (B. Chiarelli, ed.), pp. 151-181, Toronto: Rosenberg & Sellier.

Chiarelli, Brunetto 1972. The karyotypes of the gibbons, in: *Gibbon and Siamang* (D.M. Rumbaugh, ed.), Vol. 1, pp. 90-102, Basel: Karger.

Chiarelli, Brunetto 1975. The study of Primate chromosomes, in: *Primate Functional Morphology and Evolution* (R.H. Tuttle, ed.), pp. 103-127, The Hague: Mouton.

Chivers, David J. 1971a. The Malayan siamang. *Malayan Nature Journal*, 24:78-86.

Chivers, David J. 1971b. Spatial relations within the siamang group, in: *Proceedings of the Third International Congress of Primatology, Zurich 1970*, Vol. 3, pp. 14-21, Basel: Karger.

Chivers, David J. 1972. The siamang and the gibbon in the Malay Peninsula, in: *Gibbon and Siamang* (D.M. Rumbaugh, ed.), Vol. 1, pp. 103-135, Basel: Karger.

Chivers, David J. 1973. Introduction to the socio-ecology of Malayan forest primates, in: *Comparative Ecology and Behaviour of Primates* (R.P. Michael and J.H. Crook, eds.), pp. 101-146, London: Academic Press.

Chivers, David J. 1974. The siamang in Malaya. *Contributions to Primatology*, 4:1-335, Basel: Karger.

Chivers, David J. 1975. Daily patterns of ranging and feeding in siamang, in: *Contemporary Primatology* (S. Kondo, M. Kawai, and A. Ehara, eds.), pp. 362-372, Basel: Karger.

Chivers, David J. 1976. Communication within and between family groups of siamang (*Symphalangus syndactylus*). *Behaviour*, 57:116-135.

Chivers, David J. 1977a. The lesser apes, in: *Primate Conservation* (H.S.H. Prince Rainier and G.H. Bourne, eds.), pp. 539-598.

Chivers, David J. 1977b. The ecology of gibbons: some preliminary considerations based on observations in the Malay Peninsula, in: *Use of Non-human Primates in Biomedical Research* (M.R.N. Prasad and T.C. Anand Kumar, eds.), pp. 85-105, New Delhi: Indian National Science Academy.

Chivers, David J. 1977c. The feeding behavior of siamang (*Symphalangus syndactylus*), in: *Feeding Ecology: Studies of feeding and ranging behaviour in lemurs, monkeys and apes* (T.H. Clutton-Brock, ed.), pp. 355-382, London: Academic Press.

Chivers, David J. 1978. Sexual behaviour of wild siamang, in: *Recent Advances in Primatology*, Vol. 1, *Behaviour* (D.J. Chivers and J. Herbert, eds.), pp. 609-610, London: Academic Press.

Chivers, David J. 1980. *Malayan Forest Primates: Ten year's study in tropical rain forest*, New York: Plenum Press.

Chivers, David J. 1984. Feeding and ranging in gibbons: a summary, in: *The Lesser Apes: Evolutionary and Behavioural Biology* (H. Preuschoft, D.J. Chivers, W.Y. Brockelman, and N. Creel, eds.), pp. 267-281, Edinburgh: Edinburgh University Press.

Chivers, David J., and Chivers, Sarah T. 1975. Event preceding and following the birth of a wild siamang. *Primates*, 16:227-230.

Chivers, David J., and Hladik, Claude Marcel 1980. Morphology of the gastrointestinal tract in primates: comparisons with other mammals in relation to diet. *Journal of Morphology*, 166:337-386.

Chivers, David J., and MacKinnon, John 1977. On the behaviour of siamang after playback of their calls. *Primates*, 18:943-948.

Chivers, David J., and Raemaekers, Jeremy John 1980. Long-term changes in behaviour, in: *Malayan Forest Primates: Ten year's study in tropical rain forest* (D.J. Chivers, ed.), pp. 209-260, New York: Plenum Press.

Chivers, David J.; Raemaekers, Jeremy John; and Aldrich-Blake, F. Pelham G. 1975. Long-term observations of siamang behaviour. *Folia Primatologica*, 23:1-49.

Ciochon, Russell L., and Corruccini, Robert S. 1983. *New Interpretations of Ape and Human Ancestry*, New York: Plenum Press.

Clark, Cathleen B. 1977. A preliminary report on weaning among chimpanzees of the Gombe National Park, Tanzania, in: *Primate Bio-social Development: Biological, social, and ecological determinants* (Suzanne Chevalier-Skolnikoff and Frank E. Poirier, eds.), pp. 235-260, New York: Garland Publishing, Inc.

Clark, Fogle C. 1961. Avoidance conditioning in the chimpanzee. *Journal of the Experimental Analysis of Behavior*, 4:393-395.

Clark, Wilfrid E. LeGros 1968. *Chant of Pleasant Exploration*, Edinburgh: Livingstone.

Clark, Wilfrid E. LeGros 1971. *The Antecedents of Man*, 3rd ed., Chicago: Quadrangle Books.

Clark, Wilfrid E. LeGros, and Campbell, Bernard G. 1978. *The Fossil Evidence for Human Evolution*, 3rd ed., Chicago, Illinois: The University of Chicago Press.

Clutton-Brock, Timothy H., and Harvey, Paul H. 1977. Primate ecology and social organization. *Journal of Zoology, London*, 183:1-39.

Cohen, Joel E. 1975. The size and demographic composition of social groups of wild orang-utans. *Animal Behaviour*, 23:543-550.

Colbert, Edwin H., and Hooijer, Dirk Albert 1953. Pleistocene mammals from the limestone fissures of Szechwan, China. *Bulletin of the American Museum of Natural History*, 102:1-134.

Coolidge, Harold Jefferson 1929. A revision of the genus *Gorilla. Memoirs of the Museum of Comparative Zoology, Harvard*, 50:291-383.

Coolidge, Harold J. 1933. *Pan paniscus*, pygmy chimpanzee from south of the Congo River. *American Journal of Physical Anthropology*, 18:1-59.

Coolidge, Harold J. 1936. Zoological results of the George Vanderbilt African expedition of 1934. Part IV, - notes on four gorillas from the Sanga River Region. *Proceedings of the Academy of Natural Sciences of Philadelphia*, 88:479-501.

Coolidge, Harold J., and Shea, Brian T. 1982. External body dimensions of *Pan paniscus* and *Pan troglodytes* chimpanzees. *Primates*, 23:245-251.

Corruccini, Robert S. 1981. Analytical techniques for Cartesian coordinate data with reference to the relationship between *Hylobates* and *Symphalangus* (Hylobatidae; Hominoidea). *Systematic Zoology*, 30:32-40.

Cortright, Gerald W. 1976. A kinematic analysis of gibbon bipedalism. *American Journal of Physical Anthropology*, 44:172.

Cousins, Don 1972. Body measurements and weights of wild and captive gorillas, *Gorilla gorilla. Der Zoologische Garten N. F., Leipzig*, 41:261-277.

Cowey, Alan, and Weiskrantz, Lawrence 1975. Demonstration of cross-modal matching in rhesus monkeys, *Macaca mulatta. Neuropsychologia*, 13:117-120.

Cowles, John T. 1937. Food-tokens as incentives for learning by chimpanzees. *Comparative Psychology Monographs*, 14:1-96.

Cowles, John T., and Nissen, Henry W. 1937. Reward-expectancy in delayed responses of chimpanzees. *Journal of Comparative Psychology*, 24:345-358.

Cramer, Douglas L. 1977. Craniofacial morphology of *Pan paniscus*: a morphometric and evolutionary appraisal, in: *Contributions to Primatology*, Vol. 10, Basel: Karger.

Cramer, Douglas L., and Zihlman, Adrienne L. 1978. Sexual dimorphism in the pygmy chimpanzee, *Pan paniscus*, in: *Recent Advances in Primatology*, Vol. 3, *Evolution* (D.J. Chivers and K.A. Joysey, eds.), pp. 487-490, London: Academic Press.

Crawford, Meredith P. 1937. The cooperative solving of problems by young chimpanzees. *Comparative Psychology Monographs* 14:1-88.

Crawford, Meredith P. 1941. The cooperative solving by chimpanzees of problems requiring serial responses to color cues. *Journal of Social Psychology* 13:259-280.

Crawford, Meredith P., and Spence, Kenneth W. 1939. Observational learning of discrimination problems by chimpanzees. *Journal of Comparative Psychology* 27:133-147.

Creel, Norman, and Preuschoft, Holger 1976. Cranial morphology in the lesser apes, in: *Gibbon and Siamang* (D.M. Rumbaugh, ed.), Vol. 4, pp. 219-303, Basel: Karger.

Creel, Norman, and Preuschoft, Holger 1984. Systematics of the lesser apes: a quantitative taxonomic analysis of craniometric and other variables, in: *The Lesser Apes: Evolutionary and Behavioural Biology* (Holger Preuschoft, David J. Chivers, Warren Y. Brockelman, and Normal Creel, eds.), pp. 562-613, Edinburgh: Edinburgh University Press.

Cronin, John E. 1983. Apes, humans, and molecular clocks: a reappraisal, in: *New Interpretations of Ape and Human Ancestry* (Russell L. Ciochon and Robert S. Corruccini, eds.), pp. 115-135, New York: Plenum Press.

Cronin, John E.; Sarich, Vincent M.; and Ryder, O. 1984. Molecular evolution and speciation in the lesser apes, in: *The Lesser Apes: Evolutionary and Behavioural Biology* (Holger Preuschoft, David J. Chivers, Warren Y. Brockelman, and Norman Creel, eds.), pp. 467-485, Edinburgh: Edinburgh University Press.

Crook, John H., and Gartlan, J. Stephen 1966. Evolution of primate societies. *Nature* 210:1200-1203.

Crook, John H., and Goss-Custard, John D., 1972. Social ethology. *Annual Review of Psychology* 23:277-312.

Cummings, Edward Estlin 1972. *Complete Poems*, New York: Harcourt Brace Jovanovich, Inc.

Dahl, Jeremy F. 1985. The external genitalia of female pygmy chimpanzees. *The Anatomical Record* 211:24-28.

Dang, D.C.; Demars, Christian, and Goustard, Michel 1969. Étude du "grand chant" du gibbon (*Hylobates concolor*). *Annales des Sciences Naturelles Zoologie, Paris* XI:505-514.

Darga, Linda L.; Baba, Marietta L.; Weiss, Mark S.; and Goodman, Morris 1984. Molecular perspectives on the evolution of the lesser apes, in: *The Lesser Apes: Evolutionary and Behavioural Biology* (Holger Preuschoft, David J. Chivers, Warren Y. Brockelman, and Norman Creel, eds.), pp. 448-466, Edinburgh: Edinburgh University Press.

Darga, Linda L.; Goodman, Morris; and Weiss, Mark L. 1973. Molecular evidence on the cladistic relationships of the Hylobatidae, in: *Gibbon and Siamang* (D.M. Rumbaugh, ed.), Vol. 2, pp. 149-162, Basel: Karger.

Darley, Frederic L. 1967. *Brain Mechanisms Underlying Speech and Language: Proceedings of a Conference held at Princeton, New Jersey, 1965*, New York: Grune and Stratton.

Darwin, Charles 1871. *The Descent of Man—and Selection in Relation to Sex*, London: Murray.

Davenport, Richard K. 1967. The orang-utan in Sabah. *Folia Primatologica* 5:247-263.

Davenport, Richard K. 1976. Cross-modal perception in apes. *Annals of the New York Academy of Sciences* 280:143-149.

Davenport, Richard K. 1977. Cross-modal perception: a basis for language?, in: *Language Learning by a Chimpanzee: The Lana Project* (D.M. Rumbaugh, ed.), pp. 73-83, New York: Academic Press.

Davenport, Richard K., Jr., and Menzel, Emil W. 1960. Oddity preference in the chimpanzee. *Psychological Reports* 7:523-526.

Davenport, Richard K., Jr., and Rogers, Charles M. 1968. Intellectual performance of differentially reared chimpanzees: I. delayed response. *American Journal of Mental Deficiency* 72:674-680.

Davenport, Richard K., and Rogers, Charles M. 1970. Intermodal equivalence of stimuli in apes. *Science* 168:279-280.

Davenport, Richard K., and Rogers, Charles M. 1971. Perception of photographs by apes. *Behaviour* 34:318-320.

Davenport, Richard K., Jr.; Rogers, Charles M.; and Menzel, Emil W. 1969. Intellectual performance of differentially reared chimpanzees: II. discrimination learning set. *American Journal of Mental Deficiency* 73:963-969.

Davenport, Richard K.; Rogers, Charles M.; and Rumbaugh, Duane M. 1973. Long-term cognitive deficits in chimpanzees associated with early impoverished rearing. *Developmental Psychology* 9:343-347.

Davenport, Richard K.; Rogers, Charles M.; and Russell, I. Steele 1973. Cross-modal perception in apes. *Neuropsychologia* 11:21-28.

Davenport, Richard K.; Rogers, Charles M.; and Russell, I. Steele 1975. Cross-model perception in apes: altered visual cues and delay. *Neuropsychologia* 13:229-235.

Davis, Rickie R., and Markowitz, Hal 1978. Orangutan performance on a light-dark reversal discrimination in the zoo. *Primates* 19:755-759.

Davis, Roger T.; McDowell, Arnold A.; and Nissen, Henry W. 1957. Solution of bent-wire problems by monkeys and chimpanzees. *Journal of Comparative and Physiological Psychology* 50:441-444.

Delson, Eric 1977. Catarrhine phylogeny and classification: principles, methods and comments. *Journal of Human Evolution* 6:433-459.

Delson, Eric 1979. *Oreopithecus* is a cercopithecoid after all. *American Journal of Physical Anthropology* 50:431-432.

Delson, Eric 1985. *Ancestors: The Hard Evidence*, New York: Alan R. Liss, Inc.

Delson, Eric, and Andrews, Peter 1975. Evolution and interrelationships of the catarrhine primates, in: *Phylogeny of the Primates* (W.P. Luckett and F.S. Szalay, eds.), pp. 405-446, New York: Plenum Press.

Demars, Christian; Berthomier, C.; and Goustard, Michel 1978. The ontogenesis of the 'great call' of gibbons (*Hylobates concolor*), in: *Recent Advances in Primatology*, Vol. 1, *Behaviour* (D.J. Chivers and J. Herbert, eds.), pp. 827-830, London: Academic Press.

Demars, Christian, and Goustard, Michel 1972. Structure et regles de deroulement des emissions sonores des Hylobates (*Hylobates concolor*). *Bulletin Biologique de la France et de la Belgique* 106:177-191.

Dendy, Larry, B. 1973. Communication—three-way: chimpanzee, man, computer. *Computers and Automation* 22:5-8.

Denenberg, Victor H. 1981. Hemispheric laterality in animals and the effects of early experience. *The Behavioral and Brain Sciences*, 4:1-49.

Depputte, Bertrand, and Goustard, Michel 1978. Etude du répertoire vocal du gibbon a favoris blanc (*Hylobates concolor leucogenys*): analyse structurale des vocalisations. *Zeitschrift für Tierpsychologie*, 48:225-250.

Deriagina, M. 1982. Note on the manipulatory activity of apes. *Journal of Human Evolution*, 11:171-172.

Dewson, James H., III 1978. Some behavioural effects of removal of superior temporal cortex in the monkey, in: *Recent Advances in Primatology*, Vol. 1, *Behaviour* (D.J. Chivers and J. Herbert, eds.), pp. 763-768, London: Academic Press.

Dixson, Alan F. 1981. *The Natural History of the Gorilla*, New York: Columbia University Press.

Döhl, Jürgen 1970. Lernversuche mit Schimpanzen. *Bil der Wissenschaft*, 7:1107-1115.

Döhl, Jürgen 1972. Über die Möglichkeiten für psychologische Untersuchungen an Menschenaffen in Zoologischen Garten: Eindrücke eines Aussenstehenden. *Der Zoologische Garten N.F., Leipzig*, 42:288-295.

Döhl, Jürgen 1973. Gedächtnisprüfung eines Schimpansen für erlernte komplizierte Handlungsweisen. *Zeitschrift für Tierpsychologie*, 33:204-208.

Döhl, Jürgen 1975. Das Verhalten eines Zwegschimpanzen (*Pan paniscus*) bei einfachsten "Puzzzle-Spiel"-Aufgaben. *Zeitschrift für Tierpsychologie*, 38;:461-471.

Döhl, Jürgen, and Podolczak, Doris 1973. Versuche zur Manipulierfreudigkeit von zwei jungen Orang-Utans (*Pongo pygmaeus*) im Frankfurter Zoo. *Der Zoologische Garten N.F. Leipzig*, 43:81-94.

Donisthorpe, Jill H. 1958. A pilot study of the mountain gorilla (*Gorilla gorilla beringei*) in South West Uganda, February to September, 1957. *South African Journal of Science*, 54:195-217.

Dooley, Gwendolyn B., and Gill, Timothy 1977a. Mathematical capabilities of Lana chimpanzee, in: *Progress in Ape Research* (Geoffrey H. Bourne, ed.), pp. 133-142, New York: Academic Press.

Dooley, Gwendolyn B., and Gill, Timothy V. 1977b. Acquisition and use of mathematical skills by a linguistic chimpanzee, in: *Language Learning by a Chimpanzee: The Lana Project* (Duane M. Rumbaugh, ed.), pp. 247-260, New York: Academic Press.

Drumm, Patrick; Gardner, Beatrix T.; and Gardner, R. Allen 1986. Vocal and gestural responses to announcements and events by cross-fostered chimpanzees. *American Journal of Psychology*, 99:1-29.

Du Chaillu, Paul Belloni 1861. *Explorations and Adventures in Equatorial Africa*, New York: Harper & Brothers.

Du Chaillu, Paul B. 1867. *A Journey to Ashango-Land: and Further Penetration into Equatorial Africa*, New York: D. Appleton & Co.

Dupuy, André R. 1970. Sur la présence du chimpanzé dans les limites du Parc national du Niokolo-Koba (Sénégal). *Bulletin de l'Institut Fondamental d'Afrique Noire,* 32:1090-1099.

Dupuy, André R., and Verschuren, Jacques 1977. Wildlife and parks in Senegal. *Oryx,* 14:36-46.

Eckhardt, Robert B. 1972. Population genetics and human origins. *Scientific American,* 226:94-103.

Eckhardt, Robert B. 1973. *Gigantopithecus* as a hominid ancestor. *Anthropologischer Anzeiger,* 34:1-8.

Eckhardt, Robert B. 1975a. The relative body weights of Bornean and Sumatran orangutans. *American Journal of Physical Anthropology,* 42:349-350.

Eckhardt, Robert B. 1975b. *Gigantopithecus* as a hominid, in: *Paleoanthropology: Morphology and Paleoecology* (R.H. Tuttle, ed.), pp. 103-129, The Hague: Mouton.

Economo, C.v., and Horn, L. 1930. Über Windungsrelief, Masse und Rindenarchitektonik der Supratemporalfläche, ihre individuellen und ihre Seitenunterschiede. *Zeitschrift für die gesamte Neurologie und Psychiatrie,* 130:678-757.

Edwards, Sara D. 1982. Social potential expressed in captive, group-living orang utans, in: *The Orang Utan: Its biology and conservation* (Leobert E.M. De Boer, ed.), pp. 249-255, The Hague: Dr. W. Junk Publishers.

Edwards, Sara D., and Snowdon, Charles T. 1980. Social behavior of captive, group-living orang-utans. *International Journal of Primatology,* 1:39-62.

Eiseley, Loren 1958. *Darwin's Century,* Garden City, New York: Doubleday & Co.

Eisenberg, John F.; Muckenhirn, Nancy A.; and Rudran, Rasanayagam 1972. The relation between ecology and social structure in primates. *Science,* 176:863-874.

Elder, James Harlan 1934. Auditory acuity of the chimpanzee. *Journal of Comparative and Physiological Psychology,* 17:157-183.

Ellefson, John O. 1968. Territorial behavior in the common white-handed gibbon, *Hylobates lar* Linn, in: *Primates: Studies in Adaptation and Variability,* (P. Jay, ed.), pp. 180-199, New York: Holt, Rinehart & Winston.

Ellefson, John O. 1974. A natural history of white-handed gibbons, in: *Gibbon and Siamang,* (D.M. Rumbaugh, ed.), Vol. 3, pp. 1-136, Basel: Karger.

Elliott, R.C. 1976. Observations on a small group of mountain gorillas (*Gorilla gorilla beringei*). *Folia Primatologica,* 25:12-24.

Elliott, R.C. 1977. Cross-modal recognition in three primates. *Neuropsychologia,* 15:183-186.

Ellis, Jim, Jr. 1975. Orangutan tool use at Oklahoma City Zoo. *The Keeper* (AAZK), 1:5-6.

Emlen, John T. 1962. The display of the gorilla. *Proceedings of the American Philosophical Society,* 106:516-519.

Emlen, John T., Jr., and Schaller, George B. 1960a. Distribution and status of the mountain gorilla (*Gorilla gorilla beringei*)-1959. *Zoologica,* 45:41-52.

Emlen, John T., Jr., and Schaller, George B. 1960b. In the home of the mountain gorilla. *Animal Kingdom,* 63:98-108.

Emlen, Stephen T., and Oring, Lewis W. 1977. Ecology, sexual selection, and the evolution of mating systems. *Science,* 197:215-223.

Emmons, Louise H.; Gautier-Hion, Annie; and Dubost, Gerard 1983. Community structure of the frugivorous-folivorous forest mammals of Gabon. *Journal of the Zoological Society, London,* 199:209-222.

Epstein, Robert; Lanza, Robert P.; and Skinner, Burrhus Frederic 1980. Symbolic communication between two pigeons (*Columbia livia domistica*). *Science*, 207:543-545.

Epstein, Robert; Lanza, Robert P.; and Skinner, Burrhus Frederic 1981. "Self-awareness" in the pigeon. *Science*, 212:695-696.

Essock, Susan M. 1978. Colour perception and colour classification of Lana chimpanzee, in: *Recent Advances in Primatology*, Vol. 1, *Behaviour* (D.J. Chivers and J. Herbert, eds.), pp. 901-902, London: Academic Press.

Essock, Susan M.; Gill, Timothy V.; and Rumbaugh, Duane M. 1977. Object- and color-naming skills of Lana chimpanzee, in: *Progress in Ape Research* (Geoffrey H. Bourne, ed.), pp. 143-148, New York: Academic Press.

Essock, Susan M., and Rumbaugh, Duane M. 1978. Development and measurement of cognitive capabilities in captive nonhuman primates, in: *Behavior of Captive Wild Animals*, (Hal Markowitz and Victor J. Stevens, eds.), pp. 161-208 and 300-303, Chicago: Nelson-Hall.

Essock-Vitale, Susan M. 1978. Comparison of ape and monkey modes of problem solution. *Journal of Comparative and Physiological Psychology*, 92:942-957.

Estep, Daniel Q., and Bruce, Katherine E.M. 1981. The concept of rape in non-humans: a critique. *Animal Behaviour*, 29:1272-1273.

Ettlinger, George 1967. Analysis of cross-modal effects and their relationship to language, in: *Brain Mechanisms underlying Speech and Language*, (F.L. Darley, ed.), pp. 53-60, New York: Grune and Stratton.

Ettlinger, George 1973. The transfer of information between sense-modalities: a neuropsychological review, in: *Memory and Transfer of Information*, (H.P. Zippel, ed.), pp. 43-64, New York: Plenum.

Ettlinger, George 1983. A comparative evaluation of the cognitive skills of the chimpanzee and the monkey. *International Journal of Neuroscience*, 22:7-19.

Ettlinger, George 1984. Humans, apes and monkeys: the changing neuropsychological viewpoint. *Neuropsychologia*, 22:685-696.

Ettlinger, George, and Blakemore, Colin B. 1967. Cross-modal matching in the monkey. *Neuropsychologia*, 5:147-154.

Ettlinger, George, and Blakemore, Colin B. 1969. Cross-modal transfer set in the monkey. *Neuropsychologia*, 7:41-47.

Ettlinger, George, and Garcha, H.S. 1980. Cross-modal recognition by the monkey: the effects of cortical removals. *Neuropsychologia*, 18:685-692.

Ettlinger, George, and Jarvis, M.J. 1976. Cross-modal transfer in chimpanzees. *Nature*, 259:44-46.

Falk, Dean 1978. Cerebral asymmetry in Old World monkeys. *Acta Anatomica*, 101:334-339.

Falk, Dean 1980. Language, handedness, and primate brains: did the australopithecines sign? *American Anthropologist*, 82:72-78.

Farrer, Donald N. 1967. Picture memory in the chimpanzee. *Perceptual and Motor Skills*, 25:305-315.

Fenart, R., and Deblock, R. 1973. *Pan paniscus-Pan troglodytes*. Crâniométrie. Etude comparative et ontogénique selon les méthods classiques et vestibulaire. *Annales du Musée Royale de l'Afrique Centrale, Turvuren-Belgique*, I:1-593.

Fenart, R., and Deblock, R. 1974. Sexual differences in adult skulls of *Pan troglodytes*. *Journal of Human Evolution*, 3:123-133.

Ferster, Charles B. 1958a. Control of behavior in chimpanzees and pigeons by time out from positive reinforcement. *Psychological Monographs*, 72:1-38.

Ferster, Charles, B. 1958b. Intermittent reinforcement of a complex response in a chimpanzee. *Journal of the Experimental Analysis of Behavior*, 1:163-165.

Ferster, Charles B. 1964. Arithmetic behavior in chimpanzees. *Scientific American*, 210:98-106.

Ferster, Charles B., and Hammer, C.E., Jr. 1966. Synthesizing the components of arithmetic behavior, in: *Operant Behavior: Areas of Research and Application* (Werner K. Honig, ed.), pp. 634-676, Englewood Cliffs, New Jersey: Prentice-Hall, Inc.

Finch, Glen 1941a. Chimpanzee handedness. *Science*, 94:117-118.

Finch, Glen 1941b. The solution of patterned string problems by chimpanzees. *Journal of Comparative Psychology*, 32:83-90.

Finch, Glen 1942. Delayed matching-from-sample and non-spatial delayed response in chimpanzees. *Journal of Comparative Psychology*, 34:315-319.

Fischer, Gloria J. 1962. The formation of learning sets in young gorillas. *Journal of Comparative and Physiological Psychology*, 55:924-925.

Fischer, Gloria J., and Kitchener, S.L. 1965. Comparative learning in young gorillas and orangutans. *Journal of Genetic Psychology*, 107:337-348.

Fischer, Robert B.; Meunier, Gary F.; and White, Pamela J. 1982. Evidence of laterality in the lowland gorilla. *Perceptual and Motor Skills*, 54:1093-1094.

Fleagle, John G. 1974. Dynamics of a brachiating siamang (*Hylobates* (*Symphalangus*) *syndactylus*). *Nature*, 248:259-260.

Fleagle, John G. 1976. Locomotion and posture of the Malayan siamang and implications for hominid evolution. *Folia Primatologica*, 26:245-269.

Fleagle, John G. 1977. Brachiation and biomechanics: the siamang as example. *Malayan Nature Journal*, 30:45-51.

Fleagle, John G. 1984. Are there any fossil gibbons?, in: *The Lesser Apes: Evolutionary and Behavioural Biology* (H. Preuschoft, D.J. Chivers, W.Y. Brockelman, and H. Creel, eds.), pp. 431-447, Edinburgh: University of Edinburgh Press.

Fleagle, John G., and Simons, Elwyn L. 1978. *Micropithecus clarki*, a small ape from the Miocene of Uganda. *American Journal of Physical Anthropology*, 49:427-440.

Fleagle, John G.; Stern, Jack T., Jr.; Jungers, William L.; Susman, Randall L.; Vangor, Andrea K.; and Wells, James P. 1981. Climbing: a biomechanical link with brachiation and with bipedalism. *Symposia of the Zoological Society of London*, 48:359-375.

Fobes, James L., and King, James E. 1982. Measuring primate learning abilities, in: *Primate Behavior* (James L. Fobes and James E. King, eds.), pp. 289-326, New York: Academic Press, Inc.

Foley, John P., Jr. 1935. Judgement of facial expression of emotion in the chimpanzee. *Journal of Social Psychology*, 6:31-67.

Fooden, Jack 1971. Report on primates collected in western Thailand, January-April, 1967. *Fieldiana, Zoology*, 59:1-62.

Fooden, Jack, and Izor, Robert J. 1983. Growth curves, dental emergence norms, and supplementary morphological observations in known-age captive orangutans. *American Journal of Primatology*, 5:285-310.

Ford, Henry A. 1852-53. On the characteristics of the *Troglodytes gorilla*. *Proceedings of the Academy of Natural Sciences, Philadelphia*,6:30-33.

Forster, Milton C. 1935. Temporal relations of behavior in chimpanzee and man as measured by reaction time. *Journal of Comparative Psychology*, 20:361-383.

Fossey, Dian 1970. Making friends with mountain gorillas. *National Geographic Magazine*, 137:48-67.

Fossey, Dian 1971. More years with mountain gorillas. *National Geographic Magazine*, 140:574-585.

Fossey, Dian 1972. Vocalizations of the mountain gorilla (*Gorilla gorilla beringei*). *Animal Behaviour*, 20:36-53.

Fossey, Dian 1974. Observations on the home range of one group of mountain gorillas (*Gorilla gorilla beringei*). *Animal Behaviour*, 22:568-581.

Fossey, Dian 1979. Development of the mountain gorilla (*Gorilla gorilla beringei*): the first thirty-six months, in: *The Great Apes* (D.A. Hamburg and E.R. McCown, eds.), pp. 138-184, Menlo Park, California: The Benjamin/Cummings Publishing Co.

Fossey, Dian 1981. The imperiled mountain gorilla. *National Geographic Magazine*, 159:500-523.

Fossey, Dian 1982. Reproduction among free-living mountain gorillas. *American Journal of Primatology, Supplement*, 1:97-104.

Fossey, Dian 1983. *Gorillas in the Mist*, New York: Houghton Mifflin Company.

Fossey, Dian 1984. Infanticide in mountain gorillas (*Gorilla gorilla beringei*) with comparative notes on chimpanzees, in: *Infanticide: Comparative and Evolutionary Perspectives* (Glenn Hausfater and Sarah Blaffer Hrdy, eds.), pp. 217-235, New York: Aldine.

Fossey, Dian, and Harcourt, Alexander H. 1977. Feeding ecology of free-ranging mountain gorilla (*Gorilla gorilla beringei*), in : *Primate Ecology: Studies of feeding and ranging behaviour in lemurs, monkeys and apes* (T.H. Clutton-Brock, ed.), pp. 415-447, London: Academic Press.

Fouts, Roger S. 1972. Use of guidance in teaching sign language to a chimpanzee (*Pan troglodytes*). *Journal of Comparative and Physiological Psychology*, 80:515-522.

Fouts, Roger S. 1973. Acquisition and testing of gestural signs in four young chimpanzees. *Science*, 180:978-980.

Fouts, Roger S. 1974. Language: origins, definitions and chimpanzees. *Journal of Human Evolution*, 3:475-482.

Fouts, Roger S. 1975a. Capacities for language in great apes, in: *Socioecology and Psychology of Primates* (Russell H. Tuttle, ed.), pp. 371-390, The Hague: Mouton.

Fouts, Roger S. 1975b. Communication with chimpanzees, in: *Hominisation und Verhalten* (G. Kurth and I. Eibl-Eibesfeldt, eds.), pp. 137-158, Stuttgart: Gustav Fischer Verlag.

Fouts, Roger S. 1977. Ameslan in Pan, in: *Progress in Ape Research* (Geoffrey H. Bourne, ed.), pp. 117-123, New York: Academic Press.

Fouts, Roger S. 1978. Sign language in chimpanzees: implications of the visual mode and the comparative approach, in: *Sign Language and Language Acquisition in Man and Ape: New Dimensions in Comparative Pedolinguistics* (Fred C.C. Peng, ed.), pp. 121-136, Boulder, Colorado: Westview Press, Inc.

Fouts, Roger S. 1983. Chimpanzee language and elephant tails: a theoretical synthesis, in: *Language in Primates: Perspectives and Implications* (Judith de Luce and Hugh T. Wilder, eds.), pp. 63-75, New York: Springer-Verlag.

Fouts, Roger S., and Budd, Richard L. 1979. Artificial and human language acquisition in the chimpanzee, in: *The Great Apes* (David A. Hamburg and Elizabeth R. McCown, eds.), pp. 375-392, Menlo Park, California: The Benjamin/Cummings Publishing Co.

Fouts, Roger S.; Chown, Bill; and Goodin, Larry 1976. Transfer of signed responses in American Sign Language from vocal English stimuli to physical object stimuli by a chimpanzee (*Pan*). *Learning and Motivation*, 7:458-475.

Fouts, Roger S., and Couch, Joseph B. 1976. Cultural evolution of learned language in chimpanzees, in: *Communicative Behavior and Evolution* (Martin E. Hahn and Edward C. Simmel, eds.), pp. 141-161, New York: Academic Press.

Fouts, Roger S.; Hirsch, Alan D.; and Fouts, Deborah H. 1982. Cultural transmission of a human language in a chimpanzee mother-infant relationship, in: *Child Nurturance*, Vol. 3 (Hiram E. Fitzgerald, John A. Mullins, and Patricia Gage, eds.), pp. 159-193, New York: Plenum Publishing Co.

Fouts, Roger S., and Rigby, Randall L. 1977. Man-chimpanzee communication, in: *How Animals Communicate* (Thomas A. Sebeok, ed.), pp. 1034-1054, Bloomington: Indiana University Press.

Fox, Greysolynne J. 1972. Some comparisons between siamang and gibbon behaviour. *Folia Primatologica*, 18:122-139.

Fox, Michael W. 1982. Are most animals "mindless automatons"?: a reply to Gordon G. Gallup, Jr. *American Journal of Primatology*, 3:341-343.

Frayer, David W. 1973. *Gigantopithecus* and its relationship to *Australopithecus. American Journal of Physical Anthropology*, 39:413-426.

Frisch, John E. 1963. Dental variability in a population of gibbons (*Hylobates lar*), in: *Dental Anthropology* (D. Brothwell, ed.), pp. 15-28, Oxford: Pergamon.

Frisch, John E. 1973. The hylobatid dentition, in: *Gibbon and Siamang*, Vol. 2 (D.M. Rumbaugh, ed.), pp. 55-95, Basel: S. Karger.

Furness, William H. 1916. Observations on the mentality of chimpanzees and orang-utans. *Proceedings of the American Philosophical Society*, 55:281-290.

Galaburda, Albert M.; LeMay, Marjorie; Kemper, Thomas L.; and Geschwind, Norman 1978. Right-left asymmetries in the brain. *Science*, 199:852-856.

Galdikas, Biruté 1978. Orangutans and hominid evolution, in: *Spectrum: Essays presented to Sultan Takdir Alisjahbana on his seventieth birthday* (S. Udin, ed.), pp. 287-309, Jakarta: Dian Rakyat.

Galdikas, Biruté M.F. 1979. Orangutan adaptation at Tanjung Puting Reserve: mating and ecology, in: *The Great Apes* (D.A. Hamburg and E.R. McCown, eds.), pp. 194-233, Menlo Park, California: The Benjamin/Cummings Publishing Co.

Galdikas, Biruté M.F. 1980. Indonesia's orangutans: living with the great orange apes. *National Geographic Magazine*, 157:830-853.

Galdikas, Biruté M.F. 1981a. Modern adaptations in orangutans. *Nature*, 291:266.

Galdikas, Biruté M.F. 1981b. Orangutan reproduction in the wild, in: *Reproductive Biology of the Great Apes* (Charles E. Graham, ed.) pp. 281-300, New York: Academic Press.

Galdikas, Biruté M.F. 1982a. Orang-utan tool-use at Tanjung Puting Reserve, Central Indonesian Borneo (Kalimantan Tengah). *Journal of Human Evolution*, 10:19-33.

Galdikas, Biruté M.F. 1982b. Wild orangutan birth at Tanjung Puting Reserve. *Primates*, 23:500-510.

Galdikas, Biruté M.F. 1982c. Orang utans as seed dispersers at Tanjung Puting, Central Kalimantan: implications for conservation, in: *The Orang utan: Its Biology and Conservation* (L.E.M. de Boer, ed.), pp. 285-298, The Hague: Dr. W. Junk Publishers.

Galdikas, Biruté M.F. 1983. The orangutan long call and snag crashing at Tanjung Puting Reserve. *Primates*, 24:371-384.

Galdikas, Biruté M.F. 1984. Adult female sociality among wild orangutans at Tanjung Puting Reserve, in: *Female Primates: Studies by Women Primatologists* (Meredith F. Small, ed.), pp. 217-235, New York: Alan R. Liss, Inc.

Galdikas, Biruté M.F. 1985a. Subadult male orangutan sociality and reproductive behavior at Tanjung Puting. *American Journal of Primatology*, 8:87-99.

Galdikas, Biruté M.F. 1985b. Orangutan sociality at Tanjung Puting. *American Journal of Primatology*, 9:101-119.

Galdikas, Biruté M.F. 1985c. Adult male sociality and reproductive tactics among orangutans at Tanjung Puting. *Folia Primatologica*, 45:9-24.

Galdikas, Biruté M.F., and Teleki, Geza 1981. Variations in subsistence activities of female and male pongids: new perspectives on the origins of hominid labor division. *Current Anthropology*, 22:241-255 and 316-320.

Galdikas-Brindamour, Biruté 1975. Orangutans, Indonesia's "people of the forest." *National Geographic Magazine*, 148:444-473.

Galdikas-Brindamour, Biruté 1978. The intelligence of orangutans. *L.S.B. Leakey Foundation News*, 11:56-57.

Gallup, Gordon G., Jr. 1966. Mirror-image reinforcement in monkeys. *Psychonomic Science*, 5:39-40.

Gallup, Gordon G., Jr. 1968. Mirror-image stimulation. *Psychological Bulletin*, 70:782-793.

Gallup, Gordon G., Jr. 1970. Chimpanzees: self-recognition. *Science*, 167:86-87.

Gallup, Gordon G., Jr. 1971. Selbsterkennen bei Schimpansen. *Umshau in Wissenschaft und Technik, Frankfurt/Main*, 6:209.

Gallup, Gordon G., Jr. 1975. Towards an operational definition of self-awareness, in: *Socioecology and Psychology of Primates* (Russell H. Tuttle, ed.), pp. 309-341, The Hague: Mouton.

Gallup, Gordon G., Jr. 1979. Self-awareness in primates. *American Scientist*, 67:417-521.

Gallup, Gordon G., Jr. 1982. Self-awareness and the emergence of mind in primates. *American Journal of Primatology*, 2:237-248.

Gallup, Gordon G., Jr.; Boren, James L.; Gagliardi, Gregg J.; and Wallnau, Larry B. 1977. A mirror for the mind of man, or will the chimpanzee create an identity crisis for *Homo sapiens? Journal of Human Evolution*, 6:303-313.

Gallup, Gordon G., Jr., and McClure, Michael K. 1971. Preference for mirror-image stimulation in differentially reared rhesus monkeys. *Journal of Comparative and Physiological Psychology*, 75:403-407.

Gallup, Gordon G., Jr.; McClure, Michael K.; and Hill, Suzanne D. 1971. Capacity for self-recognition in differentially reared chimpanzees. *The Psychological Record*, 21:69-74.

Gandini, Gustavo, and Baldwin, Pamela J. 1978. An encounter between chimpanzees and a leopard in Senegal. *Carnivore*, 1:107-109.

Garcha, H.S., and Ettlinger, George 1979. Object sorting by chimpanzees and monkeys. *Cortex*, 15:213-224.

Gardner, Beatrix T. 1981. Project Nim: who taught whom? *Contemporary Psychology*, 26:425-426.

Gardner, Beatrice T., Gardner, R. Allen 1971. Two-way communication with an infant chimpanzee, in: *Behavior of Nonhuman Primates. Modern Research Trends*, Vol. 4 (Allan M. Schrier and Fred Stollnitz, eds.), pp. 117-184, New York: Academic Press.

Gardner, Beatrice T., and Gardner, R. Allen 1974. Comparing the early utterances of child and chimpanzee, in: *Minnesota Symposia on Child Psychology* (A. Pick, ed.), Vol. 8, pp. 3-24, Minneapolis: University of Minnesota Press.

Gardner, Beatrice T., and Gardner, R. Allen 1975. Evidence for sentence constituents in the early utterances of child and chimpanzee. *Journal of Experimental Psychology: General*, 104:244-267.

Gardner, Beatrice T., and Gardner, R. Allen 1980. Two comparative psychologists look at language acquisition, in: *Children's Language*, Vol. 2 (Keith E. Nelson, ed.), pp. 331-369, New York: Gardner Press, Inc.

Gardner, L. Pearl, and Nissen, Henry W. 1948. Simple discrimination behavior of young chimpanzees: comparisons with human aments and domestic animals. *Journal of Genetic Psychology*, 72:145-164.

Gardner, R. Allen, and Gardner, Beatrice T. 1969. Teaching sign language to a chimpanzee. *Science*, 165:664-672.

Gardner, R. Allen, and Gardner, Beatrice T. 1972. Communication with a young chimpanzee: Washoe's vocabulary. *Colloques internationaux du Centre National de la Recherche Scientifique*, 198:241-264.

Gardner, R. Allen, and Gardner, Beatrice T. 1975. Early signs of language in child and chimpanzee. *Science*, 187:752-753.

Gardner, R. Allen, and Gardner, Beatrice T. 1978. Comparative psychology and language acquisition. *Annals of the New York Academy of Sciences*, 309:37-76.

Gardner, R. Allen, and Gardner, Beatrix T. 1984. A vocabulary test for chimpanzees (*Pan troglodytes*). Journal of Comparative Psychology, 98:381-404.

Garner, R.L. 1896. *Gorillas and Chimpanzees*, London: Osgood, McIlvaine & Co.

Gartlan, J. Stephen, and Struhsaker, Thomas T. 1972. Polyspecific associations and niche separation of rain-forest anthropoids in Cameroon, West Africa. *Journal of Zoology, London*, 168:221-266.

Gaulin, Steven J.C., and Konner, Melvin 1977. On the natural diet of primates, including humans, in: *Nutrition and the Brain* (R.J. Wurtman and J.J. Wurtman, eds.), Vol, 1, pp. 1-86, New York: Raven Press.

Gaulin, Steven J.C., and Sailer, Lee Douglas 1984. Sexual dimorphism in weight among the primates: the relative impact of allometry and sexual selection. *International Journal of Primatology*, 5:515-535.

Gavan, James A. 1971. Longitudinal, postnatal growth in chimpanzee, in: *The Chimpanzee* (G.H. Bourne, ed.), Vol. 4, pp. 46-102, Basel: Karger.

Geerling, C., and Bokdam, J. 1973. Fauna of the Comoe National Park, Ivory Coast. *Biological Conservation*, 5:251-257.

Geissmann, Thomas 1983. Female capped gibbon (*Hylobates pileatus* Gray 1891) sings male song. *Journal of Human Evolution*, 12:667-671.

Gellermann, Louis W. 1933a. Form discrimination in chimpanzees and two-year-old children: I. form (triangularity) *per se. Journal of Genetic Psychology*, 42:3-27.

Gellermann, Louis W. 1933b. Form discrimination in chimpanzees and two-year-old children: II. form versus background. *Journal of Genetic Psychology*, 42:28-50.

Geschwind, Norman 1965. Disconnection syndromes in animals and man. *Brain*, 88:237-294, 585-644.

Geschwind, Norman 1971. Some differences between human and other primate brains, in: *Cognitive Processes of Nonhuman Primates* (Leonard E. Jarrard, ed.), pp. 149-154, New York: Academic Press.

Geschwind, Norman 1972. Language and the brain. *Scientific American*, 226:76-83.

Geschwind, Norman, and Levitsky, Walter 1968. Human brain: left-right asymmetries in temporal speech region. *Science*, 161:186-187.

Ghiglieri, Michael Patrick 1984a. *The Chimpanzees of Kibale Forest: A Field Study of Ecology and Social Structure*, New York: Columbia University Press.

Ghiglieri, Michael P. 1984b. Feeding ecology and sociality of chimpanzees in Kibale Forest, Uganda, in: *Adaptations for Foraging in Nonhuman Primates* (Peter S. Rodman and John G.H. Cant, eds.), pp. 161-194, New York: Columbia University Press.

Gibbons, Michael F. 1970. A trip through Korup. *Discovery*, 6:7-14.

Gijzen, Agatha 1974. Studbook of *Pan paniscus* Schwarz, 1929. *Acta Zoologica Pathologica Antverpiensia*, 61:119-164.

Gill, Timothy V. 1977. Talking to Lana: the question of conversation, in: *Progress in Ape Research* (Geoffrey H. Bourne, ed.), pp. 125-132, New York: Academic Press.

Gill, Timothy V. 1978. Conversing with Lana, in: *Recent Advances in Primatology*, Vol. 1, *Behaviour* (D.J. Chivers and J. Herbert, eds.), pp. 861-866, London: Academic Press.

Gill, Timothy V., and Rumbaugh, Duane M. 1974a. Learning processes of bright and dull apes. *American Journal of Mental Deficiency*, 78:683-687.

Gill, Timothy V., and Rumbaugh, Duane M. 1974b. Mastery of naming skills by a chimpanzee. *Journal of Human Evolution*, 3:483-492.

Gillan, Douglas J. 1981. Reasoning in the chimpanzee: II. transitive inference. *Journal of Experimental Psychology: Animal Behavior Processes*, 7:150-164.

Gillan, Douglas J.; Premack, David; and Woodruff, Guy 1981. Reasoning in the chimpanzee. I. analogical reasoning. *Journal of Experimental Psychology: Animal Behavior Processes*, 7:1-16.

Gittins, S. Paul 1978a. The species range of the gibbon *Hylobates agilis*, in: *Recent Advances in Primatology*, Vol. 3, *Evolution* (D.J. Chivers and K.A. Joysey, eds.), pp. 319-321, London: Academic Press.

Gittins, S. Paul 1978b. Hark! The beautiful song of the gibbon. *New Scientist*, 80:832-834.

Gittins, S. Paul 1980. Territorial behavior in the agile gibbon. *International Journal of Primatology*, 1:381-399.

Gittins, S. Paul 1983. Use of the forest canopy by the agile gibbon. *Folia Primatologica*, 40:134-144.

Gittins, S. Paul 1984a. The distribution and status of the hoolock gibbon in Bangladesh, in: *The Lesser Apes: Evolutionary and Behavioral Biology* (H. Preschoft, D.J. Chivers, W.Y. Brockelman, and N. Creel, eds.), pp. 13-15, Edinburgh: Edinburgh University Press.

Gittins, S. Paul 1984b. The vocal repertoire and song of the agile gibbon, in: *The Lesser Apes: Evolutionary and Behavioural Biology* (H. Preuschoft, D.J. Chivers, W.Y. Brockelman, and N. Creel, eds.), pp. 354-375, Edinburgh: Edinburgh University Press.

Gittins, S. Paul 1984c. Territorial advertisement and defence in gibbons, in: *The Lesser Apes: Evolutionary and Behavioural Biology* (H. Preuschoft, D.J. Chivers, W.Y. Brockelman, and N. Creel, eds.), pp. 420-424, Edinburgh: Edinburgh University Press.

Gittins, S. Paul, and Akonda, A.W. 1982. What survives in Bangladesh? *Oryx*, 16:275-281.

Gittins, S. Paul, and Raemaekers, Jeremy J. 1980. Siamang, lar and agile gibbons, in: *Malayan Forest Primates* (D.J. Chivers, ed.), pp. 63-105, New York: Plenum Press.

Gittins, S. Paul, and Tilson, Ronald L. 1984. Notes on the ecology and behaviour of the hoolock gibbon, in: *The Lesser Apes: Evolutionary and Behavioural Biology* (H. Preuschoft, D.J. Chivers, W.Y. Brockelman, and N. Creel, eds.), pp. 258-266, Edinburgh: Edinburgh University Press.

von Glasersfeld, Ernst 1976. The development of language as purposive behavior. *Annals of the New York Academy of Sciences*, 280:212-226.

von Glasersfeld, Ernst 1978. Les chimpanzees et le language. *La Recherche*, 9:725-732.

von Glasersfeld, Ernst; Warner, Harold; Pisani, Pier Paulo; Rumbaugh, Duane M.; Gill, Timothy V.; and Bell, Charles 1973. A computer mediates communication with a chimpanzee. *Computers and Automation*, 22:9-11.

Gonzales, Richard C.; Gentry, George V.; and Bitterman, Morton E. 1954. Relational discrimination of intermediate size in the chimpanzee. *Journal of Comparative and Physiological Psychology*, 47:385-388.

Goodall, Alan G. 1977. Feeding and ranging behavior of a mountain gorilla group (*Gorilla gorilla beringei*) in the Tshibinda-Kahuzi region (Zaire), in: *Primate Ecology: Studies of feeding and ranging behaviour in lemurs, monkeys and apes* (T.H. Clutton-Brock, ed.), pp. 449-479, London: Academic Press.

Goodall, Jane 1962. Nest building behavior in the free ranging chimpanzee. *Annals of the New York Academy of Sciences*, 102:445-467.

Goodall, Jane 1963a. Feeding behaviour of wild chimpanzees—a preliminary report. *Symposia of the Zoological Society, London*, 10:39-47.

Goodall, Jane 1963b. My life among wild chimpanzees. *National Geographic Magazine*, 124:272-308.

Goodall, Jane 1965. Chimpanzees of the Gombe Stream Reserve, in: *Primate Behavior* (I. DeVore, ed.), pp. 425-473, New York: Holt, Rinehart & Winston.

Goodall, Jane 1977. Infant killing and cannibalism in free-living chimpanzees. *Folia Primatologica*, 28:259-282.

Goodall, Jane 1979. Life and death at Gombe. *National Geographic Magazine*, 155:592-621.

Goodall, Jane 1981. The chimpanzees of Gombe. *The L.S.B. Leakey Foundation News*, 21:8-9.

Goodall, Jane 1983. Population dynamics during a 15 year period in one community of free-living chimpanzees in the Gombe National Park, Tanzania. *Zeitschrift für Tierpsychologie*, 61:1-60.

Goodall, Jane; Bandora, Adriano; Bergmann, Emilie; Busse, Curt; Matama, Hilali; Mpongo, Esilom; Pierce, Ann; and Riss, David 1979. Intercommunity interactions in the chimpanzee population of the Gombe National Park, in: *The Great Apes* (David A. Hamburg and Elizabeth R. McCown, eds.), pp. 13-53, Menlo Park, California: The Benjamin/Cummings Publishing Company.

Goodman, Morris 1962. Evolution of the immunologic species specificity of human serum proteins. *Human Biology*, 34:104-150.

Goodman, Morris 1963. Man's place in the phylogeny of the primates as reflected in serum proteins, in: *Classification and Human Evolution* (S.L. Washburn, ed.), pp. 204-234.

Goodman, Morris; Baba, Marietta L.; and Darga, Linda L. 1983. The bearing of molecular data on the cladogenesis and times of divergence of hominoid lineages, in: *New Interpretations of Ape and Human Ancestry* (Russell L. Ciochon and Robert S. Corruccini, eds.), pp. 67-68, New York: Plenum Press.

Goodman, Morris, and Cronin, John E. 1982. Molecular anthropology: its development and current directions, in: *A History of American Physical Anthropology, 1930-1980* (Frank Spencer, ed.), pp. 105-146, New York: Academic Press.

Goodwin, A.W., and Darian-Smith, Ian (eds.) 1985. *Hand Function and the Neocortex*, Berlin: Springer-Verlag.

Gossette, Robert L. 1973. Comparative analysis of serial discrimination reversal (SDR) performances of the gibbon, *Hylobates lar*, in: *Gibbon and Siamang* (D.M. Rumbaugh, ed.), Vol. 2, pp. 208-220, Basel: Karger.

Goustard, Michel 1965. Introduction à l'étude des émissions sonores des Hylobatidés. *Annales des Science Naturelle, Zoologie et Biologie Animale*, 7:359-396.

Goustard, Michel 1976. The vocalizations of *Hylobates*, in: *Gibbon and Siamang* (D.M. Rumbaugh, ed.), Vol. 4, pp. 135-166, Basel: Karger.

Goustard, Michel 1979a. Les interactions acoustiques au cours des grandes émissions sonores des gibbons (*Hylobates concolor leucogenys*). *Journal de Psychologie*, 2:133-156.

Goustard, Michel 1979b. Stéréotypie et plasticité des émissions sonores des Gibbons à favoris (*Hylobates concolor leucogenys*). *Comptes Rendus Academie des Science Paris, Series D*, 288:1615-1618.

Goustard, Michel 1979c. Les interactons acoustiques au cours des grandes émissions sonores des Gibbons (*Hylobates concolor leucogenys*). *Comptes Rendus Academie des Sciences Paris, Series D*, 288:1671-1673.

Goustard, Michel 1982. Les vocalisations des singes anthropomorphes. *Journal de Psychologie*, 1:141-166.

Goustard, Michel 1984a. Les émissions sonores territoriales des mâles de Gibbons a mains blanches (*Hylobates lar*) observés dans leur habitat naturel, en Thaïland. *Comptes Rendus Academie des Sciences Paris*, series III, 298:65-67.

Goustard, Michel 1984b. Patterns and functions of loud calls in the concolor gibbon, in: *The Lesser Apes: Evolutionary and Behavioural Biology* (H. Preuschoft, D.J. Chivers, W.Y. Brockelman, and N. Crel, eds.), pp. 404-415, Edinburgh: Edinburgh University Press.

Goustard, Michel, and Demars, Christian 1971. Sur les chants à roulades, chez le gibbon *Hylobates concolor*. *Bulletin Biologique de la France et de la Belgique*, 105:243-251.

Goustard, Michel, and Demars, Christian 1973. Structure et forme des segments du grant chant du gibbon (*Hylobates concolor*). *Bulletin Biologique de la France et de la Belgique*, 107:171-187.

Grand, Theodore I. 1972. A mechanical interpretation of terminal branch feeding. *Journal of Mammalogy*, 53:198-201.

Green, Kenneth M. 1978. Primates of Bangladesh: a preliminary survey of population and habitat. *Biological Conservation*, 13:141-160.

Greene, John C. 1959. *The Death of Adam*, Ames Iowa: The Iowa State University Press.

Greenfield, Leonard Owen 1979. On the adaptive pattern of "*Ramapithecus*." *American Journal of Physical Anthropology*, 50:527-548.

Gregory, William King, and Raven, Henry C. 1937. *In Quest of Gorillas*, New Bedford, Massachusetts: Darwin Press.

Grether, Walter F. 1940a. Chimpanzee color vision. I. hue discrimination at three spectral points. *Journal of Comparative Psychology*, 29:167-177.

Grether, Walter F. 1940b. Chimpanzee color vision. II. color mixture proportions. *Journal of Comparative Psychology*, 29:179-186.

Grether, Walter F. 1940c. Chimpanzee color vision. III. spectral limits. *Journal of Comparative Psychology*, 29:187-192.

Grether, Walter F. 1940d. A comparison of human and chimpanzee spectral hue discrimination curves. *Journal of Experimental Psychology*, 26:394-403.

Grether, Walter F. 1941. Spectral saturation curves for chimpanzee and man. *Journal of Experimental Psychology*, 28:419-427.

Grether, Walter F. 1942. The magnitude of simultaneous color contrast and simultaneous brightness contrast for chimpanzee and man. *Journal of Experimental Psychology*, 30:69-83.

Griffin, Donard R. 1976. *The Question of Animal Awareness: Evolutionary Continuity of Mental Experience*, New York: The Rockefeller University Press.

Griffin, Donald R. 1981. *The Question of Animal Awareness: Evolutionary Continuity of Mental Experience*, 2nd edn., New York: The Rockefeller University Press.

Griffin, Donald R. 1984a. Animal thinking. *American Scientist*, 72:456-464.

Griffin, Donald R. 1984b. *Animal Thinking*, Cambridge, Massachusetts: Harvard University Press.

Grilly, David M. 1975. Sex differences in delayed matching-to-sample performance of chimpanzees. *Psychological Reports*, 37:203-207.

Groves, Colin P. 1967. Ecology and taxonomy of the gorilla. *Nature*, 213:890-893.

Groves, Colin P. 1968a. The classification of the gibbons (Primates, Pongidae). *Zeitschrift für Saugetierkunde*, 33:239-246.

Groves, Colin P. 1968b. A new subspecies of gibbon from northern Thailand, *Hylobates lar carpenteri* new subspecies. *Proceedings of the Biological Society of Washington*, 81:625-627.

Groves, Colin P. 1969. *Hylobates lar* (Linnaeus, 1771), *H. entelloides* (I. Geoffroy St. Hilaire, 1842) and *H. hoolock* (Harlan, 1834) (Mammalia, Pongidae). Proposal to place these names on the official list of specific names in zoology. *Nomenclature*, 25:162-164.

Groves, Colin P. 1970a. Taxonomic and individual variation in gibbons. *Symposia of the Zoological Society, London*, 26:127-134.

Groves, Colin P. 1970b. *Gorillas*, New York: Arco Publishing Co., Inc.

Groves, Colin P. 1970c. Population systematics of the gorilla. *Journal of Zoology*,

London, 161:287-300.

Groves, Colin P. 1971a. Geographic and individual variations in Bornean gibbons, with remarks on the systematics of the subgenus *Hylobates*. *Folia Primatologica*, 14:139-153.

Groves, Colin P. 1971b. *Pongo pygmaeus*. *Mammalian Species*, 4:1-6.

Groves, Colin P. 1971c. Distribution and place of origin of the gorilla. *Man*, 6:44-51.

Groves, Colin P. 1972. Systematics and phylogeny of gibbons, in: *Gibbon and Siamang* (D.M. Rumbaugh, ed.), Vol. 1, pp. 1-89, Basel: Karger.

Groves, Colin P. 1984. A new look at the taxonomy of the gibbons, in: *The Lesser Apes: Evolutionary and Behavioural Biology* (Holger Preuschoft, David J. Chivers, Warren Y. Brockelman, and Normal Creel, eds.), pp. 542-561, Edinburgh: Edinburgh University Press.

Groves, Colin P., and Humphrey, Nicholas K. 1973. Asymmetry in gorilla skulls: evidence of lateralized brain function? *Nature*, 244:53-54.

Groves, Colin P., and Stott, Kenhelm W., Jr. 1979. Systematic relationships of gorillas from Kahuzi, Tshiaberimu and Kayonza. *Folia Primatologica*, 32:161-179.

Grzimek, Bernhard 1957. Masse und Gewichte von Flachland-Gorillas. *Zeitschrift für Saugetierkunde*, 21:192-194.

Gulik, R.H. van 1967. *The Gibbon in China*, Leiden, Holland: E.J. Brill.

Gunderson, Virginia M. 1983. Development of cross-modal recognition in infant pigtail monkeys (*Macaca nemestrina*). *Developmental Psychology*, 19:398-404.

Gur, Ruben C.; Packer, Ira K.; Hungerbuhler, Jean Pierre; Reivich, Martin; Obrist, Walter D.; Amarnek, Wayne S.; and Sackeim, Harold A. 1980. Differences in the distribution of gray and white matter in human cerebral hemispheres. *Science*, 207:1226-1228.

Gyldenstolpe, N. 1928. Zoological results of the Swedish expedition to Central Africa 1921: Vertebrata, 5. Mammals from the Birunga Volcanoes north of Lake Kivu. *Arkiv for Zoologi, Uppsala*, 20A:1-76.

Haeckel, Ernst 1866. *Generelle Morphologie*, Berlin: Reimer.

Haeckel, Ernst 1868. *Natürliche Schöpfungsgeschichte*, Berlin: Reimer.

Haeckel, Ernst 1874. *Anthropogenie oder Entwickelungsgeschichte des Menschen*, Leipzig: Engelmann.

Haimoff, Elliott H. 1981. Video analysis of siamang (*Hylobates syndactylus*) songs. *Behaviour*, 76:128-151.

Haimoff, Elliott H. 1983. Occurrence of anti-resonance in the song of the siamang (*Hylobates syndactylus*). *American Journal of Primatology*, 5:249-256.

Haimoff, Elliott H. 1984. Acoustic and organizational features of gibbon songs, in: *The Lesser Apes: Evolutionary and Behavioural Biology* (H. Preuschoft, D.J. Chivers, W.Y. Brockelman, and H. Creel, eds.), pp. 333-353, Edinburgh: Edinburgh University Press.

Haimoff, Elliott H. 1985. The organization of song in Müller's gibbon (*Hylobates muelleri*). *International Journal of Primatology*, 6:173-192.

Haimoff, Elliott H., and Gittins, S. Paul 1985. Individuality in the songs of wild agile gibbons (*Hylobates agilis*) of Peninsular Malaysia. *American Journal of Primatology*, 8:239-247.

Haimoff, Elliott H.; Gittins, S. Paul; Whitten, Anthony J.; and Chivers, David J. 1984. A phylogeny and classification of gibbons based on morphology and ethology, in: *The Lesser Apes: Evolutionary and Behavioural Biology* (H. Preuschoft, D.J. Chivers, W.Y. Brockelman, and N. Creel, eds.), pp. 614-632, Edinburgh: Edinburgh University Press.

Hall, Anthony D.; Braggio, John T.; Buchanan, James P.; and Nadler, Ronald D. 1982. Partitioning the influence of level from rate factors on the performance of children and apes on a cognitive task. *Journal of Human Evolution*, 11:335-348.

Hall, K. Ronald L. 1963. Tool-using performances as indicators of behavioral adaptability. *Current Anthropology*, 4:479-494.

Halperin, Stewart D. 1979. Temporary association patterns in free ranging chimpanzees: an assessment of individual grouping preferences, in: *The Great Apes* (David A. Hamburg and Elizabeth R. McCown, eds.), pp. 491-499, Menlo Park, California: The Benjamin/Cummings Publishing Company.

Hamburg, David A. 1971. Aggressive behavior of chimpanzees and baboons in natural habitats. *Journal of Psychiatric Research*, 8:385-398.

Hamilton, William J., and Arrowood, Patricia C. 1978. Copulatory vocalizations of chacma baboons (*Papio ursinus*), gibbons (*Hylobates hoolock*), and humans. *Science*, 200:1405-1409.

Hamilton, William Roger; Whybrow, P.J.; and Andrews, Peter 1978. Fauna of fossil mammals from the Miocene of Saudi Arabia. *Nature*, 274:248-249.

Haraway, Maury M.; Maples, Ernest G.; and Tolson, Steve 1981. Taped vocalization as a reinforcer of vocal behavior in a siamang gibbon (*Symphalangus syndactylus*). *Psychological Reports*, 49:995-999.

Harcourt, Alexander H. 1977. Virunga gorillas—the case against translocations. *Oryx*, 13:469-472.

Harcourt, Alexander H. 1978a. Activity periods and patterns of social interaction: a neglected problem. *Behaviour*, 66:121-135.

Harcourt, Alexander H. 1978b. Strategies of emigration and transfer by primates, with particular reference to gorillas. *Zeitschrift für Tierpsychologie*, 48:401-420.

Harcourt, Alexander H. 1979a. The social relations and group structure of wild mountain gorilla, in: *The Great Apes* (D.A. Hamburg and E.R. McCown, eds.), pp. 186-192, Menlo Park, California: The Benjamin/Cummings Publishing Co.

Harcourt, Alexander H. 1979b. Social relationships among adult female mountain gorillas. *Animal Behaviour*, 27:251-264.

Harcourt, Alexander H. 1979c. Social relationships between adult male and female mountain gorillas in the wild. *Animal Behaviour*, 27:325-342.

Harcourt, Alexander H. 1979d. Contrasts between male relationships in wild gorilla groups. *Behavioral Ecology and Sociobiology*, 5:39-49.

Harcourt, Alexander H. 1980-81. Can Uganda's gorillas survive?—a survey of the Bwindi Forest Reserve. *Biological Conservation*, 19:269-282.

Harcourt, Alexander H. 1981. Intermale competition and the reproductive behavior of the great apes, in: *Reproductive Biology of the Great Apes* (Charles E. Graham, ed.), pp. 301-318, New York: Academic Press.

Harcourt, Alexander H., and Curry-Lindahl, Kai 1978. The FPS mountain gorilla project—a report from Rwanda. *Oryx*, 14:316-324.

Harcourt, Alexander H., and Fossey, Dian 1981. The Virunga gorillas: decline of an 'island' population. *African Journal of Ecology*, 19:83-97.

Harcourt, Alexander H.; Fossey, Dian; Sabater Pi, Jorge 1981. Demography of *Gorilla gorilla. Journal of Zoology, London*, 195:215-233.

Harcourt, Alexander H.; Fossey, Dian; Stewart, Kelly J.; and Watts, David P. 1980. Reproduction of wild gorillas and some comparisons with chimpanzees. *Journal of Reproduction and Fertility, Supplement*, 28:59-70.

Harcourt, Alexander H., and Groom, F.G. 1972. Gorilla census. *Oryx*, 11:355-363.

Harcourt, Alexander H., and Stewart, Kelly J. 1977. Apes, sex, and societies. *New Scientist*, 76:160-162.

Harcourt, Alexander H., and Stewart, Kelly J. 1978a. Coprophagy by wild mountain gorillas. *East African Wildlife Journal*, 16:223-225.

Harcourt, Alexander H., and Stewart, Kelly J. 1978b. Sexual behaviour of wild mountain gorillas, in: *Recent Advances in Primatology*, Vol. 1, *Behaviour* (David J. Chivers and J. Herbert, eds.), pp. 611-612, London: Academic Press.

Harcourt, Alexander H., and Stewart, Kelly J. 1981. Gorilla male relationships: can differences during immaturity lead to contrasting reproductive tactics in adulthood? *Animal Behaviour*, 29:206-210.

Harcourt, Alexander H., and Stewart, Kelly J. 1983. Interactions, relationships and social structure: the great apes, in: *Primate Social Relationships: an integrated approach* (Robert A. Hinde, ed.), pp. 307-314, Sunderland, Massachusetts: Sinauer Associates, Inc.

Harcourt, Alexander H., and Stewart, Kelly J. 1984. Gorillas' time feeding: aspects of methodology, body size, competition and diet. *African Journal of Ecology*, 22:207-215.

Harcourt, Alexander H.; Stewart, Kelly J.; and Fossey, Dian 1976. Male emigration and female transfer in wild mountain gorilla. *Nature*, 263:226-227.

Harcourt, Alexander H.; Stewart, Kelly J.; and Fossey, Dian 1981. Gorilla reproduction in the wild, in: *Reproductive Biology of the Great Apes* (Charles E. Graham, ed.), pp. 265-279, New York: Academic Press.

Harlow, Harry F. 1951. Primate learning, in: *Comparative Psychology* (Calvin P. Stone, ed.), pp. 183-238, New York: Prentice-Hall.

Harrisson, Barbara 1962. *Orang-utan*, London: Collins.

Harrisson, Barbara 1969. The nesting behaviour of semi-wild juvenile orang-utans. *The Sarawak Museum Journal*, 17:336-384.

Harrison, Ross, and Nissen, Henry W. 1941a. Spatial separation in the delayed response performance of chimpanzees. *Journal of Comparative Psychology*, 31:427-435.

Harrison, Ross, and Nissen, Henry W. 1941b. The response of chimpanzees to relative and absolute positions in delayed response problems. *Journal of Comparative Psychology*, 31:447-455.

Hasegawa, Toshikazu, and Hiraiwa-Hasegawa, Mariko 1983. Opportunistic and restrictive matings among wild chimpanzees in the Mahale Mountains, Tanzania. *Journal of Ethology*, 1:75-85.

Hasegawa, Toshikazu; Hiraiwa, Mariko; Nishida, Toshisada; and Takasaki, Hiroyuki 1983. New evidence on scavenging behavior in wild chimpanzees. *Current Anthropology*, 24:231-232.

Haslerud, George M. 1938. The effect of movement of stimulus objects upon avoidance reactions in chimpanzees. *Journal of Comparative Psychology*, 25:507-528.

Hay, Richard L., and Leakey, Mary D. 1982. The fossil footprints of Laetoli. *Scientific American*, 246:50-57.

Hayes, Cathy 1951. *The Ape in Our House*, New York: Harper & Brothers.

Hayes, Keith J., and Hayes, Catherine 1951. The intellectual development of a home-raised chimpanzee. *Proceedings of the American Philosophical Society*, 95:105-109.

Hayes, Keith J., and Hayes, Catherine 1952. Imitation in a home-raised chimpanzee. *Journal of Comparative and Physiological Psychology*, 45:450-459.

Hayes, Keith J., and Hayes, Catherine 1953. Picture perception in a home-raised chimpanzee. *Journal of Comparative and Physiological Psychology*, 46:470-474.

Hayes, Keith J., and Hayes, Catherine 1954. The cultural capacity of chimpanzee. *Human Biology*, 26:288-303.

Hayes, Keith J., and Nissen, Catherine H. 1971. Higher mental functions of a home-raised chimpanzee, in: *Behavior of Nonhuman Primates: Modern Research Trends*, Vol. 4 (Allan M. Schrier and Fred Stollnitz, eds.), pp. 59-115, New York: Academic Press.

Hayes, Keith J.; Thompson, Robert; and Hayes, Catherine 1953a. Discrimination learning sets in chimpanzees. *Journal of Comparative and Physiological Psychology*, 46:99-104.

Hayes, Keith J.; Thompson, Robert; and Hayes, Catherine 1953b. Concurrent discrimination learning in chimpanzees. *Journal of Comparative and Physiological Psychology*, 46:105-107.

Hayes, Keith J., and Thompson, Robert 1953. Nonspatial delayed response to trial-unique stimuli in sophisticated chimpanzees. *Journal of Comparative and Physiological Psychology*, 46:498-500.

Heffner, R., and Masterson, Bruce 1975. Variation in form of the pyramidal tract and its relationship to digital dexterity. *Brain, Behavior and Evolution*, 12:161-200.

Heller, Wendy, and Levy, Jerre 1981. Perception and expression of emotion in right-handers and left-handers. *Neuropsychologia*, 19:263-272.

Hess, Jorg P. 1973. Some observations on the sexual behavior of captive lowland gorillas, *Gorilla gorilla*, in: *Comparative Ecology and Behaviour of Primates* (Richard P. Michael and John H. Crook, eds.), pp. 507-581, London: Academic Press.

Hewett, T.D., and Ettlinger, George 1978. Cross-modal performance: the absence of transfer in non-human primates capable of recognition. *Neuropsychologia*, 16:361-366.

Hewett, T.D., and Ettlinger, George 1979. Cross-modal performance: the nature of the failure at "transfer" in non-human primates capable of "recognition." *Neuropsychologia*, 17:511-514.

Hill, Suzanne D.; Bundy, Rosalie A.; Gallup, Gordon, G., Jr.; and McClure, Michael K. 1970. Responsiveness of young nursery reared chimpanzees to mirrors. *Proceedings of the Louisiana Academy of Sciences*, 33:77-82.

Hill, W.C. Osman 1967. The taxonomy of the genus *Pan*, in: *Neue Ergebnisse der Primatologie* (D. Starck, R. Schneider, and H.-J. Kuhn, eds.), pp. 47-54, Stuttgart: Fischer.

Hill, W.C. Osman 1969. The nomenclature, taxonomy and distribution of chimpanzees, in: *The Chimpanzee* (G.H. Bourne, ed.), Vol. 1, pp. 22-49, Basel: Karger.

Hiraiwa-Hasegawa, Mariko; Hasegawa, Toshikazu; and Nishida, Toshisada 1984. Demographic study of a large-sized unit-group of chimpanzees in the Mahale Mountains, Tanzania: a preliminary report. *Primates*, 25:401-413.

Hirschler, P. 1942. Anthropoid and human endocranial casts. *Natuurkunde Verhandligen van de Noord-Hollandsche*, Amsterdam: Uitgevers Maatschappij.

Hladik, Claude Marcel 1973. Alimentation et activité d'un groupe de chimpanzés reintroduits en forêt gabonaise. *La Terre et la Vie*, 27:343-413.

Hladik, Claude Marcel 1974. La vie d'un groupe de chimpanzés dans la forêt du Gabon. *Science et Nature*, 121:5-14.

Hladik, Claude Marcel 1975. Ecology, diet, and social patterning in Old and New World Primates, in: *Socioecology and Psychology of Primates* (R.H. Tuttle, ed.), pp. 3-35, The Hague: Mouton.

Hladik, Claude Marcel 1977. Chimpanzees of Gabon and chimpanzees of Gombe: some comparative data on diet, in: *Primate Ecology: Studies of feeding and ranging behaviour in lemurs, monkeys and apes* (T.H. Clutton-Brock, ed.), pp. 481-501, London: Academic Press.

Hladik, Claude Marcel, and Gueguen, Leon 1974. Géophagie et nutrition minerale chez les Primates sauvages. *Comptes Rendus Academie des Sciences Paris*, serie D, 279:1393-1396.

Hladik, Claude Marcel, and Viroben, Gerard 1974. L'alimentation protéique du Chimpanzé dans son environment forestier naturel. *Comptes Rendus Academie des Sciences Paris*, Serie D, 279:1475-1478.

Hofer, Helmut 1972. Uber den Gesang des Orang-Utan (*Pongo pygmaeus*). *Der Zoologisch Garten, N.F., Leipzig*, 41:299-302.

Hollihn, Uwe 1984. Bimanual suspensory behaviour: morphology, selective advantages and phylogeny, in: *The Lesser Apes: Evolutionary and Behavioural Biology* (H. Preuschoft, D.J. Chivers, W.Y. Brockelman, and N. Creel, eds.), pp. 85-95, Edinburgh: Edinburgh University Press.

Hollihn, Uwe, and Jungers, William L. 1984. Kinesiologische Untersuchungen zur Brachiation bei Weisshandgibbons (*Hylobates lar*). *Zeitschrift für Morphologie und Anthropologie*, 74:275-293.

Holloway, Ralph L., Jr. 1968. The evolution of the primate brain: some aspects of quantitative relations. *Brain Research*, 7:121-172.

Holloway, Ralph L. 1978. Problems of brain endocast interpretation and African hominid evolution, in: *Early Hominids of Africa* (C. Jolly, ed.), pp. 379-401, New York: St. Martin's Press.

Holloway, Ralph L. 1979. Brain size, allometry, and reorganization: toward a synthesis, in: *Development and Evolution of Brain Size* (M.E. Hahn, C. Jensen, and B.C. Dudek, eds.), pp. 59-88, New York: Academic Press.

Holsi, P., and Lang, E.M. 1970. Die Chromosomen des Zwergsiamang (*Symphalangus klossi*). *Schweizer Archiv für Tierheilkunde*, 6:296-297.

van Hooff, Johan Antoon Reinier Alex Maria 1962. Facial expressions in higher primates *Symposia of the Zoological Society, London*, 8:97-125.

van Hooff, Johan A.R.A.M. 1963. Facial expressions in higher primates. *Symposia of the Zoological Society, London*, 10:103-104.

van Hooff, Johan A.R.A.M. 1967. The facial displays of the catarrhine monkeys and apes, in: *Primate Ethology* (Desmond Morris, ed.), pp. 7-68, London: Weidenfeld and Nicolson.

van Hooff, Johan A.R.A.M. 1971. *Aspecten van het Sociale Gedrag en de Communicatie bij Humane en Hogere Niet-humane Primaten*, Rotterdam: Bronder-Offset n.v.

van Hooff, Johan A.R.A.M. 1972. A comparative approach to the phylogeny of laughter and smiling, in: *Non-verbal Communication* (R.A. Hinde, ed.), pp. 209-241, Cambridge: Cambridge University Press.

van Hooff, Johan A.R.A.M. 1973. A structural analysis of the social behaviour of a semi-captive group of chimpanzees, in: *Social Communication and Movement, Studies in Interaction and Expression in Man and Chimpanzee* (Mario von Cranach and Ian Vine, eds.), pp. 75-162, London: Academic Press.

van Hooff, Johan A.R.A.M. 1976. The comparison of facial expression in man and higher primates, in: *Methods of Inference from Animal to Human Behaviour* (Mario von Cranach, ed.), pp. 165-196, Chicago: Aldine.

van Hooff, Johan A.R.A.M. 1981. Facial expressions, in: *The Oxford Companion to Animal Behaviour* (D. McFarland, ed.), pp. 165-176, Oxford University Press.

Hooijer, Dirk Albert 1948. Prehistoric teeth of man and the orang-utan from central Sumatra, with notes on the fossil orang-utan from Java and Southern China. *Zoologische Mededlingen Museum (Leiden)*, 29:175-301.

Hooijer, Dirk Albert 1960. Quaternary gibbons from the Malay archipelago. *Zoologische Verhandlingen, Leiden* 46:1-41.

Hooijer, Dirk Albert 1961. The orang-utan in Niah Cave prehistory. *Sarawak Museum Journal*, 10:408-421.

Hooijer, Dirk Albert 1962. Prehistoric bone: the gibbons and monkeys of Niah Great Cave. *Sarawak Museum Journal*, 11:428-449.

Hooton, Earnest Albert 1931; 2nd ed. 1946. *Up from the Ape*, New York: Macmillan Co.

Hooton, Earnest Albert, 1937. *Apes, Men, and Morons*, New York: G.P. Putnam's Sons.

Hooton, Earnest Albert 1939. *Crime and the Man*, Cambridge, Mass.: Harvard University Press.

Hooton, Earnest Albert 1940. *Why Men Behave like Apes and Vice Versa; Or Body and Behavior*, Princeton, New Jersey: Princeton University Press.

Hooton, Earnest Albert 1942. *Man's Poor Relations*, New York: Doubleday, Doran and Comapny.

Horn, Arthur 1975. Adaptations of the pygmy chimpanzee (*Pan paniscus*) to the forests of the Zaire Basin. *American Journal of Physical Anthropology*, 42:307.

Horn, Arthur D. 1980. Some observations on the ecology of the bonobo chimpanzee (*Pan paniscus*, Schwarz 1929) near Lake Tumba, Zaire. *Folia Primatologica*, 34:145-169.

Hornaday, William T. 1879. On the species of Bornean orangs, with notes on their habits. *Proceedings of the American Association for the Advancement of Science, Salem*, 28:438-455.

Hornaday, William T. 1885. *Two Years in the Jungle*, New York: Charles Scribner & Sons.

Horr, David Agee 1972. The Borneo orang-utan. *Borneo Research Bulletin*, 4:46-50.

Horr, David Agee 1975. The Borneo Orang-utan: population structure and dynamics in relationship to ecology and reproductive strategy, in: *Primate Behavior* (L.A. Rosenblum, ed.), Vol. 4, pp. 307-323, New York: Academic Press.

Horr, David Agee 1977. Orang-utan maturation: growing up in a female world, in: *Primate Bio-Social Development* (S. Chevalier-Skolnikoff and F.E. Poirier, eds.), pp. 289-321, New York: Garland Press.

Howell, F. Clark 1978. Hominidae, in: *Evolution of African Mammals* (V.J. Maglio and H.B.S. Cooke, eds.), pp. 154-248, Cambridge, Mass.: Harvard University Press.

Hürzeler, Johannes 1958. *Oreopithecus bambolii* Gervais, a preliminary report. *Verhandlungen der Naturforschenden Gesellschaft in Basel*, 69:1-48.

Hürzeler, Johannes 1968. Questions et reflexions sur l'histoire des anthropomorphes. *Annales de Paleontologie*, 54:195-233.

Huxley, Thomas Henry 1863. *Evidence as to Man's Place in Nature*, London: Williams & Norgate.

Hylander, William L. 1975. Incisor size and diet in anthropoids with special reference to Cercopithecidae. *Science*, 189:1095-1098.

International Anatomical Nomenclature Committee 1977. *Nomina Anatomica, Fourth Edition*, Amsterdam: Excerpta Medica.

Ishida, Hidemi; Kimura, Tasuku; and Okada, Morihiko 1975. Patterns of bipedal walking in anthropoid primates, in *Symposia of the Fifth Congress of the International Primatological Society (1974)* (S. Kondo, M. Kawai, A. Ehara, and S. Kawamura, eds.), pp. 287-301, Tokyo: Japan Science Press.

Ishida, Hidemi; Kimura, Tasuku; Okada, Morihiko; and Yamazaki, Nobutoshi 1984. Kinesiological aspects of bipedal walking in gibbons, in: *The Lesser Apes: Evolutionary and Behavioural Biology* (H. Preuschoft, D.J. Chivers, W.Y. Brockelman, and N. Creel, eds.), pp. 135-145, Edinburgh: Edinburgh University Press.

Ishida, Hidemi; Kumakura, Hiroo; and Kondo, Shiro 1985. Primate bipedalism and quadrupedalism: comparative electromyography, in: *Primate Morphophysiology, Locomotor Analyses, and Human Bipedalism* (S. Kondo, ed.), pp. 59-79, Tokyo: University of Tokyo Press.

Ishida, Hidemi; Okada, Morihiko; Tuttle, Russell; and Kimura, Tasuku 1978. Activities of hindlimb muscles in bipedal gibbons, in: *Recent Advances in Primatology* (D.J. Chivers and K.A. Joysey, eds.), Vol. 3, pp. 459-462, London: Academic Press.

Itani, Junichiro 1979. Distribution and adaptation of chimpanzees in an arid area, in: *The Great Apes* (D.A. Hamburg and E.R. McCown, eds.), pp. 54-71, Menlo Park, California: The Benjamin/Cummings Publishing Co.

Itani, Junichiro 1980. Social structure of African great apes. *Journal of Reproduction & Fertility, Supplement No. 28*:33-41.

Itani, Junichiro 1982. Intraspecific killing among non-human primates. *Journal of Social and Biological Structures*, 5:361-368.

Itani, Junichiro, and Suzuki, Akira 1967. The social unit of wild chimpanzees. *Primates*, 8:355-381.

Izawa, Kohsei 1970. Unit-groups of chimpanzees and their nomadism in the savannah woodland. *Primates*, 11:1-46.

Izawa, Kohsei, and Itani, Junichiro 1966. Chimpanzees in Kasakati Basin, Tanzania, (I) ecological study in the rainy season 1963-1964. *Kyoto University African Studies*, 1:73-156.

Jackson, William J.; Reite, Martin L.; and Buxton, Donald F. 1969. The chimpanzee central nervous system: a comparative review, in: *Primates in Medicine* (H.H. Reynolds, ed.), Vol. 4, pp. 1-51, Basel: S. Karger.

Janzen, Daniel H. 1979. How to be a fig. *Annual Review of Ecology and Systematics*, 10:13-52.

Jarvik, Murray E. 1953. Discrimination of colored food and food signs by primates. *Journal of Comparative and Physiological Psychology*, 46:390-392.

Jarvik, Murray E. 1956. Simple color discrimination in chimpanzees: effect of varying contiguity between cue and incentive. *Journal of Comparative and Physiological Psychology*, 49:492-495.

Jarvis, M.J., and Ettlinger, George 1977. Cross-modal recognition in chimpanzees and monkeys. *Neuropsychologia*, 15:499-506.

Jarvis, M.J., and Ettlinger, George 1978. Cross-modal performance in monkeys and apes: is there a substantial difference?, in: *Recent Advances in Primatology*, Vol. 1, *Behaviour* (D.J. Chivers and J. Herbert, eds.), pp. 953-956, London: Academic Press.

Jenkins, William Oliver 1943. A spatial factor in chimpanzee learning. *Journal of Comparative Psychology*, 35:81-84.

Jerison, Harry J. 1973. *Evolution of the Brain and Intelligence*, New York: Academic Press.

Johanson, Donald Carl 1974. Some metric aspects of the permanent and deciduous dentition of the pygmy chimpanzee (*Pan paniscus*). *American Journal of Physical Anthropology*, 41: 39-48.

Johanson, Donald C., and White, Timothy D. 1979. A systematic assessment of early African hominids. *Science*, 203:321-330.

Johanson, Donald C.; White, Timothy D.; and Coppens, Yves 1978. A new species of the genus *Australopithecus* (Primates: Hominidae) from the Pliocene of eastern Africa. *Kirtlandia*, 28:1-14.

Jolicoeur, Pierre; Pirlot, Paul; Baron, Georg; and Stephan, Heinz 1984. Brain structure and correlation patterns in Insectivora, Chiroptera, and Primates. *Systematic Zoology*, 33:14-29.

Jones, Clyde, and Sabater Pi, Jorge 1971. Comparative ecology of *Gorilla gorilla* (Savage and Wyman) and *Pan troglodytes* (Blumenbach) in Rio Muni, West Africa. *Bibliotheca Primatologica*, 13:1-96.

Jones, Marvin L. 1969. The geographical races of orang-utan. *Proceedings of the Second International Congress of Primatology, Atlanta, Georgia, 1968*, 2:217-223, Basel: Karger.

Jordan, Claudia 1982. Object manipulation and tool-use in captive pygmy chimpanzees (*Pan paniscus*). *Journal of Human Evolution*, 11:35-39.

Jordan, Claudia, and Jordan, Heimo 1977. Versuche zur Symbol-Ereignis-Verknupfung bei einer Zwergschimpanzen (*Pan paniscus* Schwarz, 1929). *Primates*, 18:515-529.

Jungers, William L., and Stern, Jack T., Jr. 1980. Telemetered electromyography of forelimb muscle chains in gibbons (*Hylobates lar*). *Science*, 208:617-619.

Jungers, William L., and Stern, Jack T., Jr. 1984. Kinesiological aspects of brachiation in lar gibbons, in: *The Lesser Apes: Evolutionary and Behavioural Biology* (H. Preuschoft, D.J. Chivers, W.Y. Brockelman, and N. Creel, eds.), pp. 119-134, Edinburgh: Edinburgh University Press.

Jungers, William L., and Susman, Randall L. 1984. Body size and skeletal allometry in African apes, in: *The Pygmy Chimpanzee. Evolutionary Biology and Behavior* (Randall L. Susman, ed.), pp. 131-177, New York: Plenum Publishing Corporation.

Kano, Takayoshi 1971a. The chimpanzee of Filabanga, western Tanzania. *Primates*, 12:229-246.

Kano, Takayoshi 1971b. Distribution of the primates on the eastern shore of Lake Tanganyika. *Primates*, 12:281-304.

Kano, Takayoshi 1972. Distribution and adaptation of the chimpanzee on the eastern shore of Lake Tanganyika. *Kyoto University African Studies*, 7:37-129.

Kano, Takayoshi 1979. A pilot study on the ecology of pygmy chimpanzees, *Pan paniscus*, in: *The Great Apes* (D.A. Hamburg and E.R. McCown, eds.), pp. 122-135, Menlo Park, California: The Benjamin/Cummings Publishing Co.

Kano, Takayoshi 1980. Social behavior of wild pygmy chimpanzees (*Pan paniscus*) of Wamba: a preliminary report. *Journal of Human Evolution*, 9:243-260.

Kano, Takayoshi 1982a. The social group of pygmy chimpanzees (*Pan paniscus*) of Wamba, *Primates*, 23:171-188.

Kano, Takayoshi 1982b. The use of leafy twigs for rain cover by the pygmy chimpanzees of Wamba. *Primates*, 23:453-457.

Kano, Takayoshi 1984a. Distribution of pygmy chimpanzees (*Pan paniscus*) in the Central Zaire Basin. *Folia Primatologica*, 43:36-52.

Kano, Takayoshi 1984b. Observations of physical abnormalities among the wild bonobos (*Pan paniscus*) of Wamba, Zaire. *American Journal of Physical Anthropology*, 63:1-11.

Kano, Takayoshi, and Mulavwa, Mbangi 1984. Feeding ecology of the pygmy chimpanzees (*Pan paniscus*) of Wamba, in: *The Pygmy Chimpanzee: Evolutionary Biology and Behavior* (Randall L. Susman, ed.), pp. 233-274, New York: Plenum Press.

Kappeler, Markus 1984a. The gibbon in Java, in: *The Lesser Apes: Evolutionary and Behavioural Biology* (H. Preuschoft, D.J. Chivers, W.Y. Brockelman, and N. Creel, eds.), pp. 19-31, Edinburgh: Edinburgh University Press.

Kappeler, Markus 1984b. Vocal bouts and territorial maintenance in the moloch gibbon, in: *The Lesser Apes: Evolutionary and Behavioural Biology* (H. Preuschoft, D.J. Chivers, W.Y. Brockelman, and N. Creel, eds.), pp. 376-389, Edinburgh: Edinburgh University Press.

Kappeler, Markus 1984c. Diet and feeding behaviour of the moloch gibbon, in: *The Lesser Apes: Evolutionary and Behavioural Biology* (H. Preuschoft, D.J. Chivers, W.Y. Brockelman, and N. Creel, eds.), pp. 228-241, Edinburgh: Edinburgh University Press.

Kawabe, Munemi 1966. One observed case of hunting behavior among wild chimpanzees living in the savanna woodland of western Tanzania. *Primates*, 7:393-396.

Kawabe, Munemi 1970. A preliminary study of the wild siamang gibbon (*Hylobates syndactylus*) at Fraser's Hill, Malaysia. *Primates*, 11:285-291.

Kawai, Masao, and Mizuhara, Hiroki 1959-1960. An ecological study on the wild mountain gorilla (*Gorilla gorilla beringei*). *Primates*, 2:1-42.

Kawanaka, Kenji 1981. Infanticide and cannibalism in chimpanzees, with special reference to the newly observed case in the Mahale Mountains. *African Study Monographs*, 1:69-99.

Kawanaka, Kenji 1982a. Further studies on predation by chimpanzees in the Mahale Mountains. *Primates*, 23:364-384.

Kawanaka, Kenji 1982b. A case of inter-unit-group encounter in chimpanzees of the Mahale Mountains. *Primates*, 23:558-562.

Kawanaka, Kenji 1984. Association, ranging, and the social unit in chimpanzees of the Mahale Mountains, Tanzania. *International Journal of Primatology*, 5:411-434.

Kawanaka Kenji, and Nishida, Toshisada 1974. Recent advances in the study of inter-unit-group relationships and social structure of wild chimpanzees of the Mahali Mountains, in: *Proceedings from the Symposia of the Fifth Congress of the International Primatological Society* (S. Kondo, M. Kawai, A. Ehara, and S. Kawamura, eds.), pp. 173-186, Tokyo: Japan Science Press.

Kay, Richard F. 1975. The functional adaptations of primate molar teeth. *American Journal Physical Anthropology*, 43:195-216.

Kay, Richard F. 1981. The nut-crackers—a new theory of the adaptations of the Ramapithecinae. *American Journal of Physical Anthropology*, 55:141-151.

Kay, Richard F., and Hylander, William L. 1978. The dental structure of mammalian folivores with special reference to Primates and Phalangeroidea (Marsupialia), in: *The Ecology of Arboreal Folivores* (G.G. Montgomery, ed.), pp. 173-191, Washington, D.C.: Smithsonian Institute Press.

Kay, Richard F., and Simons, Elwyn L. 1983. A reassessment of the relationship between later Miocene and subsequent Hominoidea, in: *New Interpretations of Ape and Human Ancestry* (R.L. Ciochon and R.S. Corruccini, eds.), pp. 577-624, New York: Plenum Press.

Keith, Arthur 1891. Anatomical notes on Malay apes. *Journal of the Straits Branch of the Royal Asiatic Society*, 23:77-94.

Keith, Arthur 1903. The extent to which the posterior segments of the body have been transmuted and suppressed in the evolution of man and allied primates. *Journal of Anatomy and Physiology*, 37:18-40.

Keith, Arthur 1912. Certain phases in the evolution of man. *British Medical Journal*, 1:734-737, 788-790.

Keith, Arthur 1923. Man's posture: its evolution and disorders. *British Medical Journal*, 1:451-454, 499-502, 545-548, 587-590, 624-626, 669-672.

Keith, Arthur 1927. *Concerning Man's Origin*, London: Watts.

Keith, Arthur 1934. *The Construction of Man's Family Tree*, London: Watts.

Keith, Arthur 1940. Fifty years ago. *American Journal of Physical Anthropology*, 26:251-267.

Kelleher, Roger T. 1957a. Conditioned reinforcement in chimpanzees. *Journal of Comparative and Physiological Psychology*, 50:571-575.

Kelleher, Roger T. 1957b. A multiple schedule of conditioned reinforcement with chimpanzees. *Psychological Reports*, 3:485-491.

Kelleher, Roger T. 1958. Concept formation in chimpanzees. *Science*, 128:777-778.

Kelleher, Roger T. 1965. Operant conditioning, in: *Behavior of Nonhuman Primates, Modern Research Trends* (Allan M. Schrier, Harry F. Harlow, and Fred Stollnitz, eds.), Vol. 1, pp. 211-247, New York: Academic Press.

Kellogg, Winthrop N. 1968. Chimpanzees in experimental homes. *The Psychological Record*, 18:489-498.

Kellogg, Winthrop N. 1969. Research on the home-raised chimpanzee, in: *The Chimpanzee* (G.H. Bourne, ed.), Vol. 1, pp. 369-392, Basel: S. Karger.

Kellogg, Winthrop N., and Kellogg, L.A. 1933. *The Ape and the Child: A Study of Environmental Influence upon Early Behavior*, New York: McGraw-Hill.

Kimura, Tasuku, 1985. Bipedal and quadrupedal walking of primates: comparative dynamics, in: *Primate Morphophysiology, Locomotor Analyses and Human Bipedalism* (S. Kondo, ed.), pp. 81-104, Tokyo: University of Tokyo Press.

Kimura, Tasuku; Okada, Morihiko; and Ishida, Hidemi 1977. Dynamics of primate bipedal walking as viewed from the force of foot. *Primates*, 18:137-147.

Kimura, Tasuku; Okada, Morihiko; and Ishida, Hidemi 1979. Kinesiological characteristics of primate walking: its significance in human walking, in: *Environment, Behavior and Morphology: Dynamic Interactions in Primates* (M.E. Morbeck, H. Preuschoft, and N. Gomberg, eds.), pp. 297-311, New York: Gustav Fischer.

Kimura, Tasuku; Okada, Morihiko; Yamazaki, Nobutoshi; and Ishida, Hidemi 1983. Speed of the bipedal gaits of man and nonhuman primates. *Annales des Sciences Naturelles, Zoologie, Paris*, Series 13, 5:145-158.

King, James E. 1973. Learning and generalization of a two-dimensional sameness-difference concept by chimpanzees and orang-utans. *Journal of Comparative and Physiological Psychology*, 84:140-148.

King, James E., and Fobes, James L. 1982. Complex learning by primates, in: *Primate Behavior* (James L. Fobes and James E. King, eds.), pp. 327-360, New York: Academic Press.

Kintz, B.L.; Foster, Marilyn S.; Hart, James O.; O'Malley, John J.; Palmer, Edward L.; and Sullivan, Sharon L. 1969. A comparison of learning sets in humans, primates, and subprimates. *The Journal of General Psychology*, 80:189-204.

Kinzey, Warren G. 1984. The dentition of the pygmy chimpanzee, *Pan paniscus*, in: *The Pygmy Chimpanzee: Evolutionary Biology and Behavior* (R.L. Susman, ed.), pp. 65-88, New York: Plenum Press.

Kitahara-Frisch, Jean 1971. Evolution of the siamang (*Symphalangus syndactylus*) in Southeast Asia during the Pleistocene. *Proceedings of the Third International Congress of Primatology, Zurich 1970*, 1:67-73, Basel: Karger.

Kitahara-Frisch, Jean, and Norikoshi, Kohshi 1982. Spontaneous sponge-making in captive chimpanzees. *Journal of Human Evolution*, 11:41-47.

Kitamura, Koji 1983. Pygmy association patterns in ranging. *Primates*, 24:1-12.

Kleiman, Devra G. 1977. Monogamy in mammals. *The Quarterly Review of Biology*, 52:39-69.

Kluge, Arnold G. 1983. Cladistics and the classification of the great apes, in: *New Interpretations of Ape and Human Ancestry* (R.L. Ciochon and R.S. Corruccini, eds.), pp. 151-177, New York: Plenum Press.

von Koenigswald, Gustav Heinrich Ralph 1952. *Gigantopithecus blacki* von Koenigswald , a giant fossil hominoid from the Pleistocene of Southern China. *Anthropological Papers of the American Museum of Natural History*, 43:293-325.

Kohlbrügge, J.H.F. 1890-91. Versuch einer Anatomie des Genus *Hylobates*, in: *Zoolische Ergebnisse einer Reise in Niederländisch Oest-Indien* (Max Weber, ed.), Vol. 2, pp. 139-254, Leiden: E.J. Brill.

Köhler, Wolfgang 1921. Zur Psychologie des Schimpansen (Aus der Anthropoiden-station auf Teneriffa). *Psychologische Forschung, Berlin*, I:2-46.

Köhler, Wolfgang 1925. *The Mentality of Apes*, New York: Humanities Press, Inc.

Kortlandt, Adriaan 1962. Chimpanzees in the wild. *Scientific American*, 206:128-138.

Kortlandt, Adriaan 1968. Handgebrauch bei freilebenden Schimpansen, in: *Handgebrauch und Verstandigung bei Affen und Fruhmenschen* (B. Rensch, ed.), pp. 59-102, Bern: Hans Huber.

Kortlandt, Adriaan 1972. *New Perspectives on Ape and Human Evolution*, Amsterdam: Stichting voor Psychobiologie.

Kortlandt, Adriaan 1981. "Siabon" or "gibbang"? *Laboratory Primate Newsletter*, 20:12.

Kortlandt, Adriaan, and Kooij, M. 1963. Protohominid behaviour in primates (preliminary communication). *Symposia of the Zoological Society, London*, 10:61-88.

Kots, Nadezhda Nikolaevna (Ladygina) 1921. Report of the Zoopsychology Laboratory of the Darwinian Museum Moscow (in Russian).

Koyama, Naoki 1971. Observations on mating behavior of wild siamang gibbons at Fraser's Hill, Malaysia. *Primates*, 2:183-189.

Kretzoi, Miklos 1975. New ramapithecines and *Pliopithecus* from the lower Pliocene of Rudabanya in north-eastern Hungary. *Nature*, 257:578-581.

Kuroda, Suehisa 1979. Grouping of the pygmy chimpanzees. *Primates*, 20:161-183.

Kuroda, Suehisa 1980. Social behavior of the pygmy chimpanzees. *Primates*, 21:181-197.

Kuroda, Suehisa 1984. Interactions over food among pygmy chimpanzees, in: *The Pygmy Chimpanzee: Evolutionary Biology and Behavior* (Randall L. Susman, ed.), pp. 301-324, New York: Plenum Press.

Ladas, Alice Kahn; Whipple, Beverly; and Perry, John D. 1982. *The G Spot and Other Recent Discoveries about Human Sexuality*, New York: Holt, Rinehart and Winston.

Ladygina-Kots, Nadia N. 1935. Ditia shimpanze i ditia cheloveka. *Scientific Memoirs of the Museum Darwinianum Moscow*.

Laidler, Keith 1978. Language in the orang-utan, in: *Action, Gesture and Symbol: The Emergence of Language* (Andrew Lock, ed.), pp. 133-155, London: Academic Press.

Laitman, Jeffrey T., and Heimbuch, Raymond C. 1984. A measure of basicranial flexion in *Pan paniscus*, the pygmy chimpanzee, in: *The Pygmy Chimpanzee: Evolutionary Biology and Behavior* (R.L. Susman, ed.), pp. 49-63, New York: Plenum Press.

Lamarck, Jean de 1809. *Philosophie zoologique*, Paris: Dentu.

Lamprecht, Jürg 1970. Duettgesang beim Siamang, *Symphalangus syndactylus* (Hominoidea, Hylobatinae). *Zeitschrift für Tierpsychologie*, 27:186-204.

Lancaster, Jane B. 1968. On the evolution of tool-using behavior. *American Anthropologist*, 70:56-66.

Lancaster, Jane B. 1968. Primate communication systems and the emergence of human language, in: *Primates: Studies in Adaptation and Variability* (Phyllis C. Jay, ed.), pp. 439-457, New York: Holt, Rinehart and Winston.

Langdale-Brown, I.; Osmaston, H.A.; and Wilson, James G. 1964. *The Vegetation of Uganda and its Bearing on Land-use*, Entebbe: Uganda Government Printers.

van Lawick-Goodall, Jane 1965. New discoveries among Africa's chimpanzees. *National Geographic Magazine*, 128:802-831.

van Lawick-Goodall, Baroness Jane 1967. *My Friends the Wild Chimpanzees*, Washington, D.C.: National Geographic Society.

van Lawick-Goodall, Jane 1968a. The behaviour of free-living chimpanzees in the Gombe Stream area. *Animal Behaviour Monographs*, 1:161-311.

van Lawick-Goodall, Jane 1968b. A preliminary report on expressive movements and communication in the Gombe Stream chimpanzees, in: *Primates: Studies in Adaptation and Variability* (Phyllis C. Jay, ed.), pp. 313-374, New York: Holt, Rinehart & Winston.

van Lawick-Goodall, Jane 1971. *In the Shadow of Man*, Boston: Houghton Mifflin Co.

van Lawick-Goodall, Jane 1973a. The behavior of chimpanzees in their natural habitat. *American Journal of Psychiatry*, 130:1-12.

van Lawick-Goodall, Jane 1973b. Cultural elements in a chimpanzee community. *Symposia of the Fourth International Congress of Primatology*, Vol. 1, *Precultural Primate Behavior* (Emil W. Menzel, Jr., ed.), pp. 144-184), Basel: Karger.

van Lawick-Goodall, Jane 1975. The behaviour of the chimpanzee, in: *Hominisation und Verhalten* (G. Kurth and I. Eibl-Eibesfeldt, eds.), pp. 74-136, Stuttgart: Gustav Fischer Verlag.

Leakey, Louis S.B. 1962. A new lower Pliocene fossil primate from Kenya. *Annals and Magazine of Natural History*, 4:689-696.

Leakey, Mary D. 1979. 3.6 million year old footprints in the ashes of time. *National Geographic Magazine*, 155:446-457.

Leakey, Mary D. 1981. Tracks and tools. *Philosophical Transactions of the Royal Society, London*, B-292:95-102.

Leakey, Mary D., and Hay, Richard L. 1979. Pliocene footprints in the Laetolil Beds at Laetoli, northern Tanzania. *Nature*, 278:317-323.

Lefebvre, Louis 1982. Food exchange strategies in an infant chimpanzee. *Journal of Human Evolution*, 11:195-204.

LeMay, Marjorie 1976. Morphological cerebral asymmetries of modern man, fossil man, and nonhuman primate. *Annals of the New York Academy of Sciences*, 280:349-366.

LeMay, Marjorie; Billing, Michael S.; and Geschwind, Norman 1982. Asymmetries of the brains and skulls of nonhuman primates, in: *Primate Brain Evolution: Methods and Concepts* (E. Armstrong and D. Falk, eds.), pp. 263-277, New York: Plenum Press.

LeMay, Marjorie, and Culebras, Antonio 1972. Human brain: morphologic differences in the hemispheres demonstrable by carotid arteriography. *The New England Journal of Medicine*, 287:168-170.

LeMay, Marjorie, and Geschwind, Norman 1975. Hemispheric differences in the brains of great apes. *Brain, Behavior and Evolution*, 11:48-52.

Lethmate, Jürgen, 1976a. Gebrauch und Herstellung von Trinkwerkzeugen bei Orang-Utans. *Zoologischer Anzeiger, Jena*, 197:251-263.

Lethmate, Jürgen 1977d. Versuche zum Schlagstockverfahren mit zwei jungen Orang-Utan. *Zoologischer Anzeiger, Jena*, 197:264-271.

Lethmate, Jürgen 1976c. Werkzeugintelligenz von Orang-Utans. *UMSCHAU*, 76:782-784.

Lethmate, Jürgen 1976d. Werkzeugverhalten von Orang-Utans. *Biologie in unserer Zeit*, 6:33-40.

Lethmate, Jürgen 1977a. Instrumentelles Verhalten zoolebender Orang-Utans. *Zeitschrift für Morphologie und Anthropologie*, 68:57-87.

Lethmate, Jürgen 1977b. Versuche zum Schlagstockverfahren mit zwei jungen Orang-Utans. *Zoologischer Anzeiger, Jena*, 199:209-226.

Lethmate, Jürgen 1977c. Weitere Versuche zum Manipulier-und Werkzeugverhalten junger Orang-Utans. *Primates*, 18:531-543.

Lethmate, Jürgen 1977d. Nestbauverhalten eines isoliert aufgezogenen, jungen Orang-Utans. *Primates*, 18:545-554.

Lethmate, Jürgen 1977e. Versuche zum "vorbedingten" Handeln mit einem jungen Orang-Utan. *Primates*, 19:727-736.

Lethmate, Jürgen 1977f. Werkzeugherstellung eines jungen Orang-Utans. *Behaviour*, 62:174-189.

Lethmate, Jürgen 1977g. Problemlöseverhalten von Orang-Utans (*Pongo pygmaeus*). *Fortschritte der Verhaltensforschung*, No. 19, Berlin: Paul Parey.

Lethmate, Jürgen 1979. Instrumental behaviour of zoo orang-utans. *Journal of Human Evolution*, 8:741-744.

Lethmate, Jürgen 1982. Tool-using skills of orang-utans. *Journal of Human Evolution*, 11:49-64.

Lethmate, Jürgen, and Dücker, Gerti 1973. Untersuchungen zum Selbsterkennen im Spiegel bei Orang-Utans einigen anderen Affenarten. *Zeitschrift für Tierpsychologie*, 33:248-269.

Levy, Jerre 1980. Cerebral asymmetry and the psychology of man, in: *The Brain and Psychology* (M.C. Wittrock, ed.), pp. 259-321, New York: Academic Press.

Levy, Jerre 1982. Performance, capacity, and hemispheric dominance for processing and behavior: metacontrolling programs in the central nervous system. *Cognition and Brain Theory*, 5:199-210.

Levy, Jerre 1983. Is cerebral asymmetry of function a dynamic process? Implications for specifying degree of lateral differentiation. *Neuropsychologia*, 21:3-11.

Levy, Jerre; Heller, Wendy; Banich, Marie T.; Burton, Leslie A. 1983. Are variations among right-handed individuals in perceptual asymmetries caused by characteristic arousal difference between hemispheres? *Journal of Experimental Psychology: Human Perception and Performance*, 9:329-359.

Levy, Jerre, and Reid, Marylou 1978. Variations in cerebral organization as a function of handedness, hand posture in writing, and sex. *Journal of Experimental Psychology: General*, 107:119-144.

Leyton, A.S.F., and Sherrington, Charles S. 1917. Observations on the excitable cortex of the chimpanzee, orang-utan, and gorilla. *Quarterly Journal of Experimental Physiology*, 11:135-222.

Linden, Eugene 1974. *Apes, Men and Language*, New York: Saturday Review Press.

Linden, Eugene 1981. *Apes, Men and Language*, 2nd edn., Harmondsworth, Middlesex, England: Penguin Books Ltd.

Lipson, Susan, and Pilbeam, David 1982. *Ramapithecus* and hominoid evolution. *Journal of Human Evolution*, 11:545-548.

Lockard, Joan S. 1984. Handedness in a captive group of lowland gorillas. *International Journal of Primatology*, 5:356.

Lynch, Gary, Hechtel, S., and Jacobs, D. 1983. Neonate size and evolution of brain size in the anthropoid primates. *Journal of Human Evolution*, 12:519-522.

MacKinnon, John 1971. The orang-utan in Sabah today. *Oryx*, 11:141-191.

MacKinnon, John 1973. Orang-utans in Sumatra. *Oryx*, 12:234-242.

MacKinnon, John 1974a. The behaviour and ecology of wild orang-utans (*Pongo pygmaeus*). *Animal Behaviour*, 22:3-74.

MacKinnon, John 1974b. *In Search of the Red Ape*, London: Collins.

MacKinnon, John 1975. Distinguishing characters of the insular forms of orang-utan. *International Zoo Yearbook*, 15:195-197.

MacKinnon, John 1976. Mountain gorillas and bonobos. *Oryx*, 13:372-382.

MacKinnon, John 1977. A comparative ecology of Asian apes. *Primates*, 18:747-772.

MacKinnon, John 1978. *The Ape within Us*, London: William Collins Sons.

MacKinnon, John 1979. Reproductive behavior in wild orangutan populations, in: *The Great Apes* (D.A. Hamburg and E.R. McCown, eds.), pp. 256-273, Menlo Park, California: Benjamin/Cummings Publishing Co.

MacKinnon, John R. 1984. The distribution and status of gibbons in Indonesia, in: *The Lesser Apes: Evolutionary and Behavioural Biology* (H. Preuschoft, D.J. Chivers, W.Y. Brockelman, and N. Creel, eds.), pp. 16-18, Edinburgh: Edinburgh University Press.

MacKinnon, John, and MacKinnon, Kathleen 1977. The formation of a new gibbon group. *Primates*, 18:701-708.

MacKinnon, John R., and MacKinnon, Kathleen S. 1978. Comparative feeding ecology of six sympatric primates in West Malaysia, in: *Recent Advances in Primatology*, Vol. 1, *Behaviour* (D.J. Chivers and J. Herbert, eds.), pp. 305-321, London: Academic Press.

MacKinnon, John R., and MacKinnon, Kathleen S. 1980. Niche differentiation in a primate community, in: *Malayan Forest Primates* (D.J. Chivers, ed.), pp. 167-190, New York: Plenum Press.

Mai, Larry L. 1980. A model of chromosome evolution and its bearing on cladogenesis in the Hominoidea, in: *New Interpretations of Ape and Human Ancestry* (Russell L. Ciochon and Robert S. Corruccini, eds.), pp. 87-114, New York: Plenum Press.

Malone, Daniel R.; Tolan, James C.; and Rogers, Charles M. 1980. Cross-modal matching of objects and photographs in the monkey. *Neuropsychologia*, 18:693-697.

Maple, Terry L. 1980. *Orang-utan Behavior*, New York: Van Nostrand Reinhold Co.

Maple, Terry L., and Hoff, Michael P. 1982. *Gorilla Behavior*, New York: Van Nostrand Reinhold Company.

Marks, Jon 1983. Hominoid cytogenetics and evolution. *Yearbook of Physical Anthropology*, 26:131-159.

Marler, Peter 1965. Communication in monkeys and apes, in: *Primate Behavior* (I. DeVore, ed.), pp. 544-584, New York: Holt, Rinehart & Winston.

Marler, Peter 1969. Vocalizations of wild chimpanzees: an introduction. *Proceedings of the Second International Congress of Primatology, Atlanta, GA 1968* (C.R. Carpenter, ed.), Vol. 1, pp. 94-100, Basel: Karger.

Marler, Peter 1976. Social organization, communication and graded signals: the chimpanzee and the gorilla, in: *Growing Points in Ethology* (P.P.G. Bateson and R.A. Hinde, eds.), pp. 239-280, Cambridge, England: Cambridge University Press.

Marler, Peter 1983. Monkey calls: how are they perceived and what do they mean?, in: *Advances in the Study of Mammalian Behavior* (J.F. Eisenberg and D.G. Kleiman, eds.), pp. 343-356, *Special Publication of the American Society of Mammalogists*, No. 7.

Marler, Peter, and Hobbett, Linda 1975. Individuality in a long-range vocalization of wild chimpanzees. *Zeitschrift für Tierpsychologie*, 38:97-109.

Marler, Peter, and Tenaza, Richard 1977. Signaling behavior of apes with special reference to vocalization, in: *How Animals Communicate* (Thomas A. Sebeok, ed.), pp. 965-1033, Bloomington: Indiana University Press.

Marshall, Joe T. 1981. The agile gibbon in South Thailand. *Natural History Bulletin of the Siam Society*, 29:129-136.

Marshall, Joe T., and Marshall, E.R. 1976. Gibbons and their territorial songs. *Science*, 193:235-237.

Marshall, Joe T., Jr.; Ross, Bruce A.; and Chantharojvong, Swart 1972. The species of gibbons in Thailand. *Journal of Mammalogy*, 53:479-486.

Marshall, Joe T., Sugardjito, Jito; and Markaya, M. 1984. Gibbons of the lar group: relationships based on voice, in: *The Lesser Apes: Evolutionary and Behavioural Biology* (H. Preuschoft, D.J. Chivers, W.Y. Brockelman, and N. Creel, eds.), pp. 533-541, Edinburgh: Edinburgh University Press.

Mason, William A. 1979. Environmental models and mental modes: representational processes in the great apes, in: *The Great Apes* (David A. Hamburg and Elizabeth R. McCown, eds.), pp. 277-293, Menlo Park, California: The Benjamin/Cummings Publishing Company.

Matsuzawa, Tetsuro 1985a. Use of numbers by a chimpanzee. *Nature*, 315:57-59.

Matsuzawa, Tetsuro 1985b. Colour naming and classification in a chimpanzee. *Journal of Human Evolution*, 14:283-291.

Matthew, William Diller, and Granger, W. 1923. New fossil mammals from the Pliocene of Sze-Chuan, China. *Bulletin of the American Museum of Natural History*, 48:563-598.

McBeath, Norman Milne, and McGrew, William C. 1982. Tools used by wild chimpanzees to obtain termites at Mt. Assirik, Senegal: the influence of habitat. *Journal of Human Evolution*, 11:65-72.

McCann, C. 1933. Notes on the colouration and habits of the white-browed gibbon or hoolock (*Hylobates hoolock* Harl.). *Journal of the Bombay Natural History Society*, 36:395-405.

McClure, H. Elliott 1964. Some observations of primates in climax dipterocarp forest near Kuala Lumpur, Malaya. *Primates*, 5:39-58.

McClure, Michael K., and Helland, Joan 1979. A chimpanzee's use of dimensions in responding same and different. *The Psychological Record*, 29:371-378.

McCulloch, Thomas L., and Nissen, Henry W. 1937. Equated and non-equated stimulus situations in discrimination learning by chimpanzees. II. comparison with limited response. *Journal of Comparative Psychology*, 23:365-376.

McDowell, Arnold A., and Nissen, Henry W. 1959. Solution of a bi-manual coordination problem by monkeys and chimpanzees. *Journal of Genetic Psychology*, 94:35-42.

McGinnis, Patrick R. 1979. Sexual behavior in free-living chimpanzees: consort relationships, in: *The Great Apes* (David A. Hamburg and Elizabeth R. McCown, eds.), pp. 429-439, Menlo Park, California: The Benjamin/Cummings Publishing Company.

McGlone, J. 1980. Sex differences in human brain asymmetry: a critical survey. *The Behavioral and Brain Sciences*, 3:215-263.

McGrew, William C. 1974. Tool use by wild chimpanzees in feeding upon driver ants. *Journal of Human Evolution*, 3:501-508.

McGrew, William C. 1975. Patterns of plant food sharing by wild chimpanzees, in: *Contemporary Primatology* (S. Kondo, M. Kawai and A. Ehara, eds.), pp. 304-309, Basel: Karger.

McGrew, William C. 1979. Evolutionary implications of sex differences in chimpanzee predation and tool use, in: *The Great Apes* (David A. Hamburg and Elizabeth R. McCown, eds.), pp. 441-463, Menlo Park, California: The Benjamin/Cummings Publishing Company.

McGrew, William C. 1981. The female chimpanzee as a human evolutionary prototype, in: *Woman the Gatherer* (Frances Dahlberg, ed.), pp. 35-73, New Haven: Yale University Press.

McGrew, William C. 1982. Recent advances in the study of tool-use by nonhuman primates, in: *Advanced Views in Primate Biology* (A.B. Chiarelli and R.S. Corruccini, eds.), pp. 177-183, Berlin: Springer-Verlag.

McGrew, William C. 1983. Animal foods in the diets of wild chimpanzees (*Pan troglodytes*): why cross-cultural variation? *Journal of Ethology*, 1:46-61.

McGrew, William C.; Baldwin, Pamela J.; and Tutin, Caroline E.G. 1981. Chimpanzees in a hot, dry and open habitat: Mt. Assirik, Senegal, West Africa. *Journal of Human Evolution*, 10:227-244.

McGrew, William C.; Baldwin, Pamela J.; and Tutin, Caroline E.G. 1982. Recherches scientifiques dans les parcs nationaux du Sénégal. XXV. Observations préliminaires sur les chimpanzés (*Pan troglodytes verus*) du Parc National du Niokolo-Koba. *Mémoires de l'Institut Fondamental d'Afrique Noire*, Ifan-Dakar, 92:335-340.

McGrew, William C., and Collins, David Anthony 1985. Tool use by wild chimpanzees (*Pan troglodytes*) to obtain termites (*Macrotermes herus*) in the Mahale Mountains, Tanzania. *American Journal of Primatology*, 9:47-62.

McGrew, William C., and Rogers, Mary Elizabeth 1983. Chimpanzees, tools, and termites: new record from Gabon. *American Journal of Primatology*, 5:171-174.

McGrew, William C.; Sharman, Martin John; Baldwin, Pamela J.; and Tutin, Caroline E.G. 1982. On early hominid plant-food niches. *Current Anthropology*, 23:213-214.

McGrew, William C., and Tutin, Caroline E.G. 1972. Chimpanzee dentistry. *JADA*, 85:1198-1204.

McGrew, William C., and Tutin, Caroline E.G. 1973. Chimpanzee tool use in dental grooming. *Nature*, 41:477-478.

McGrew, William C., and Tutin, Caroline E.G. 1978. Evidence for a social custom in wild chimpanzees? *Man* (N.S.), 13:234-251.

McGrew, William C.; Tutin, Caroline E.G.; and Baldwin, P.J. 1979a. New data on meat eating by wild chimpanzees. *Current Anthropology*, 20:238-239.

McGrew, William C.; Tutin, Caroline E.G.; and Baldwin, P.J. 1979b. Chimpanzees, tools and termites: cross-cultural comparisons of Senegal, Tanzania, and Rio Muni. *Man*, 14:185-214.

McGrew, William C.; Tutin, Caroline E.G.; and Midgett, Palmer S., Jr. 1975. Tool use in a group of captive chimpanzees I. escape. *Zeitschrift für Tierpsychologie*, 37:145-162.

McGrew, William C.; Tutin, Caroline E.G.; Baldwin, Pamela J.; Sharman, Martin John; and Whitten, Anthony 1978. Primates preying upon vertebrates: new records from West Africa. *Carnivore*, 1:41-45.

Medway, Lord 1966. Animal remains from Lobang Angus, Niah. *Sarawak Museum Journal*, 14:185-216.

Medway, Lord 1972. The Gunong Benom expedition, 1967. 6. the distribution and latitudinal zonation of birds and mammals on Gunong Benom. *Bulletin of the British Museum of Natural History, Zoology*, 23:103-154.

Menzel, Emil W., Jr. 1964. Responsiveness to object-movement in young chimpanzees. *Behaviour*, 24:147-160.

Menzel, Emil W., Jr. 1969. Chimpanzee utilization of space and responsiveness to objects: age differences and comparison with macaques. *Proceedings of the Second International Congress of Primatology*, Atlanta, GA 1968, Vol. 1, *Behavior* (C.R. Carpenter, ed.), pp. 72-80, Basel: Karger.

Menzel, Emil W., Jr. 1970. Menzel reporting on spontaneous use of poles as ladders. *Delta Primate Report* (December).

Menzel, Emil W., Jr. 1971a. Group behavior in young chimpanzees: responsiveness to cumulative novel changes in a large outdoor enclosure. *Journal of Comparative and Physiological Psychology*, 74:46-51.

Menzel, Emil W., Jr. 1971b. Communication about the environment in a group of young chimpanzees. *Folia Primatologica*, 15:220-232.

Menzel, Emil W., Jr. 1972. Spontaneous invention of ladders in a group of young chimpanzees. *Folia Primatologica*, 17:87-106.

Menzel, Emil W., Jr. 1973a. Further observations on the use of ladders in a group of young chimpanzees. *Folia Primatologica*, 19:450-457.

Menzel, Emil W., Jr. 1973b. Leadership and communication in young chimpanzees. *Symposia of the Fourth International Congress of Primatology, Portland, Ore.*, 1972, Vol. 1, *Precultural Primate Behavior* (Emil W. Menzel, Jr., ed), pp. 192-225, Basel: Karger.

Menzel, Emil W., Jr. 1973c. Chimpanzee spatial memory organization. *Science*, 182:943-945.

Menzel, Emil W., Jr. 1974. A group of young chimpanzees in a one-acre field, in: *Behavior of Nonhuman Primates: Modern Research Trends* (A.M. Schrier and F. Stollnitz, eds.), pp. 83-153, New York: Academic Press.

Menzel, Emil W., Jr. 1978. Cognitive mapping in chimpanzees, in: *Cognitive Processes in Animal Behavior* (Stewart A. Hulse, Harry Fowler, and Werner K. Honig, eds.), pp. 375-422, Hillsdale, New Jersey: Lawrence Erlbaum Assoc., Publishers.

Menzel, Emil W., Jr. 1979. Communication of object-locations in a group of young chimpanzees, in: *The Great Apes* (David A. Hamburg and Elizabeth R. McCown, eds.), pp. 359-371, Menlo Park, California: The Benjamin/Cummings Publishing Company.

Menzel, Emil W., Jr., and Davenport, Richard K., Jr. 1961. Preference for clear versus distorted viewing in the chimpanzee. *Science*, 134:1531.

Menzel, Emil W., Jr.; Davenport, Richard K., Jr.; and Rogers, Charles M. 1961. Some aspects of behavior toward novelty in young chimpanzees. *Journal of Comparative and Physiological Psychology*, 54:16-19.

Menzel, Emil W., Jr.; Davenport, Richard K.; and Rogers, Charles M. 1970. The development of tool using in wild-born and restriction-reared chimpanzees. *Folia Primatologica*, 12:273-283.

Menzel, Emil W., Jr.; Davenport, Richard K.; and Rogers, Charles M. 1972. Protocultural aspects of chimpanzees' responsiveness to novel objects. *Folia Primatologica*, 16:161-170.

Menzel, Emil E., and Halperin, Stewart 1975. Purposive behavior as a basis for objective communication between chimpanzees. *Science*, 189:652-654.

Menzel, Emil E., Jr.; Premack, David; and Woodruff, Guy 1978. Map reading in chimpanzees. *Folia Primatologica*, 29:241-249.

Menzel, Emil W., Jr.; Savage-Rumbaugh, E. Sue; and Lawson, Janet 1985. Chimpanzee (*Pan troglodytes*) spatial problem solving with the use of mirrors and televised equivalents of mirrors. *Journal of Comparative Psychology*, 99:211-217.

Merfield, Fred G., and Miller, Harry 1956. *Gorilla Hunter*, New York: Farrar, Straus and Cudahy.

Meyer, Donald R.; Treichler, F. Robert; and Meyer, Patricia M. 1965. Discrete-trial training techniques and stimulus variables, in: *Behavior of Nonhuman Primates, Modern Research Trends* (Allan M. Schrier, Harry F. Harlow, and Fred Stollnitz, eds.), Vol. 1, pp. 1-49, New York: Academic Press.

Michael, Jack 1984. Verbal behavior. *Journal of the Experimental Analysis of Behavior*, 42:363-376.

Miles, H. Lyn 1976. The communicative competence of child and chimpanzee. *Annals of the New York Academy of Sciences*, 280:592-597.

Miles, H. Lyn 1978. Language acquisition in apes and children, in: *Sign Language and Language Acquisition in Man and Ape: New Dimensions in Comparative Pedolinguistics* (Fred C.C. Peng, ed.), pp. 103-120, Boulder, Colorado: Westview Press, Inc.

Miles, H. Lyn 1983. Apes and language: the search for communicative competence, in: *Language in Primates: Perspectives and Implications* (Judith de Luce and Hugh T. Wilder, eds.), pp. 43-61, New York: Springer-Verlag.

Miles, Raymond C. 1965. Discrimination-learning sets, in: *Behavior of Nonhuman Primates: Modern Research Trends* (Allan Schrier, Harry F. Harlow, and Fred Stollnitz, eds.), Vol. 1, pp. 51-95, New York: Academic Press.

Miller, Gerrit S., Jr. 1942. Zoological results of the George Vanderbilt Sumatran Expedition, 1936-1939. Part V. Mammals collected by Frederick A. Ulmer, Jr. on Sumatra and Nias. *Proceedings of the Academy of Natural Sciences of Philadelphia*, 94:107-165.

Milner, A. David, and Ettlinger, George 1970. Cross-modal transfer of serial reversal learning in the monkey. *Neuropsychologia*, 8:251-258.

Mistler-Lachman, Janet L., and Lachman, Roy 1974. Language in man, monkeys, and machines. *Science*, 186:871-872.

Mitani, John C. 1984. The behavioral regulation of monogamy in gibbons (*Hylobates muelleri*). *Behavioral Ecology and Sociobiology*, 15:225-229.

Mitani, John C. 1985a. Responses of gibbons (*Hylobates muelleri*) to self, neighbor, and stranger song duets. *International Journal of Primatology*, 6:193-200.

Mitani, John C. 1985b. Gibbon song duets and intergroup spacing. *Behaviour*, 92:59-96.

Mitani, John C. 1985c. Mating behaviour of male orangutans in the Kutai Game Reserve, Indonesia. *Animal Behaviour*, 33:392-402.

Mitani, John C. 1985d. Sexual selection and adult male orangutan long calls. *Animal Behaviour*, 33:272-283.

Monfort, A., and Monfort, N. 1973. Quelques observations sur les grands mammifères du Parc national de Tai (Côte d'Ivoire). *La Terre et la Vie*, 27:499-506.

Mori, Akio 1982. An ethological study on chimpanzees at the artificial feeding place in the Mahale Mountains, Tanzania—with special reference to the booming situation. *Primates*, 23:45-65.

Mori, Akio 1983. Comparison of the communicative vocalizations and behaviors of group ranging in eastern gorillas, chimpanzees and pygmy chimpanzees. *Primates*, 24:486-500.

Mori, Akio 1984. An ethological study of pygmy chimpanzees in Wamba, Zaire: a comparison with chimpanzees. *Primates*, 25:255-278.

Morris, Kathryn, and Goodall, Jane 1977. Competition for meat between chimpanzees and baboons of the Gombe National Park. *Folia Primatologica*, 28:109-121.

Muncer, Steven J. 1983a. "Conservations" with a chimpanzee. *Developmental Psychobiology*, 16:1-11.

Muncer, Steven J. 1983b. Is Nim, the chimpanzee, problem-solving? *Perceptual and Motor Skills*, 57:132-134.

Muncer, Steven J., and Ettlinger, George 1981. Communication by a chimpanzee: first-trial mastery of word order that is critical for meaning, but failure to negate conjunctions. *Neuropsychologia*, 19:73-78.

Murnyak, Dennis F. 1981. Censusing the gorillas in Kahuzi-Biega National Park. *Biological Conservation*, 21:163-176.

Myers, B.J.; Kunz, R.E.; and Kamara, J.A. 1973. Parasites and commensals of chimpanzees captured in Sierra Leone, West Africa. *Proceedings of the Helminthological Society of Washington*, 40:298-299.

Myers, Richard H., and Shafer, David A 1979. Hybrid ape offspring of a mating of gibbon and siamang. *Science*, 205:308-310.

Myers, Ronald E. 1967. Cerebral connectionism and brain function, in: *Brain Mechanisms Underlying Speech and Language*, (F.L. Darley, ed.). Proceedings of a conference held at Princeton, New Jersey, 1965, pp. 61-72, New York: Grune and Stratton.

Napier, John R. 1960. Studies on the hands of living primates. *Proceedings of the Zoological Society, London*, 134:647-656.

Napier, John R. 1961. Prehensility and opposability in the hands of primates. *Symposia of the Zoological Society, London*, 5:115-132.

Napier, John R., and Napier, Prunella H. 1967. *A Handbook of Living Primates*, New York: Academic Press.

Nash, Victoria J. 1982. Tool use by captive chimpanzees at an artificial termite mound. *Zoo Biology*, 1:211-221.

Nicolson, Nancy A. 1977. A comparison of early behavioral development in wild and captive chimpanzees, in: *Primate Bio-social Development: Biological, Social, and Ecological Determinants* (Suzanne Chevalier-Skolnikoff and Frank E. Poirier, eds.), pp. 529-560, New York: Garland Publishing, Inc.

Niemitz, Carsten, and Kok, Daniel 1976. Observations on the vocalization of a captive infant Orang utan (*Pongo pygmaeus*). *Sarawak Museum Journal*, 24:237-250.

Nishida, Toshisada 1968. The social group of wild chimpanzees in the Mahali Mountains. *Primates*, 9:167-224.

Nishida, Toshisada 1970. Social behavior and relationship among wild chimpanzees of the Mahali Mountains. *Primates*, 11:47-87.

Nishida, Toshisada 1972a. Preliminary information on the pygmy chimpanzees (*Pan paniscus*) of the Congo Basin. *Primates*, 13:415-425.

Nishida, Toshisada 1972b. A note on the ecology of the red-colobus monkeys (*Colobus badius tephrosceles*) living in the Mahali Mountains. *Primates*, 13:57-64.

Nishida, Toshisada 1973. The ant-gathering behaviour by the use of tools among wild chimpanzees of the Mahali Mountains. *Journal of Human Evolution*, 2:357-370.

Nishida, Toshisada 1976. The bark-eating habits in Primates, with special reference to their status in the diet of wild chimpanzees. *Folia Primatologica*, 25:277-287.

Nishida, Toshisada 1979a. The social structure of chimpanzees of the Mahale Mountains, in: *The Great Apes* (D.A. Hamburg and E.R. McCown, eds.), pp. 72-121, Menlo Park, California: The Benjamin/Cummings Publishing Co.

Nishida, Toshisada 1980. Local differences in responses to water among wild chimpanzees. *Folia Primatologica*, 33:189-209.

Nishida, Toshisada 1983a. Alpha status and agonistic alliance in wild chimpanzees (*Pan troglodytes*). *Primates*, 24:318-336.

Nishida, Toshisada 1983b. Alloparental behavior in wild chimpanzees of the Mahale Mountains, Tanzania. *Folia Primatologica*, 41:1-33.

Nishida, Toshisada, and Hiraiwa, Mariko 1982. Natural history of a tool-using behavior by wild chimpanzees in feeding upon wood-boring ants. *Journal of Human Evolution*, 11:73-99.

Nishida, Toshisada, and Hiraiwa-Hasegawa, Mariko 1984. Behavior of an adult male in one-male unit-group of chimpanzees in the Mahale Mountains, Tanzania. *International Journal of Primatology*, 5:367.

Nishida, Toshisada, and Hiraiwa-Hasegawa, Mariko 1985. Responses to a stranger mother-son pair in the wild chimpanzee: a case report. *Primates*, 26:1-13.

Nishida, Toshisada; Hiraiwa-Hasegawa, Mariko; Hasegawa, Toshikazu; and Takahata, Yukio 1985. Group extinction and female transfer in wild chimpanzees in the Mahale National Park, Tanzania. *Zeitschrift für Tierpsychologie*, 67:284-301.

Nishida, Toshisada, and Kawanaka, Kenji 1972. Inter-unit-group relationships among wild chimpanzees of the Mahali Mountains. *Kyoto University African Studies*, 7:131-169.

Nishida, Toshisada, and Kawanaka, Kenji 1985. Within-group cannibalism by adult male chimpanzees. *Primates*, 26:274-284.

Nishida, Toshisada, and Uehara, Shigeo 1983. Natural diet of chimpanzees (*Pan troglodytes schweinfurthii*): long-term record from the Mahale Mountains, Tanzania. *African Study Monographs*, 3:109-130.

Nishida, Toshisada; Uehara, Shigeo; and Nyundo, Ramadhani 1979. Predatory behavior among wild chimpanzees of the Mahale Mountains. *Primates*, 20:1-20.

Nishida, Toshisada; Wrangham, Richard W.; Goodall, Jane; and Uehara, Shigeo 1983. Local differences in plant-feeding habits of chimpanzees between the Mahale Mountains and Gombe National Park, Tanzania. *Journal of Human*

Evolution, 12:467-480.

Nissen, Henry W. 1931. A field study of the chimpanzee. *Comparative Psychology Monographs*, 8:1-122.

Nissen, Henry W. 1942. Ambivalent cues in discriminative behavior of chimpanzees. *Journal of Psychology*, 14:3-33.

Nissen, Henry W. 1951a. Analysis of a complex conditional reaction in chimpanzee. *Journal of Comparative and Physiological Psychology*, 44:9-16.

Nissen, Henry W. 1951b. Phylogenetic comparison, in: *Handbook of Experimental Psychology* (Stanley Smith Stevens, ed.), pp. 347-386, New York: Wiley.

Nissen, Henry W.; Blum, Josephine Semmes, and Blum, Robert A. 1948. Analysis of matching behavior in chimpanzee. *Journal of Comparative and Physiological Psychology*, 41:62-74.

Nissen, Henry W.; Blum, Josephine Semmes; and Blum, Robert A. 1949. Conditional matching behavior in chimpanzees; implications for the comparative study of intelligence. *Journal of Comparative and Physiological Psychology*, 42:339-356.

Nissen, Henry W.; Carpenter, Clarence Ray; and Cowles, J.T. 1936. Stimulus-versus-response-differentiation in delayed reactions of chimpanzees. *Journal of Genetic Psychology*, 48:112-136.

Nissen, Henry W., and Crawford, Meredith P. 1936. A preliminary study of food-sharing behavior in young chimpanzees. *Journal of Comparative and Physiological Psychology*, 22:383-419.

Nissen, Henry W., and Elder, James H. 1935. The influence of amount of incentive on delayed response performances of chimpanzees. *Journal of Genetic Psychology*, 47:49-72.

Nissen, Henry W., and Harrison, Ross 1941. Visual and positional cues in the delayed responses of chimpanzees. *Journal of Comparative Psychology*, 31:437-445.

Nissen, Henry W., and Jenkins, William Oliver 1943. Reduction and rivalry of cues in the discrimination behavior of chimpanzees. *Journal of Comparative Psychology*, 35:85-95.

Nissen, Henry W., and McCulloch, Thomas L. 1937a. Equated and non-equated stimulus situations in discrimination learning by chimpanzees. I. Comparison with unlimited response. *Journal of Comparative Psychology*, 23:165-189.

Nissen, Henry W., and McCulloch, Thomas L. 1937b. Equated and non-equated stimulus situations in discrimination learning by chimpanzees. III. prepotency of response to oddity through training. *Journal of Comparative Psychology*, 23:377-381.

Nissen, Henry W.; Riesen, Austin H.; and Nowlis, Vincent 1938. Delayed response and discrimination learning by chimpanzees. *Journal of Comparative Psychology*, 26:361-386.

Nissen, Henry W., and Taylor, F.V. 1939. Delayed alteration to non-positional cues in chimpanzee. *Journal of Psychology*, 7:323-332.

Norikoshi, Kohshi 1982. One observed case of cannibalism among wild chimpanzees of the Mahale Mountains. *Primates*, 23:66-74.

Norikoshi, Kohshi 1983. Prevalent phenomenon of predation observed among wild chimpanzees of the Mehale Mountains. *Journal of the Anthropological Society, Nippon*, 91:475-480.

Okada, Morihiko 1985. Primate bipedal walking: comparative kinematics, in:

Primate Morphophysiology, Locomotor Analyses and Human Bipedalism (S. Kondo, ed.), pp. 47-58, Tokyo: University of Tokyo Press.

Okada, Morihiko, and Kondo, Shiro 1982. Gait and EMGs during bipedal walk of a gibbon (*Hylobates agilis*) on flat surface. *The Journal of the Anthropological Society of Nippon*, 90:325-330.

Okada, Morihiko; Yamazaki, Nobutoshi; Ishida, Hidemi; Kimura, Tasuku; and Kondo, Shiro 1983. Biomechanical characteristics of hylobatid bipedal walking on flat surfaces. *Annales des Sciences Naturelles, Zoologie, Paris*, Series 13, 5:137-144.

Okano, T. 1965. Preliminary survey of the orangutan in North Borneo (Sabah). *Primates*, 6:123-128.

Osborn, Rosalie M. 1963. Observations on the behaviour of the mountain gorilla. *Symposia of the Zoological Society, London*, 10:29-37.

Oxnard, Charles E. 1983. Sexual dimorphisms in the overall proportions of primates. *American Journal of Primatology*, 4:1-22.

Pakkenberg, H., and Voigt, J. 1964. Brain weight of the Danes. *Acta anatomica*, 56:297-307.

Palmans, M. 1956. Un chimpanzé pas comme les autres: *Pan paniscus. Zoo*, 21:80-84.

Papaiouannou, J. 1973. Observations on locomotor and general behavior of the siamang. *Malayan Nature Journal*, 26:46-52.

Parker, Christopher E. 1968. The use of tools by apes. *Zoonooz*, 41:10-13.

Parker, Christopher E. 1969. Responsiveness, manipulation, and implementation behavior in chimpanzees, gorillas, and orang-utans, in: *Proceedings of the Second International Congress of Primatology, Atlanta, Ga., 1968*, Vol. 1, *Behavior* (C.R. Carpenter, ed.), pp. 160-166, Basel: Karger.

Parker, Christopher E. 1973. Manipulatory behavior and responsiveness, in: *Gibbon and Siamang*, (D.M. Rumbaugh, ed.), Vol. 2, pp. 185-207, Basel: Karger.

Parker, Christopher E. 1974. Behavioral diversity in ten species of nonhuman primates. *Journal of Comparative and Physiological Psychology*, 87:930-937.

Parker, Christopher E. 1978 Opportunism and the rise of intelligence. *Journal of Human Evolution*, 7:597-608.

Parker, Sue Taylor, and Gibson, Kathleen R. 1977. Object manipulation, tool use and sensorimotor intelligence as feeding adaptations in cebus monkeys and great apes. *Journal of Human Evolution*, 6:623-641.

Passingham, Richard E. 1973 Anatomical differences between the neocortex of man and other primates. *Brain, Behavior and Evolution*, 7:337-359.

Passingham, Richard E. 1975a. The brain and intelligence. *Brain, Behavior and Evolution*, 11:1-5.

Passingham, Richard E. 1975b. Changes in the size and organization of the brain in man and his ancestors. *Brain, Behavior and Evolution*, 11:73-90.

Passingham, Richard E. 1978. Brain size and intelligence in primates, in: *Recent Advances in Primatology*, Vol. 3, *Evolution* (D.J. Chivers and K.A. Joysey, eds.), pp. 85-86, London: Academic Press.

Passingham, Richard E. 1981a. Primate specialization in brain and intelligence. *Symposia of the Zoological Society, London*, 46:361-388.

Passingham, Richard E. 1981b. Brocas's area and the origins of human vocal skill. *Philosophical Transactions of the Royal Society, London* B, 292:167-175.

Passingham, Richard E. 1982. *The Human Primate*, Oxford, England: W.H. Freeman and Company.

Passingham, Richard E., and Ettlinger, George 1974. A comparison of cortical functions in man and the other primates. *International Review of Neurobiology*, 16:233-299.

Pate, James L., and Rumbaugh, Duane M. 1983. The language-like behavior of Lana chimpanzee: is it merely discrimination and paired-associate learning? *Animal Learning and Behavior*, 11:134-138.

Patterson, Francine G. 1978a. The gestures of a gorilla: language acquisition in another pongid. *Brain and Language*, 5:72-97.

Patterson, Francine 1978b. Conversations with a gorilla. *The National Geographic Magazine*, 154:438-465.

Patterson, Francine 1978c. Lingustic capabilities of a lowland gorilla, in: *Sign Language and Language Acquisitions in Man and Ape: New Dimensions in Comparative Pedolinguistics* (Fred C.C. Peng, ed.), pp. 161-201, Boulder, Colorado: Westview Press, Inc.

Patterson, Francine Grace Penelope 1978d. Francine Grace Penelope Patterson, in: *In the Spirit of Enterprise: From the Rolex Awards* (Gregory B. Stone, ed.), pp. 182-186, San Francisco: W.H. Freeman and Company.

Patterson, Francine G. 1980. Innovative uses of language by a gorilla: a case study, in: *Children's Language*, Vol. 2 (Keith E. Nelson, ed.), pp. 497-561, New York: Gardner Press, Inc.

Patterson, Francine G. 1981. Ape language. *Science*, 211: 86-87.

Patterson, Francine 1983. Child, chimp and gorilla: a comparison of early language samples. *Gorilla*, 7:2-3.

Patterson, Francine, and Linden, Eugene 1981. *The Education of Koko*, New York: Holt, Rinehart and Winston.

Patterson, Thomas L., and Tzeng, Ovid J.L. 1979. Long-term memory for abstract concepts in the lowland gorilla (*Gorilla g. Gorilla*). *Bulletin of the Psychonomic Society*, 13:279-282.

de Pelham, Alison, and Burton, Frances D. 1976. More on predatory behaviour in nonhuman primates. *Current Anthropology*, 17:512-513.

de Pelham, Alison, and Burton, Frances D. 1977. Reply. *Current Anthropology*, 18:108-109.

Penfield, Wilder, and Roberts, Lamar 1959. *Speech and Brain-mechanisms*, Princeton, New Jersey: Princeton University Press.

Peterson, Michael R.; Beecher, Michael D.; Zoloth, Stephen R.; Moody, David B.; and Stebbins, William C. 1978. Neural lateralization of species-specific vocalizations by Japanese macaques (*Macaca fuscata*). *Science*, 202:324-327.

Peterson, Michael R.; Beecher, Michael D.; Zoloth, Stephen R.; Green, Steven; Marler, Peter R.; Moody, David B.; and Stebbins, William C. 1984. Neural lateralization of vocalizations by Japanese macaques: communicative significance is more important than acoustic structure. *Behavioral Neuroscience*, 98: 779-790.

Petitto, Laura A., and Seidenberg, Mark S. 1979. On the evidence for linguistic abilities in signing apes. *Brain and Language*, 8:162-183.

Pfeifer, Richard Arwed 1936. Pathologie der Horstrahlung und der corticalen Horsphare, in: *Handbuch der Neurologie*, Vol. 6 (O. Bumke and O. Foerster, eds.), pp. 533-626, Berlin: Springer.

Phillips, J. 1959. *Agriculture and Ecology in Africa*, London: Faber & Faber.

Piaget, Jean 1973. *The Psychology of Intelligence*, Totowa, New Jersey: Littlefield.

Pilbeam, David R. 1969a. Tertiary Pongidae of East Africa: evolutionary relationships and taxonomy. *Bulletin of the Peabody Museum of Natural History*, 21:1-185.

Pilbeam, David R. 1969b. Newly recognized mandible of *Ramipithecus. Nature*, 222:1093-1094.

Pilbeam, David R. 1972. *The Ascent of Man*, New York: The MacMillan Co.

Pilbeam, David 1976. Neogene hominoids of South Asia and the origins of Hominidae, in: *Les Plus Anciens Hominidés* (P.V. Tobias and Y. Coppens, eds.), pp. 39-59, Paris: Centre National de la Recherche Scientifique.

Pilbeam, David 1982. New hominoid skull material from the Miocene of Pakistan. *Nature*, 295:232-234.

Pilbeam, David; Barry, John; Meyer, Grant E.; Ibrahim Shah, S.M.; Pickford, Martin H.L.; Bishop, William W.; Thomas, H.; and Jacobs, Louis L. 1977. Geology and palaeontology of Neogene strata of Pakistan. *Nature*, 270:684-689.

Pilbeam, David; Meyer, Grant E.; Badgley, Catherine; Rose, Michael D.; Pickford, Martin H.L.; Behrensmeyer, A.K.; and Ibrahim Shah, S.M. 1977. New hominoid primates from the Siwaliks of Pakistan and their bearing on hominoid evolution. *Nature*, 270:689-695.

Pilbeam, David R.; Rose, Michael D.; Badgley, Catherine; and Lipschutz, Bonnie 1980. Miocene hominoids from Pakistan. *Postilla*, 181:1-94.

Pilbeam, David, and Smith, Richard 1981. New skull remains of *Sivapithecus* from Pakistan. *Memoir Geological Survey of Pakistan*, 11:1-13.

Pitman, C.R. 1931. *A Game Warden among His Charges*, London: Nisbet.

Pitman, Capt. C.R.S. 1935. The gorillas of the Kayonsa Region, Western Kigezi, S.W. Uganda. *Proceedings of the Zoological Society, London*, 105:477-494.

Plooij, Frans X. 1978. Some basic traits of language in wild chimpanzees?, in: *Action, Gesture and Symbol, the Emergence of Language* (Andrew Lock, ed.), pp. 111-113, London: Academic Press.

Plooij, Frans X. 1979. How wild chimpanzee babies trigger the onset of mother-infant play—and what the mother makes of it, in: *Before Speech, the Beginnings of Human Communication* (Margaret Bullowa, ed.), pp. 223-243, Cambridge, England: Cambridge University Press.

Pocock, Reginald I. 1905. Observations upon a female specimen of Hainan gibbon (*Hylobates hainanus*), now living in the Society's Gardens. *Proceedings of the Zoological Society, London* 1905, (II):169-180.

Pohl, Peter 1983. Central auditory processing V. ear advantages for acoustic stimuli in baboons. *Brain and Language*, 20:44-53.

Polis, Gary A.; Myers, Christopher A.; and Hess, William R. 1984. A survey of intraspecific predation within the class Mammalia. *Mammal Review*, 14:187-198.

Prasad, K. 1969. Critical observations on the fossil anthropoids from the Siwaliks of India. *Folia Primitologica*, 10:288-317.

Premack, Ann James 1976. *Why Chimps Can Read*, New York: Harper and Row.

Premack, Ann James, and Premack, David 1972. Teaching language to an ape. *Scientific American*, 227:92-99.

Premack, Ann, and Premack, David 1975. Le pouvoir du mot chez les chimpanzés.

La Recherche, 6:918-925.

Premack, David 1970. A functional analysis of language. *Journal of the Experimental Analysis of Behavior*, 14:107-125.

Premack, David 1971a. On the assessment of language competence in the chimpanzee, in: *Behavior of Nonhuman Primates: Modern Research Trends*, Vol. 4 (Allan M. Schrier and Fred Stollnitz, eds.), pp. 185-228, New York: Academic Press.

Premack, David 1971b. Language in chimpanzees? *Science*, 172:808-822.

Premack, David 1971c. Some general characteristics of a method for teaching language to organisms that do not ordinarily acquire it, in: *Cognitive Processes of Nonhuman Primates* (L.E. Jarrard, ed.), pp. 47-82, New York: Academic Press.

Premack, David 1975. Putting a face together. *Science*, 188:228-236.

Premack, David 1976a. *Intelligence in Ape and Man*, Hillsdale, New Jersey: Lawrence Erlbaum Associates, Publishers.

Premack, David 1976b. Language and intelligence in ape and man. *American Scientist*, 64:674-683.

Premack, David 1976c. On the study of intelligence in chimpanzees. *Current Anthropology*, 17:516-521.

Premack, David 1977. The human ape? *The Sciences*, 17:20-23.

Premack, David 1978. Comparison of language-related factors in ape and man, in: *Recent Advances in Primatology*, Vol. 1, *Behaviour* (D.J. Chivers and J. Herbert, eds.), pp. 867-881, London: Academic Press.

Premack, David 1983. Animal cognition. *Annual Review of Psychology*, 34:351-362.

Premack, David, and Premack, Ann James 1983. *The Mind of an Ape*, New York: W.W. Norton & Co.

Premack, David, and Schwartz, A. 1966. Preparations for discussing behaviourism with chimpanzee, in: *The Genesis of Language: a Psycholingustic Approach* (F. Smith and G.A. Miller, eds.), pp. 295-325, Cambridge, Massachusetts: MIT Press.

Premack, David, and Woodruff, Guy 1978a. Chimpanzee problem-solving: a test for comprehension. *Science*, 202:532-535.

Premack, David, and Woodruff, Guy 1978b. Does the Chimpanzee have a theory of mind? *The Behavioral and Brain Sciences*, 1:515-526.

Premack, David; Woodruff, Guy; and Kennel, Keith 1978. Paper-marking test for chimpanzee: simple control for social cues. *Science*, 202:903-905.

Prestrude, Albert M. 1970. Sensory capacities of the chimpanzee: a review. *Psychological Bulletin*, 74:47-67.

Preuschoft, Holger, and Demes, Brigitte 1984. Biomechanics of brachiation, in: *The Lesser Apes: Evolutionary and Behavioural Biology* (H. Preuschoft, D.J. Chivers, W.Y. Brockelman, and N. Creel, eds.), pp. 96-118, Edinburgh: Edinburgh University Press.

Preuss, Todd M. 1982. The face of *Sivapithecus indicus*: description of a new, relatively complete specimen from the Siwaliks of Pakistan. *Folia Primitologica*, 38:141-157.

Prost, Jack H. 1965. A definitional system for the classification of primate loco-motion. *American Anthropologist*, 67:1198-1214.

Prost, Jack H. 1967. Bipedalism of man and gibbon compared using estimates of joint motion. *American Journal of Physical Anthropology*, 36:135-148.

Prouty, Leonard A.; Buchanan, Philip D.; Pollitzer, William S.; and Mootnick, Alan R. 1983a. A presumptive new hylobatid subgenus with 38 chromosomes. *Cytogenetics and Cell Genetics*, 35:141-142.

Prouty, Leonard A.; Buchanan, Philip D.; Pollitzer, William S.; and Mootnick, Alan R. 1983b. *Bunopithecus*: a genus-level taxon for the hoolock gibbon (*Hylobates hoolock*). *American Journal of Primatology*, 5:83-87.

Pusey, Anne E. 1979. Intercommunity transfer of chimpanzees in Gombe National Park, in: *The Great Apes* (David A. Hamburg and Elizabeth R. McCown, eds.), pp. 465-479, Menlo Park, California: The Benjamin/Cummings Publishing Company.

Pusey, Anne E. 1980. Inbreeding avoidance in chimpanzees. *Animal Behaviour*, 28:543-552.

Putney, R. Thompson 1985. Do willful apes know what they are aiming at? *The Psychological Record*, 35:49-62.

Radinsky, Leonard 1972. Endocasts and studies of primate brain evolution, in: *The Functional and Evolutionary Biology of Primates* (R. Tuttle, ed.), pp. 175-184, Chicago: Aldine.

Raemaekers, Jeremy 1978a. Changes through the day in the food choice of wild gibbons. *Folia Primatologica*, 30:194-205.

Raemaekers, Jeremy John 1978b. The sharing of food sources between two gibbon species in the wild. *Malayan Nature Journal*, 31:181-188.

Raemaekers, Jeremy 1984. Large versus small gibbons: relative roles of bioenergetics and competition in their ecological segregation in sympatry, in: *The Lesser Apes: Evolutionary and Behavioural Biology* (H. Preuschoft, D.J. Chivers, W.Y. Brockelman, and N. Creel, eds.), pp. 209-218, Edinburgh: Edinburgh University Press.

Raemaekers, Jeremy J., and Raemaekers, Patricia M. 1985. Field playback of loud calls to gibbons (*Hylobates lar*): territorial, sex-specific and species-specific responses. *Animal Behaviour*, 33:481-493.

Rahm, Urs 1967. Observations during chimpanzee captures in the Congo, in: *Neue Ergebnisse der Primatologie* (D. Starck, R. Schneider, and H.J. Kuhn, eds.), pp. 195-207, Stuttgart: Fischer.

Razran, Gregory 1961. The observable unconscious and inferrable conscious in current Soviet psychophysiology: interoceptive conditioning, semantic conditioning, and orienting reflex. *Psychological Review*, 68:81-147.

Reade, William Winwood 1864. *Savage Africa; being the narrative of a tour in equatorial, southwestern, and northwestern Africa; with notes on the habits of the gorilla; on the existence of unicorns and tailed men; on the slave-trade; on the origin, character, and capabilities of the Negro, and of the future civilization of Western Africa*. New York: Harper.

Redshaw, Margaret 1978. Cognitive development in human and gorilla infants. *Journal of Human Evolution*, 7:133-141.

Rempe, U. 1961. Einige Beobachtungen an Bonobos, *Pan paniscus* Schwarz 1929. *Zeitschrift für Wissenschaftliche Zoologie, Leipzig*, 165:81-87.

Rensch, Bernhard, and Döhl, Jürgen 1968. Wahlen zwischen zwei Uberschaubaren Labyrinthwegen durch einen Schimpansen. *Zeitschrift für Tierpsychologie*, 25:216-231.

Rensch, Bernhard, and Ducker, Gertrude 1966. Manipulierfahigkeit eines jungen

Orang-Utans und eines jungen Gorillas. Mit Anmerkungen uber das Spielverhalten. *Zeitschrift für Tierpsychologie*, 23:874-892.

Reynolds, Herbert H., and Farrer, Donald N. 1969. Chimpanzee psychobiology, in: *Primates in Medicine*, Vol. 4 (E.I. Goldsmith and J. Moor-Jankowski, ed.), pp. 62-93, Basel: Karger.

Reynolds, Vernon 1964. Ecologie et comportement social des chimpanzé de la foret de Budongo, Ouganga. *La Terre et al Vie*, 2:155-166.

Reynolds, Vernon 1965. *Budongo, a Forest and Its Chimpanzees*, London: Metheun and Co. Ltd.

Reynolds, Vernon 1975. How wild are the Gombe chimpanzees? *Man*, 10:123-125.

Reynolds, Vernon, and Luscombe, Gillian P. 1969a. Chimpanzee rank order and the function of displays, in: *Social Behavior of Chimpanzees in an Open Environment, IV*, Holloman Air Force Base, New Meixco: 6571st Aeromedical Research Laboratory.

Reynolds, Vernon, and Luscombe, Gillian P. 1969b. Chimpanzee social behavior, in: *Social Behavior of Chimpanzees in an Open Environment, IV*, Holloman Air Force Base, New Mexico: 6571st Aeromedical Research Laboratory.

Reynolds, Vernon, and Luscombe, Gillian 1969c. Chimpanzee rank order and the function of displays. *Proceedings of the Second International Congress of Primatology, Atlanta, Ga, 1968*, Vol. 1, *Behavior* (C.R. Carpenter, ed.), pp. 81-86, Basel: Karger.

Reynolds, Vernon, Luscombe, Gillian 1976. Greeting behaviour, displays and rank order in a group of free-ranging chimpanzees, in: *The Social Structure of Attention* (Michael R.A. Chance and Ray R. Larsen, eds.), pp. 105-115, London: John Wiley & Sons.

Reynolds, Vernon, and Reynolds, Frances 1965. Chimpanzee of the Budongo Forest, in: *Primate Behavior* (I. DeVore, ed.), pp. 368-424, New York: Holt, Rinehart & Winston.

Richards, Paul W. 1952. *The Tropical Rain Forest*, Cambridge, England: Cambridge University Press.

Riesen, Austin H. 1940. Delayed reward in discrimination learning by chimpanzees. *Comparative Psychology Monographs*, 15:1-54.

Riesen, Austin H. 1970. Chimpanzee visual perception, in: *The Chimpanzee*, Vol. 2, *Physiology, Behavior, Serology and Diseases of Chimpanzees* (G.H. Bourne, ed.), pp. 1-15, Basel: Karger.

Riesen, Austin H.; Greenberg, Bernard; Granston, Arthur S.; and Frantz, Robert L. 1953. Solutions of patterned string problems by young gorillas. *Journal of Comparative and Physiological Psychology*, 46:19-22.

Riesen, Austin H., and Kinder, Elaine F. 1952. *The Postural Development of Infant Chimpanzees*, New Haven: Yale University Press.

Riesen, Austin H., and Nissen, Henry W. 1942. Nonspatial delayed response by the matching technique. *Journal of Comparative Psychology*, 34:307-313.

Rijksen, Herman D. 1975. Social structure in a wild orangutan population in Sumatra, in: *Comtemporary Primatology* (S. Kondo, M. Kawai, and A. Ehara, eds.), pp. 373-379, Basel: Karger.

Rijksen, Herman D. 1978. *A Field Study on Sumatran Orang Utans* (Pongo pygmaeus abelii Lesson 1827), Wageningen, Nederland: H. Veenman & Zonen.

Rijksen, Herman D., and Rijksen-Graatsma, Ans G. 1975. Orang utan rescue work in North Sumatra. *Oryx*, 13:63-73.

Riopelle, Arthur J. 1963. Growth and behavioral changes in chimpanzees. *Zeitschrift für Morphologie und Anthropologie*, 53:53-61.

Riopelle, Arthur J., and Hill, Charles W. 1973. Complex processes, in: *Comparative Psychology: A Modern Survey* (Donald A. Dewsbury and Dorothy A. Rethling-schafer, eds.), pp. 510-546, New York: McGraw-Hill Book Company.

Riopelle, Arthur J., and Rogers, Charles M. 1963. Behavior of chimpanzees of different ages. *Activitas Nervosa Superior*, 5:260-263.

Riopelle, Arthur J., and Rogers, Charles M. 1965. Age changes in chimpanzees, in: *Behavior of Nonhuman Primates, Modern Research Trends* (Allan M. Schrier, Harry F. Harlow, and Fred Stollnitz, eds.), Vol. 2, pp. 449-462, New York: Academic Press.

Riss, David C., and Busse, Curt D. 1977. Fifty-day observation of a free-ranging adult male chimpanzee. *Folia Primatologica*, 28:283-297.

Riss, David, and Goodall, Jane 1977. The recent rise to the alpha-rank in a population of free-living chimpanzees. *Folia Primatologica*, 27:134-151.

Ristau, Carolyn A. 1983a. Symbols and indication in apes and other species?: comment on Savage-Rumbuagh et al. *Journal of Experimental Psychology: General*, 112:498-507.

Ristau, Carolyn A. 1983b. Language, cognition, and awareness in animals? *Annals of the New York Academy of Sciences*, 406:170-186.

Ristau, Carolyn A., and Robbins, Donald 1982a. Cognitive aspects of ape language experiments, in: *Animal Mind - Human Mind* (Donald R. Griffin, ed.), pp.299-331, Berlin: Springer Verlag.

Ristau, Carolyn A., and Robbins, Donald 1982b. Language in the great apes: a critical review. *Advances in the Study of Behavior*, 12:141-255.

Robbins, Donald, and Bush, Carol T. 1973. Memory in great apes. *Journal of Experimental Psychology*, 97:344-348.

Robbins, Donald, Compton, Phillip; and Howard, Stephen 1978. Subproblem analysis of shift behavior in the gorilla: a transition from independent to cognitive behavior. *Primates*, 19:231-236.

Robinson, John S. 1955. The sameness-difference discrimination problem in chimpanzee. *Journal of Comparative and Physiological Psychology*, 48:195-197.

Robinson, John S. 1960. The conceptual basis of the chimpanzee's performance on the sameness-difference discrimination problem. *Journal of Comparative and Physiological Psychology*, 53:368-370.

Robinson, John T. 1972. *Early Hominid Posture and Locomotion*, Chicago: The University of Chicago Press.

Robinson, P.T. 1971. Wildlife trends in Liberia and Sierra Leone. *Oryx*, 11:117-122.

Rodman, Peter S. 1973a. Population composition and adaptive organization among orang-utans of the Kutai Reserve, in: *Comparative Ecology and Behaviour of Primates* (R.P. Michael and J.H. Crook, eds.), pp. 171-209, London: Academic Press.

Rodman, Peter S. 1973b. Synecology of Bornean primates. I. a test for interspecific interactions in spatial distribution of five species. *American Journal of Physical Anthropology*, 38:655-659.

Rodman, Peter S. 1977. Feeding behaviour of orang-utans of the Kutai Nature Reserve, East Kalimantan, in: *Primate Behaviour: Studies of feeding and ranging behaviour in lemurs, monkeys and apes* (T.H. Clutton-Brock, ed.), pp. 383-413, London: Academic Press.

Rodman, Peter S. 1978. Diets, densities, and distributions of Bornean primates, in: *The Ecology of Arboreal Folivores* (G.G. Montgomery, ed.), pp. 465-478, Washington, D.C.: Smithsonian Institution Press.

Rodman, Peter S. 1979. Individual activity pattern and the solitary nature of orangutans, in: *The Great Apes* (D.A. Hamburg and E.R. McCown, eds.), pp. 234-255, Menlo Park, California: The Benjamin/Cummings Publishing Company.

Rodman, Peter S. 1984. Foraging and social systems of orangutans and chimpanzees, in: *Adaptations for Foraging in Nonhuman Primates* (Peter S. Rodman and John G.H. Cant, eds.), pp. 134-160, New York: Columbia University Press.

Rogers, Charles M., and Davenport, Richard K. 1971. Intellectual performance of differentially reared chimpanzees: III. oddity. *American Journal of Mental Deficiency*, 75:526-530.

Rogers, Charles M., and Davenport, Richard K. 1975. Capacities of nonhuman primates for perceptual integration across sensory modalities, in: *Socioecology and Psychology of Primates* (R.H. Tuttle, ed.), pp. 343-365, The Hague: Mouton.

Rohles, Frederick H., Jr. 1961. The development of an instrumental skill sequence in the chimpanzee. *Journal of the Experimental Analysis of Behavior*, 4:323-325.

Rohles, Frederick H., Jr. 1966. Operant methods in space technology, in: *Operant Behavior: Areas of Research and Application* (Werner K. Honig, ed.), pp. 677-717, Englewood Cliffs, New Jersey: Prentice-Hall, Inc.

Rohles, Frederick H., Jr. 1970. Operant conditioning research with the chimpanzee, in: *The Chimpanzee* (G.H Bourne, ed.), Vol. 2, pp. 289-317, Basel: Karger.

Rohles, Frederick H., Jr.; Belleville, R.E.; and Grunzke, Marvin E. 1961.Measurement of higher intellectual functioning in the chimpanzee—its relevance to the study of behavior in space environments. *Aerospace Medicine*, 32:121-125.

Rohles, Frederick H., and Devine, James V. 1966. Chimpanzee performance on a problem involving the concept of middleness. *Animal Behaviour*, 14:159-162.

Rohles, Frederick H., and Devine, James V. 1967. Further studies of the middleness concept with the chimpanzee. *Animal Behaviour*, 15:107-112.

Rohles, Frederick H., Jr.; Grunzke, Marvin E.; and Reynolds, Herbert H. 1963. Chimpanzee performance during the ballistic and orbital Project Mercury flights. *Journal of Comparative and Physiological Psychiatry*, 56:2-10.

Röhrer-Ertl, Olav 1982. Über Subspezies bei *Pongo pygmaeus*, Linnaeus, 1760. *Spixiana*, 5:317-321.

Röhrer-Ertl, Olav 1983. Zur Erforschungsgeschichte und Namengebung beim Orang-Utan, *Pongo satyrus* (Linnaeus, 1758); Synon. *Pongo pygmaeus* (Hoppius, 1763) (Mit Kurzbibliographie). *Spixiana*, 6:301-332.

Röhrer-Ertl, Olav 1984. *Orang-utan-studien*, Neuried: Hieronymus Verlag.

Rosenthal, Mark A. 1981. Chimpanzee triplets born in captivity. *Primates*, 22:137-138.

Rossi, Gian Franco, and Rosadini, Guido 1967. Experimental analysis of cerebral dominance in man, in: *Brain Mechanisms underlying Speech and Language* (F.L. Darley, ed.), pp. 167-184, Proceedings of a conference held at Princeton, New Jersey, 1965, New York: Grune and Stratton.

Rothblat, Lawrence A., and Wilson, William A., Jr. 1968. Intradimensional and extradimensional shifts in the monkey within and across sensory modalities. *Journal of Comparative and Physiological Psychology*, 66:549-553.

Rumbaugh, Duane M. 1968. The learning and sensory capacities of the squirrel monkey in phylogenetic perspective, in: *The Squirrel Monkey* (Leonard A. Rosenblum and Robert W. Cooper, eds.), pp. 255-317, New York: Academic Press.

Rumbaugh, Duane M. 1969. The transfer index: an alternative measure of learning set. *Proceedings of the Second International Congress of Primatology, Atlanta, GA 1968* (C.R. Carpenter, ed.), Vol. 1, pp. 267-273, Basel: Karger.

Rumbaugh, Duane M. 1970. Learning skills of anthropoids, in: *Primate Behavior: Developments in Field and Laboratory* (L.A. Rosenblum, ed.), Vol. 1, pp. 1-70, New York: Academic Press.

Rumbaugh, Duane M. 1971a. Chimpanzee intelligence, in: *The Chimpanzee* (G.H. Bourne, ed.), Vol. 4, pp. 19-45, Basel: Karger.

Rumbaugh, Duane M. 1971b. Evidence of qualitative differences in learning processes among primates. *Journal of Comparative and Physiological Psychology*, 76:250-255.

Rumbaugh, Duane M. 1974. Comparative primate learning and its contributions to understanding development, play, intelligence, and language, in: *Perspectives in Primate Biology* (B. Chiarelli, ed.), Vol. 9, pp. 253-281, New York: Plenum Press.

Rumbaugh, Duane M. 1974. Lana learns language. *Yerkes Newsletter*, 11:2-7.

Rumbaugh, Duane M. 1975. The learning and symbolizing capacities of apes and monkeys, in: *Socioecology and Psychology of Primates* (Russell H. Tuttle, ed.), pp. 353-365, The Hague: Mouton.

Rumbaugh, Duane M. 1977a. *Language Learning by a Chimpanzee: The Lana Project*, New York: Academic Press.

Rumbaugh, Duane M. 1977b. Language behavior of apes, in: *Behavioral Primatology: Advances in Research and Theory* (Allan M. Schrier, ed.), Vol. 1, pp. 105-138, Hillsdale, New Jersey: Lawrence Erlbaum Associates, Inc.

Rumbaugh, Duane M. 1977c. The emergence and state of ape language research, in: *Progress in Ape Research* (Geoffrey H. Bourne, ed.), pp. 75-83, New York: Academic Press.

Rumbaugh, Duane M. 1978. Ape language projects: a perspective, in: *Recent Advances in Primatology*, Vol. 1, *Behaviour* (D.J. Chivers and J. Herbert, eds.), pp. 855-859, London: Academic Press.

Rumbaugh, Duane M. 1981a. "Siabon" or "Gibbang"?: a reply. *Laboratory Primate Newsletter*, 20:13.

Rumbaugh, Duane M. 1981b. Who feeds Clever Hans? *Annals of the New York Academy of Sciences*, 364:26-34.

Rumbaugh, Duane M., and Gill, Timothy V. 1973. The learning skills of great apes. *Journal of Human Evolution*, 2:171-179.

Rumbaugh, Duane M., and Gill, Timothy V. 1975. Language, apes, and the apple which-is orange, please, in: *Symposia of the Fifth International Primatological Society* (S. Kondo, M. Kawai, E. Ehara, and S. Kawamura, eds.), pp. 247-257, Tokyo: Japan Science Press.

Rumbaugh, Duane M., and Gill, Timothy V. 1976a. Language and the acquisition of language-type skills by a chimpanzee (*Pan*). *Annals of the New York Academy of Sciences*, 270:90-123.

Rumbaugh, Duane M., and Gill, Timothy V. 1976b. The mastery of language-type skills by the chimpanzee (*Pan*). *Annals of the New York Academy of Sciences*, 280:562-578.

Rumbaugh, Duane M., and Gill, Timothy V. 1977. Language and language-type communication: studies with a chimpanzee, in: *Interaction, Conversation, and the Development of Language* (Michael Lewis and Leonard A. Rosenblum, eds.), pp. 115-131, New York: John Wiley & Sons.

Rumbaugh, Duane M.; Gill, Timothy V.; Brown, Josephine V.; Glasersfeld, Ernst C. von; Pisani, Pier; Warner, Harold; and Bell Charles L. 1973. A computer-controlled language training system for investigating the language skills of young apes. *Behavioral Research Methods and Instrumentation*, 5:385-392.

Rumbaugh, Duane M.; Gill, Timothy V.; Glasersfeld, Ernst C. von 1973. Reading and sentence completion by a chimpanzee (*Pan*). *Science*, 182:731-733.

Rumbaugh, Duane M.; Gill, Timothy V.; and Glasersfeld, Ernst C. von 1974. A rejoinder to language in man, monkeys, and machines. *Science*, 185:871-872.

Rumbaugh, Duane M.; Gill, Timothy V.; Glasersfeld, Ernst von; Warner, Harold; and Pisani, Pier 1975. Conversations with a chimpanzee in a computer-controlled environment. *Biological Psychiatry*, 10:627-641.

Rumbaugh, Duane M.; Gill, Timothy V.; and Wright, Sue C. 1973. Readiness to attend to visual foreground cues. *Journal of Human Evolution*, 2:181-188.

Rumbaugh, Duane M.; Glasersfeld, Ernst C. von; Gill, Timothy V.; Warner, Harold; Pisani, Pier; Brown, Josephine V.; Bell, Charles L. 1975. The language skills of a young chimpanzee in a computer-controlled training situation, in: *Socioecology and Psychology of Primates* (Russell H. Tuttle, ed.), pp. 391-401, The Hague: Mouton.

Rumbaugh, Duane M.; Glasersfeld, Ernst von; Warner, Harold; Pisani, Pier; and Gill, Timothy V. 1974. Lana (chimpanzee) learning language: a progress report. *Brain and Language*, 1:205-212.

Rumbaugh, Duane M.; Glasersfeld, Ernst C. von; Warner, Harold; Pisani, Pier; Gill, Timothy V.; Brown, Josephine V.; and Bell, Charles L. 1973. Exploring the language skills of Lana chimpanzee. *International Journal of Symbology*, 4:1-9.

Rumbaugh, Duane M., and McCormack, Carol 1967. The learning skills of primates: a comparative study of apes and monkeys, in: *Neue Ergebnisse der Primatologie* (D. Starck, R. Schneider, and H.J. Kuhn, eds.), pp. 289-306, Stuttgart: Gustav Fischer.

Rumbaugh, Duane M., and McCormack, Carol 1969. Attentional skills of great apes compared with those of gibbons and squirrel monkeys. *Proceedings of the Second International Congress of Primatology, Atlanta, GA 1968* (C.R. Carpenter, ed.), Vol. 1, pp. 167-172, Basel: Karger.

Rumbaugh, Duane M., and Pate, James L. 1984a. The evolution of cognition in primates: a comparative perspective, in: *Animal Cognition* (H.L. Roitblat, T.G. Bever, and H.S. Terrace, eds.), pp. 569-587, Hillsdale, New Jersey: Erlbaum Associates.

Rumbaugh, Duane M., and Pate, James L. 1984b. Primates' learning by levels, in: *Behavioral Evolution and Integrative Levels* (G. Greenberg and E. Tobach, eds.), pp. 221-240, Hillsdale, New Jersey: Erlbaum Associates.

Rumbaugh, Duane M., and Rice, Carol P. 1962. Learning-set formation in young great apes. *Journal of Comparative and Physiological Psychology*, 55:866-868.

Rumbaugh, Duane M.; Riesen, Austin H.; and Wright, Sue C. 1972. Creative responsiveness to objects: a report of a pilot study with young apes. *Folia Primatologica*, 17:397-403.

Rumbaugh, Duane M., and Savage-Rumbaugh, Sue 1978. Chimpanzee language research: status and potential. *Behavior Research Methods and Instrumentation*, 10:119-131.

Rumbaugh, Duane M.; Savage-Rumbaugh, E. Sue; and Gill, Timothy V. 1978. Language skills, cognition, and the chimpanzee, in: *Sign Language and Language Acquisition in Man and Ape: New Dimensions in Comparative Pedolinguistics* (Fred C.C. Peng, ed.), pp. 137-159, Boulder, Colorado: Westview Press, Inc.

Rumbaugh, Duane M.; Savage-Rumbaugh, E. Sue; and Scanlon, John L. 1982. The relationship between language in apes and human beings, in: *Primate Behavior* (James L. Fobes and James E. King, eds.), pp. 361-385, New York: Academic Press.

Rumbaugh, Duane M., and Steinmetz, Gerald T. 1971. Discrimination reversal skills of the lowland gorilla (*Gorilla g. gorilla*). *Folia Primatologica*, 16:144-152.

Rutberg, Allen T. 1983. The evolution of monogamy in primates. *Journal of Theoretical Biology*, 104:93-112.

Sabater Pi, Jorge 1960. Beitrag zur Biologie des Flachlandgorillas. *Zeitschrift für Saugetierkunde*, 25:133-141.

Sabater Pi, Jorge 1966. Rapport preliminaire sur l'alimentation dans la nature des gorilles du Rio Muni (ouest africain). *Mammalia*, 30:235-240.

Sabater Pi, Jorge 1974. Consideraciones y comentarios sobre la alimentacion de los gorilas del Africa occidental en la naturaleza. *Revista Zoo*, 21:13-15.

Sabater Pi, Jorge 1976. Aportación a una ecologia de la alimentación en estado natural de los gorilas de costa, *Gorilla gorilla gorilla*, Savage y Wyman, 1874, de Rio Muni, Republica de Guinea Ecuatorial (Africa occidental). *Ethnica*, 10:199-230.

Sabater Pi, Jorge 1977a. Observaciones y comentarios a un inventario de los alimentos consumidos en la naturaleza por los chimpancés (*Pan troglodytes troglodytes*) de Rio Muni, Africa Occidental Ecuatorial. *Revista Zoo*, Barcelona, pp. 15-17.

Sabater Pi, Jorge 1977b. Contribution to the study of alimentation of lowland gorillas in the natural state, in Rio Muni, Republic of Equatorial Guinea (West Africa). *Primates*, 18:183-204.

Sabater Pi, Jorge 1979. Feeding behavior and diet of chimpanzees (*Pan troglodytes troglodytes*) in the Okorobiko Mountains of Rio Muni (West Africa). *Zeitschrift für Tierpsychologie*, 50:265-281.

Sabater Pi, Jorge, and Groves, Colin 1972. The importance of higher primates in the diet of the Fang of Rio Muni. *Man*, 7:241-243.

Sabater Pi, Jorge, and de Lassaletta, L. 1958. Beitrag zur Kenntnis des Flachland-gorillas. *Zeitschrift für Saugetierkunde*, 23:108-114.

Sanders, Richard J. 1985. Teaching apes to ape language: explaining the imitative and nonimitative signing of a chimpanzee (*Pan troglodytes*). *Journal of Comparative Psychology*, 99:197-210.

Sarich, Vincent M. 1984. Pygmy chimpanzee systematics, in *The Pygmy Chimpanzee: Evolutionary Biology and Behavior* (R.L. Susman, ed.), pp. 43-48, New York: Plenum Press.

Sarker, Sohrab Udding, and Saker, Noor Jahan 1984. Mammals of Bangladesh: their status, distribution and habitat. *Tigerpaper*, 11:8-13.

Savage, Sue and Bakeman, Roger 1978. Sexual morphology and behavior in *Pan paniscus*, in: *Recent Advances in Primatology*, Vol. 1, *Behaviour* (D.J. Chivers and J. Herbert, eds.), pp. 613-616, London: Academic Press.

Savage-Rumbaugh, E. Sue 1979. Symbolic communication—its origins and early development in the chimpanzee. *New Directions for Child Development*, 3:1-15.

Savage-Rumbaugh, E. Sue 1981. Can apes use symbols to represent their world? *Annals of the New York Academy of Sciences*, 364:35-59.

Savage-Rumbaugh, E. Sue 1984a. *Pan paniscus* and *Pan troglodytes*: Contrasts in preverbal communicative competence, in: *The Pygmy Chimpanzee: Evolutionary Biology and Behavior* (Randall L. Susman, ed.), pp. 395-413, New York: Plenum Press.

Savage-Rumbaugh, E. Sue 1984b. Verbal behavior at a procedural level in the chimpanzee. *Journal of the Experimental Analysis of Behavior*, 41:223-250.

Savage-Rumbaugh, E. Sue; Pate, James L.; Lawson, Janet; Smith, S. Tom; and Rosenbaum, Steven 1983. Can a chimpanzee make a statement? *Journal of Experimental Psychology: General*, 112:457-492.

Savage-Rumbaugh, E. Sue, and Rumbaugh, Duane M. 1978. Symbolization, language, and chimpanzees: a theoretical reevaluation based on initial language acquisition processes in four young *Pan troglodytes*. *Brain and Language*, 6:265-300.

Savage-Rumbaugh, E. Sue, and Rumbaugh, Duane M. 1980. Language analogue project, phase II; theory and tactics, in: *Children's Language* (Keith E. Nelson, ed.), Vol. 2, pp. 267-307, New York: Gardner Press, Inc.

Savage-Rumbaugh, E. Sue, and Rumbaugh, Duane M. 1982. Ape language research is alive and well: a reply. *Anthropos*, 77:568-573.

Savage-Rumbaugh, E. Sue; Rumbaugh, Duane M.; and Boysen, Sally 1978a. Symbolic communication between two chimpanzees (*Pan troglodytes*). *Science*, 201:641-644.

Savage-Rumbaugh, E. Sue; Rumbaugh, Duane M.; and Boysen, Sally 1978b. Linguistically mediated tool use and exchange by chimpanzees (*Pan troglodytes*). *The Behavioral and Brain Sciences*, 4:539-554.

Savage-Rumbaugh, E. Sue; Rumbaugh, Duane M.; and Boysen, Sarah 1980. Do apes have language? *American Scientist*, 68:49-61.

Savage-Rumbaugh, E. Sue; Rumbaugh, Duane M.; Smith, S. Tom; and Lawson, James 1980. Reference: the linguistic essential. *Science*, 210:922-925.

Savage-Rumbaugh, E. Sue, and Wilkerson, Beverly J. 1978. Socio-sexual behavior in *Pan paniscus* and *Pan troglodytes*: a comparative study. *Journal of Human Evolution*, 7:327-344.

Schaller, George B. 1961. The orang-utan in Sarawak. *Zoologica*, 46:73-82.

Schaller, George G. 1963. *The Mountain gorilla: Ecology and Behavior*, Chicago: The University of Chicago Press.

Schaller, George B. 1965a. The behavior of the mountain gorilla, in: *Primate Behavior, Field Studies of Monkeys and Apes* (I. DeVore, ed.), pp. 324-367, New York: Holt, Rinehart and Winston.

Schaller, George B. 1965b. Behavioral comparisons of the apes, in: *Primate Behavior, Field Studies of Monkeys and Apes* (I. DeVore, ed.), pp. 474-481, New York: Holt, Rinehart and Winston.

Schaller, George B. 1970. Mountain gorilla displays, in: *Field Studies in Natural History* (Marston Bates, ed.), pp. 193-201, New York: Van Nostrand Reinhold Co.

Schaller, George B., and Emlen, John T., Jr. 1963. Observations on the ecology and social behavior of the mountain gorilla, in: *African Ecology and Human Evolution* (F.C. Howell and F. Bourlière, eds.), pp. 368-384, *Viking Fund Publications in Anthropology*, No. 36, New York: Wenner-Gren Foundation for Anthropological Research, Inc.

Schiller, Paul H. 1952. Innate constituents of complex responses in primates. *The Psychological Review*, 59:177-191.

Schiller, Paul H. 1957. Innate motor action as a basis of learning, in: *Instinctive Behavior* (C.H. Schiller, ed.), pp. 264-287, New York: International Universities Press, Inc.

Schilling, Detlef 1984. Song bouts and duetting in the concolor gibbon, in: *The Lesser Apes: Evolutionary and Behavioural Biology* (H. Preuschoft, D.J. Chivers, W.Y. Brockelman, and N. Creel, eds.), pp. 390-403, Edinburgh: Edinburgh University Press.

Schlegel, Hermann, and Müller, Salomon 1839-1844. Bijdragen tot de natuurlijke historie van den Orang-oetan (*Simia satyrus*). In Verbandelingen over de natuurlijke geschiedenis der Nederlandische overzeesche bizittingen, door de leden der Natuurkundige commissee in Indië en andere Schrijvers. Uitgegeven op last van den Koning door C.J. Temminck. *Zoologie*, 2:1-28, Leide.

Schrier, Allan M. 1984. Learning how to learn: the significance and current status of learning set formation. *Primates*, 25:95-102.

Schultz, Adolph H. 1933. Observations on the growth, classification and evolutionary specialization of gibbons and siamangs. *Human Biology*, 5:202-255, 385-428.

Schultz, Adolph H. 1934. Some distinguishing characters of the mountain gorilla. *Journal of Mammalogy*, 34:51-61.

Schultz, Adolph H. 1940. Growth and development of the chimpanzee. *Contributions to Embryology*, 28:1-63, Washington, D.C.: Carnegie Institute Publ.

Schultz, Adolph H. 1941. The relative size of the cranial capacity in primates. *American Journal of Physical Anthropology*, 28:273-287.

Schultz, Adolph H. 1944. Age changes and variability in gibbons: a morphological study on a population sample of man-like apes. *American Journal of Physical Anthropology*, 2:1-129.

Schultz, Adolph H. 1956. Postembryonic age changes. *Primatologia I* (H. Hofer, A.H. Schultz, and D. Starck, eds.), pp. 887-964, Basel: Karger.

Schultz, Adolph H. 1962a. Metric age changes and sex differences in primate skulls. *Zeitschrift für Morphologie und Anthropologie*, 52:239-255.

Schultz, Adolph H. 1962b. Die Schädelkapazität männlicher Gorillas und ihr Höchstwert. *Anthropologischer Anzeiger*, 25:197-203.

Schultz, Adolph H. 1965. The cranial capacity and the orbital volume of hominoids according to age and sex, in: *Homenaje a Juan Comas en su 65 aniversario*, Vol. II, pp. 337-357, Editorial Libros de Mexico.

Schultz, Adolph H. 1969. The skeleton of the chimpanzee, in: *The Chimpanzee* (G.H. Bourne, ed.), Vol. 1, pp. 50-103, Basel: Karger.

Schultz, Adolph H. 1973. The skeleton of the Hylobatidae and other observations on their morphology, in: *Gibbon and Siamang* (D.M. Rumbaugh, ed.), Vol. 2, pp. 1-54, Basel: Karger.

Schürmann, Cris L. 1981. Courtship and mating behavior of wild orangutans in Sumatra, in: *Primate Behavior and Sociobiology* (A.B. Chiarelli and R.S. Corruccini, eds.), pp. 130-135, Berlin: Springer-Verlag.

Schürmann, Cris 1982. Mating behaviour of wild orang utans, in: *The Orang Utan: Its biology and conservation* (Leobert E.M. De Boer, ed.), pp. 269-294, The Hague: Dr. W. Junk Publishers.

Schusterman, Ronald J. 1962. Transfer effects of successive discrimination-reversed training in chimpanzees. *Science*, 137:422-423.

Schusterman, Ronald J. 1963. The use of strategies in two-choice behavior of children and chimpanzees. *Journal of Comparative and Physiological Psychology*, 56:96-100.

Schusterman, Ronald J. 1964. Successive discrimination-reversal training and multiple discrimination training in one-trial learning by chimpanzees. *Journal of Comparative and Physiological Psychology*, 58:153-156.

Schusterman, Ronald J., and Bernstein, Irwin S. 1962. Response tendencies of gibbons in single and double alternation tasks. *Psychological Reports*, 11:521-522.

Scott, John Paul 1973. The organization of comparative psychology. *Annals of the New York Academy of Sciences*, 223:7-40.

Sebeok, Thomas A., and Rosenthal, Robert (eds.) 1981. *The Clever Hans Phenomenon: Communication with Horses, Whales, Apes, and People. Annals of the New York Academy of Sciences*, 364:1-311.

Sebeok. Thomas A., and Umiker-Sebeok, Jean (eds.), *Speaking of Apes: A Critical Anthology of Two-way Communication with Man*, New York: Plenum Press.

Segall, Marshall; Campbell, Donald T.; and Herskovits, Melville J. 1966. *The Influence of Culture on Visual Perception*, Indianapolis: Bobbs-Merrill.

Seidenberg, Mark S., and Petitto, Laura A. 1979. Signing behavior in apes: a critical review. *Cognition*, 7:177-215.

Seidenberg, Mark S., and Petitto, Laura A. 1981. Ape signing: problems of method and interpretation. *Annals of the New York Academy of Sciences*, 364:115-129.

Shafer, David A.; Myers, Richard H.; and Saltzman, David 1984. Biogenetics of the siabon (gibbon-siamang hybrids), in: *The Lesser Apes: Evolutionary and Behavioural Biology* (H. Preuschoft, D.J. Chivers, W.Y. Brockelman, and N. Creel, eds.), pp. 486-497, Edinburgh: Edinburgh University Press.

Shapiro, Gary L. 1982. Sign acquisition in a home-reared/free-ranging orangutan; comparisons with other signing apes. *American Journal of Primatology*, 3:121-129.

Shariff, Ghouse Ahmed 1953. Cell counts in the primate cerebral cortex. *The Journal of Comparative Neurology*, 98:381-400.

Shea, Brian T. 1981. Relative growth of the limbs and trunk in the African apes. *American Journal of Physical Anthropology*, 56:179-201.

Shea, Brian T. 1983a. Allometry and heterochrony in the African apes. *American Journal of Physical Anthropology*, 62:275-289.

Shea, Brian T. 1983b. Paedomorphosis and neoteny in the pygmy chimpanzee. *Science*, 222:521-522.

Shea, Brian T. 1984a. Between the gorilla and the chimpanzee: a history of debate concerning the existence of the *koolookamba* or gorilla-like chimpanzee. *Journal of Ethnobiology*, 4:1-13.

Shea, Brian T. 1984b. An allometric perspective on the morphological and evolutionary relationships between pygmy (*Pan paniscus*) and common (*Pan troglodytes*) chimpanzees, in: *The Pygmy Chimpanzee: Evolutionary Biology and Behavior* (R.L. Susman, ed.), pp. 89-130, New York: Plenum Press.

Shea, Brian T. 1985. The ontogeny of sexual dimorphism in the African apes. *American Journal of Primatology*, 8:183-188.

Short, Roger V. 1979. Sexual selection and its component parts, somatic and genital selection, as illustrated by man and the great apes. *Advances in the Study of Behavior*, 9:131-158.

Short, Roger V. 1980. The great apes of Africa. *Journal of Reproduction and Fertility*, Supplement No. 28:3-11.

Short, Roger V. 1981. Sexual selection in man and the great apes, in: *Reproductive Biology of the Great Apes* (C.E. Graham, ed.), pp. 319-341, New York: Academic Press.

Sigmon, Becky A., and Cybulski, Jerome S. 1981. *Homo erectus: Papers in Honor of Davidson Black*, Toronto: University of Toronto Press.

Silk, Joan B. 1978. Patterns of food sharing among mother and infant chimpanzees at Gombe National Park, Tanzania. *Folia Primatologica*, 29:129-141.

de Silva, G.S. 1971. Notes on the orang-utan rehabilitation project in Sabah. *Malayan Nature Journal*, 24:50-77.

de Silva, G.S. 1972. The birth of an orang-utan (*Pongo pygmaeus*) at Sepilok Game Reserve. *International Zoo Yearbook*, 12:104-105.

Simonetta, A. 1957. Catalogo e sinonimia annotata delgi ominoidi fossili ed attuali (1758-1955). *Atti della Societa Toscana di Scienze Naturali, Processi Verbalie e Memorie*, 64B:53-112.

Simons, Elwyn L. 1961. The phyletic position of *Ramapithecus*. *Postilla*, 57:1-9.

Simons, Elwyn L. 1963. A critical reappraisal of Tertiary primates, in: *Evolutionary and Genetic Biology of Primates* (J. Buettner-Janusch, ed.), Vol. 1, pp. 65-129, New York: Academic Press.

Simons, Elwyn L. 1964. On the mandible of *Ramapithecus*. *Proceedings of the National Academy of Sciences*, 51:528-535.

Simons, Elwyn L. 1968. A source for dental comparison of *Ramapithecus* with *Australopithecus* and *Homo*. *South African Journal of Science*, 64:92-112.

Simons, Elwyn L. 1972. *Primate Evolution*, New York: The MacMillan Co.

Simons, Elwyn L. 1976. Relationships between *Dryopithecus, Sivapithecus* and *Ramapithecus* and their bearing on hominid origins, in: *Les Plus Anciens Hominidés* (P.V. Tobias and Y. Coppens, eds.), pp. 60-67, Paris: Centre National de la Recherche Scientifique.

Simons, Elwyn L. 1978. Diversity among the early hominids: a vertebrate palaeontologist's viewpoint, in: *Early Hominids of Africa* (C. Jolly, ed.), pp. 543-566, New York: St. Martin's Press.

Simons, Elwyn L., and Chopra, S.R.K. 1969. *Gigantopithecus* (Pongidae, Hominoidea): a new species from North India. *Postilla*, 138:1-18.

Simons, Elwyn L., and Fleagle, John 1973. The history of extinct gibbon-like primates, in: *Gibbon and Siamang* (D.M. Rumbaugh, ed.), Vol. 2, pp. 121-148, New York: Karger.

Simons, Elwyn L., and Pilbeam, David R. 1965. Preliminary revision of the Dryopithecinae (Pongidae, Anthropoidea). *Folia Primatologica*, 3:81-152.

Simons, Elwyn L., and Pilbeam, David R. 1972. Hominoid paleoprimatology, in: *The Functional and Evolutionary Biology of Primates* (R.H. Tuttle, ed.), pp. 36-62, Chicago: Aldine-Atherton.

Simpson, George G. 1961. *Principles of Animal Taxonomy*, New York: Columbia University Press.

Simpson, George G. 1963. The meaning of taxonomic statements, in: *Classification and Human Evolution* (S.L. Washburn, ed.), pp. 1-31, New York: Wenner-Gren Foundation for Anthropological Research, Inc.

Simpson, Michael J.A. 1973. The social grooming of male chimpanzees, in: *Comparative Ecology and Behaviour of Primates* (Richard P. Michael and John H. Crook, eds.), pp. 411-505, London: Academic Press.

Skinner, Burrhus Frederic 1938. *The Behavior of Organisms: An experimental analysis*, New York: Appleton-Century-Crofts.

Smith, Fred H., and Spencer, Frank 1984. *The Origins of Modern Humans: A world survey of the fossil evidence*, New York: Alan R. Liss, Inc.

Smith, G. Elliot 1924. *The Evolution of Man*, London: Oxford University Press.

Smith, Harriet J.; King, James E.; Witt, Edwin D.; and Rickel, John E. 1975. Sameness-difference matching from sample by chimpanzees. *Bulletin of the Psychonomic Society*, 6:469-471.

Snyder, Daniel R.; Birchette, L. Muriel; and Aschenbach, Thomas S. 1978. A comparison of developmentally progressive intellectual skills between *Hylobates lar, Cebus apella,* and *Macaca mulatta,* in: *Recent Advances in Primatology,* Vol. 1, *Behaviour* (D.J. Chivers and J. Herbert eds.), pp. 945-948, London: Academic Press.

Socha, Wladyslaw W. 1984. Blood groups of pygmy and common chimpanzees, in: *The Pygmy Chimpanzee: Evolutionary Biology and Behavior* (R.L. Susman, ed.), pp. 13-41, New York: Plenum Press.

Southwick, Charles H., and Southwick, Karen L. 1985. Rhesus monkeys in Burma. *Primate Conservation*, 5:35-36.

Spence, Kenneth W. 1937. Analysis of the formation of visual discrimination habits in chimpanzees. *Journal of Comparative Psychology*, 23:77-100.

Spence, Kenneth W. 1938. Gradual versus sudden solution of discrimination problems by chimpanzees. *Journal of Comparative Psychology*, 25:213-224.

Spence, Kenneth W. 1939. The solution of multiple choice problems by chimpanzees. *Comparative Psychology Monographs*, 15:1-54.

Spence, Kenneth W. 1941. Failure of transposition in size-discrimination of chimpanzees. *American Journal of Psychology*, 54:223-229.

Spence, Kenneth W. 1942. The basis of solution by chimpanzees of the intermediate size problem. *Journal of Experimental Psychology*, 31:257-271.

Spencer, Frank; Boaz, Noel T.; Allen, Mel; and McGrew, William C. 1982. Biochemical detection of fecal hematin as a test for meat eating in chimpanzees (*Pan troglodytes*). *American Journal of Primatology*, 3:327-332.

Spinage, C.A. 1969. *Report of the Ecologist of the Rwanda National Parks*, Ministry of Overseas Development.

Spinage, C.A. 1981. Some faunal isolates of the Central African Republic. *African Journal of Ecology*, 19:125-132.

Spragg, S.D.S. 1936. Anticipatory responses in serial learning by chimpanzees. *Comparative Psychology Monographs*, 13:1-72.

Srikosamatara, Sompoad 1984. Ecology of pileated gibbons in south-east Thailand, in: *The Lesser Apes: Evolutionary and Behavioural Biology* (H. Preuschoft, D.J. Chivers, W.Y. Brockelman, and N. Creel, eds.), pp. 242-257, Edinburgh: Edinburgh University Press.

Steklis, Horst D. 1985. Primate communication, comparative neurology, and the origin of language re-examined. *Journal of Human Evolution*, 14:157-173.

Stephan, Heinz 1972. Evolution of primate brains: a comparative anatomical investigation, in: *The Functional and Evolutionary Biology of Primates* (R. Tuttle, ed.), pp. 155-174, Chicago: Aldine.

Stewart, Kelly J. 1977. The birth of a wild mountain gorilla (*Gorilla gorilla beringei*). *Primates*, 18:965-976.

Stokoe, William C. 1960. Sign language structure: an outline of the visual communication systems of the American deaf. *Studies in Linguistics*, Occasional Papers 8, New York: University of Buffalo.

Stokoe, William C. 1970. *The Study of Sign Language*, Washington, D.C.: Center for Applied Linguistics (CAL-ERIC, ED 037 719).

Stott, Kenhelm W. 1981. A suitable vernacular name for an intermediate gorilla. *Mammalia*, 45:261.

Straus, William L. Jr. 1963. The classification of *Oreopithecus*, in: *Classification and Human Evolution* (S.L. Washburn, ed.), pp. 146-177, Chicago: Aldine-Atherton.

Strong, Paschal N., Jr. 1967. Comparative studies in oddity learning: III. apparatus transfer in chimpanzees and children. *Psychonomic Science*, 7:43.

Strong, Paschal N., Jr., and Hedges, Monnie 1966. Comparative studies in simple oddity learning: I. Cats, racoons, monkeys, and chimpanzees. *Psychonomic Science*, 5:13-14.

Struhsaker, Thomas T. 1970. Phylogenetic implications of some vocalizations of *Cercopithecus* monkeys, in: *Old World Monkeys* (J.R. Napier and P.H. Napier, eds.), pp. 365-444, New York: Academic Press.

Struhsaker, Thomas T. 1975. *The Red Colobus Monkey*, Chicago: The University of Chicago Press.

Struhsaker, Thomas, and Hunkeler, P. 1971. Evidence of tool-making by chimpanzees in the Ivory Coast. *Folia Primatologica*, 15:212-219.

Suarez, Susan D., and Gallup, Gordon G., Jr. 1981. Self-recognition in chimpanzees and orangutans, but not gorillas. *Journal of Human Evolution*, 10:175-188.

Sugardjito, Jito 1982. Locomotor behaviour of the orang utan (*Pongo pygmaeus abelii*) at Ketambe, Gunung Leuser National Park. *Malayan Nature Journal*, 35:57-64.

Sugardjito, Jito 1983. Selecting nest-sites of Sumatran orang-utans, *Pongo pygmaeus abelii* in the Gunung Leuser National Park, Indonesia. *Primates*, 24:467-474.

Sugardjito, Jito, and Nurhuda, Nur 1981. Meat-eating behaviour in wild orang utans, *Pongo pygmaeus*. *Primates*, 22:414-416.

Sugiyama, Yukimaru 1968. Social organization of chimpanzees in the Budongo Forest, Uganda. *Primates*, 9:225-258.

Sugiyama, Yukimaru 1969. Social behavior of chimpanzees in the Budongo Forest, Uganda. *Primates*, 10:197-225.

Sugiyama, Yukimaru 1973a. Social organization of wild chimpanzees, in: *Behavioral Regulators of Behavior in Primates* (C.R. Carpenter, ed.), pp. 68-80, Lewisburg, Pennsylvania: Bucknell University Press.

Sugiyama, Yukimaru 1973b. The social structure of wild chimpanzees—a review of field studies, in: *Comparative Ecology and Behaviour of Primates* (Richard P. Michael and John H. Crook, eds.), pp. 376-410, London: Academic Press.

Sugiyama, Yukimaru 1981. Observations on the population dynamics and behavior of wild chimpanzees at Bossou, Guinea, in 1979-1980. *Primates*, 22:435-444.

Sugiyama, Yukimaru 1984. Population dynamics of wild chimpanzees at Bossou, Guinea, between 1976 and 1983. *Primates*, 25:391-400.

Sugiyama, Yukimaru, and Koman, Jeremy 1979a. Social structure and dynamics of wild chimpanzees at Bossou, Guinea. *Primates*, 20:323-339.

Sugiyama, Yukimaru, and Koman, Jeremy 1979b. Tool-using and -making behavior in wild chimpanzees at Bossou, Guinea. *Primates*, 20:513-524.

Susman, Randall L. 1979. Comparative and functional morphology of hominoid fingers. *American Journal of Physical Anthropology*, 50:215-236.

Susman, Randall 1980. Acrobatic pygmy chimpanzee. *Natural History*, 89:32-39.

Susman, Randall L. 1984. The locomotor behavior of *Pan paniscus* in the Lomako Forest, in: *The Pygmy Chimpanzee: Evolutionary Biology and Behavior* (R.L. Susman, ed.), pp. 369-393, New York: Plenum Press.

Susman, Randall L.; Badrian, Noel L.; and Badrian, Alison J. 1980. Locomotor behavior of *Pan paniscus* in Zaire. *American Journal of Physical Anthropology*, 53:69-80.

Sussman, Robert W., and Kinzey, Warren G. 1984. The ecological role of the Callithricidae: a review. *American Journal of Physical Anthropology*, 64:419-449.

Suzuki, Akira 1966. On the insect-eating habits among wild chimpanzees living in the savanna woodland of western Tanzania. *Primates*, 7:481-487.

Suzuki, Akira 1969. An ecological study of chimpanzees in a savanna woodland. *Primates*, 10:103-148.

Suzuki, Akira 1971. Carnivority and cannibalism observed among forest-living chimpanzees. *Journal of the Anthropological Society, Nippon*, 79:30-48.

Suzuki, Akira 1975. The origin of hominid hunting: a primatological perspective, in: *Socioecology and Psychology of Primates* (R.H. Tuttle, ed.), pp. 259-278, The Hague: Mouton.

Szalay, Frederick S., and Delson, Eric 1979. *Evolutionary History of the Primates*, New York: Academic Press.

Takahata, Yukio; Hasegawa, Toshikazu; and Nishida, Toshidada 1984. Chimpanzee predation in the Mahale Mountains from August 1979 to May 1982. *International Journal of Primatology*, 5:213-233.

Takasaki, Hiroyuki 1983a. Mahale chimpanzees taste mangoes—toward acquisition of a new food item? *Primates*, 24:273-275.

Takasaki, Hiroyuki 1983b. Seed dispersal by chimpanzees: a preliminary note. *African Study Monographs*, 3:105-108.

Takasaki, Hiroyuki, and Uehara, Shigeo 1984. Seed dispersal by chimpanzees: supplementary note 1. *African Study Monographs*, 5:91-92.

Tekkaya, Ibrahim 1974. A new species of Tortonian anthropoid (Primates, Mammalia) from Anatolia. *Bulletin of the Mineral Research and Exploration Institute of Turkey, Ankara*, 83:148-165.

Teleki, Geza 1973a. *The Predatory Behavior of Wild Chimpanzees*, Lewisburg, Pennsylvania: Bucknell University Press.

Teleki, Geza 1973b. The omnivorous chimpanzee. *Scientific American*, 228:32-42.

Teleki, Geza 1973c. Notes on chimpanzee interactions with small carnivores in Gombe National Park, Tanzania. *Primates*, 14:407-411.

Teleki, Geza 1973d. Group response to the accidental death of a chimpanzee in Gombe National Park, Tanzania. *Folia Primatologica*, 20:81-94.

Teleki, Geza 1977. Still more on predatory behavior in nonhuman primates. *Current Anthropology*, 18:107-108.

Teleki, Geza 1981. The omnivorous diet and eclectic feeding habits of chimpanzees in Gombe National Park, Tanzania, in: *Omnivorous Primates, Gathering and Hunting in Human Evolution* (R.S.O. Harding and G. Teleki, eds.), pp. 303-343, New York: Columbia University Press.

Teleki, Geza; Hunt, Edward E., Jr.; Pfifferling, John H. 1976. Demographic observations (1963-1973) in chimpanzees of Gombe National Park, Tanzania. *Journal of Human Evolution*, 5:559-598.

Tembrock, G. 1974. Sound production of *Hylobates* and *Symphalangus*, in: *Gibbon and Siamang* (D.M. Rumbaugh, ed.), Vol. 3, pp. 176-205, Basel: Karger.

Temerlin, Maurice K. 1975. *Lucy: Growing Up Human*, Palo Alto, California: Science and Behavior Books, Inc.

Tenaza, Richard R. 1975. Territory and monogamy among Kloss' gibbons (*Hylobates klossii*) in Siberut Island, Indonesia. *Folia Primatologica*, 24:60-80.

Tenaza, Richard R. 1976. Songs, choruses and countersinging of Kloss' gibbons (*Hylobates klossii*) in Siberut Island, Indonesia. *Zeitschrift für Tierpsychologie*, 40:37-52.

Tenaza, Richard 1985. Songs of hybrid gibbons (*Hylobates lar* × *H. muelleri*). *American Journal of Primatology*, 8:249-253.

Tenaza, Richard R., and Hamilton, William J., III 1971. Preliminary observations of the Mentawai Islands gibbons, *Hylobates klossii*. *Folia primiatologica*, 15:201-211.

Tenaza, Richard R., and Tilson, Ronald L. 1977. Evolution of long-distance alarm calls in Kloss's gibbons. *Nature*, 268:233-235.

Tenaza, Richard, and Tilson, Ronald L. 1985. Human predation and Kloss's gibbon (*Hylobates klossii*) sleeping trees in Siberut Island, Indonesia. *American Journal of Primatology*, 8:299-308.

Terrace, Herbert S. 1979a. *Nim*, New York: Alfred A. Knopf.

Terrace, Herbert S. 1979b. Is problem-solving language? *Journal of the Experimental Analysis of Behavior*, 31:161-175.

Terrace, Herbert S. 1981. A report to an academy, 1980. *Annals of the New York Academy of Sciences*, 364:94-114.

Terrace, Herbert S. 1982. Why Koko can't talk. *The Sciences*, 22:8-10.

Terrace, Herbert S. 1983. Apes who "talk": language or projection of language by their teachers?, in: *Language in Primates: Perspectives and Implications* (Judith de Luce and Hugh T. Wilder, eds.), pp. 19-42, New York: Springer-Verlag.

Terrace, Herbert S., and Bever, Thomas G. 1976. What might be learned from studying language in the chimpanzee?: the importance of symbolizing oneself. *Annals of the New York Academy of Sciences*, 280:579-588.

Terrace, Herbert S.; Petitto, Laura A.; Sanders, Richard J.; and Bever, Thomas G. 1979. Can an ape create a sentence? *Science*, 206:891-902.

Terrace, Herbert S.; Petitto, Laura A.; Sanders, Richard J.; and Bever, Thomas G. 1980. On the grammatical capacity of apes, in: *Children's Language* (Keith E. Nelson, ed.), Vol. 2, pp. 371-495, New York: Gardner Press, Inc.

Thenius, Erich 1981. Zur systematischen Stellung der Gibbons oder Langarmaffen (Hylobatidae, Primates). *Sitzungsberichten der Österreich Akademie der Wissenschaften Mathematik-naturwissenschaften Klasse*, Part I, 190:1-5.

Thompson, Claudia R., and Church, Russell M. 1980. An explanation of the language of a chimpanzee. *Science*, 208:313-314.

Thompson, Robert 1954. Approach versus avoidance in an ambiguous-cue discrimination problem in chimpanzees. *Journal of Comparative and Physiological Psychology*, 47:133-135.

Thompson-Handler, Nancy; Malenky, Richard K.; and Badrian, Noel 1984. Sexual behavior of *Pan paniscus* under natural conditions in the Lomako Forest, Equateur, Zaire, in: *The Pygmy Chimpanzee: Evolutionary biology and behavior* (Randall L. Susman, ed.), pp. 347-368, New York: Plenum Press.

Tien, Dao Van 1983. On the north Indochinese gibbons (*Hylobates concolor*) (Primates: Hylobatidae) in North Vietnam. *Journal of Human Evolution*, 12:367-372.

Tilney, Frederick 1928. *The Brain from Ape to Man*. New York: Paul B. Hoeber, Inc.

Tilson, Ronald L. 1979. Behaviour of hoolock gibbon (*Hylobates hoolock*) during different seasons in Assam, India. *Journal of the Bombay Natural History Society*, 76:1-16.

Tilson, Ronald L. 1981. Family formation strategies of Kloss' gibbons. *Folia Primatologica*, 35:259-287.

Tilson, Ronald L., and Tenaza, Richard R. 1982. Interspecific spacing between gibbons (*Hylobates klossii*) and langurs (*Presbytis potenziani*) on Siberut Island, Indonesia. *American Journal of Primatology*, 2:355-361.

Tinklepaugh, Otto Leif 1932. Multiple delayed reaction with chimpanzees and monkeys. *Journal of Comparative Psychology*, 13:207-243.

Tobias, Phillip V. 1971. *The Brain in Hominid Evolution*, New York: Columbia University Press.

Tobias, Phillip V. 1975. Brain evolution in the Hominoidea, in: *Primate Functional Morphology and Evolution* (R.H. Tuttle, ed.), pp. 353-392, The Hague: Mouton.

Tolman, Edward Chance 1948. Cognitive maps in rats and men. *Psychological Review*, 55:189-208.

Tomasello, Michael; George, Barbara L.; Kruger, Ann Cale; Farrar, Michael Jeffrey; and Evans, Andrea 1985. The development of gestural communication in young chimpanzees. *Journal of Human Evolution*, 14:175-186.

Tomilin, Michael I., and Yerkes, Robert M. 1935. Mother-infant relationships in chimpanzees. *Journal of Comparative Psychology*, 20:321-358.

Tóth, Tibor 1965. The variability of the weight of the brain of *Homo*, in: *Homenaje a Juan Comas en su 65 aniversario*, Vol. II, pp. 391-402, Editorial Libros de México.

Treesucon, U., and Raemaekers, Jeremy J. 1984. Group formation in gibbon through displacement of an adult. *International Journal of Primatology*, 5:387.

Trivers, Robert L. 1972. Parental investment and sexual selection, in: *Sexual Selection and the Descent of Man 1871-1971* (Bernard Campbell, ed.), pp. 136-179, Chicago: Aldine Publishing Co.

Tuinen, Peter van, and Ledbetter, David H. 1983. Cytogenetic comparison and phylogeny of three species of Hylobatidae. *American Journal of Physical Anthropology*, 61:453-466.

Turleau, Catherine; Créau-Goldberg, Nicole; Cochet, Chantal; and de Grouchy, J. 1983. Gene mapping of the gibbon: its position in primate evolution. *Human Genetics*, 64:65-72.

Tutin, Caroline E. 1975. Exceptions to promiscuity in a feral chimpanzee community, in: *Contemporary Primatology* (S. Kondo, M. Kawai, and A. Ehara, eds.), pp. 445-449, Basel: Karger.

Tutin, Caroline E.G. 1979a. Mating patterns and reproductive strategies in a community of wild chimpanzees (*Pan troglodytes schweinfurthii*). *Behavioral Ecology and Sociobiology*, 6:29-38.

Tutin, Caroline E.G. 1979b. Responses of chimpanzees to copulation, with special reference to interference by immature individuals. *Animal Behaviour*, 27:845-854.

Tutin, Caroline E.G. 1980. Reproductive behaviour of wild chimpanzees in the Gombe National Park, Tanzania. *Journal of Reproduction and Fertility*, Supplement No. 28:43-57.

Tutin, Caroline E.G., and Fernandez, Michel 1983. Gorillas feeding on termites in Gabon, West Africa. *Journal of Mammalogy*, 64:530-531.

Tutin, Caroline E.G., and Fernandez, Michel 1984a. Nationwide census of gorilla (*Gorilla g. gorilla*) and chimpanzee (*Pan t. troglodytes*) populations in Gabon. *American Journal of Primatology*, 6:313-336.

Tutin, Caroline E.G., and Fernandez, Michel 1984b. Ape census in Gabon. *Primate Eye*, 21:16-17.

Tutin, Caroline, E.G., and Fernandez, Michel 1985. Foods consumed by sympatric populations of *Gorilla g. gorilla* and *Pan t. troglodytes* in Gabon: some preliminary data. *International Journal of Primatology*, 6:27-43.

Tutin, Caroline E.G.; Fernandez, Michel; Pierce, A.H.; and Williamson, E.A. 1984. Foods consumed by sympatric populations of *Gorilla g. gorilla* and *Pan t. troglodytes* in Gabon. *International Journal of Primatology*, 5:389.

Tutin, Caroline E.G., and McGinnis, Patrick R. 1981. Chimpanzee reproduction in the wild, in: *Reproductive Biology of the Great Apes* (Charles E. Graham, ed.), pp. 230-264, New York: Academic Press.

Tutin, Caroline E.G., and McGrew, William C. 1973. Chimpanzee copulatory behaviour. *Folia Primatologica*, 19:237-256.

Tutin, Caroline E.G.; McGrew, William C.; and Baldwin, Pamela J. 1981. Responses of wild chimpanzees to potential predators, in: *Primate Behavior and Sociobiology* (A.B. Chiarelli and R.S. Corruccini, eds.), pp. 136-141, Berlin: Springer-Verlag.

Tutin, Caroline E.G.; McGrew, William C.; and Baldwin, Pamela J. 1983. Social organization of savanna-dwelling chimpanzees, *Pan troglodytes verus*, at Mt. Assirik, Senegal. *Primates*, 24:154-173.

Tuttle, Russell H. 1967. Knuckle-walking and the evolution of hominoid hands. *American Journal of Physical Anthropology*, 26:171-206.

Tuttle, Russell H. 1969a. Knuckle-walking and the problem of human origins. *Science*, 166:953-961.

Tuttle, Russell H. 1969b. Quantitative and functional studies on the hands of the Anthropoidea. I. the Hominoidea. *Journal of Morphology*, 128:309-364.

Tuttle, Russell H. 1970. Postural, propulsive and prehensile capabilities in the cheiridia of chimpanzees and other great apes, in: *The Chimpanzee* (G.H. Bourne, ed.), Vol. 2, pp. 167-253, Basel: Karger.

Tuttle, Russell H. 1972. Functional and evolutionary biology of hylobatid hands and feet, in: *Gibbon and Siamang* (D.M. Rumbaugh, ed.), Vol. 1, pp. 136-206, Basel: Karger.

Tuttle, Russell H. 1974. Darwin's apes, dental apes, and the descent of man: normal science in evolutionary anthropology. *Current Anthropology*, 15:389-426.

Tuttle, Russell H. 1975. Parallelism, brachiation and hominoid phylogeny, in: *Phylogeny of the Primates: A multidisciplinary approach* (W.P. Luckett and F.S. Szalay, eds.), pp. 447-480, New York: Plenum Publishing Co.

Tuttle, Russell H. 1976. Quantitative sociobiology. *Science*, 191:939-940.

Tuttle, Russell H. 1977. Naturalistic positional behavior of apes and models of hominid evolution, 1929-1976, in: *Progress in Ape Research* (G.H. Bourne, ed.), pp. 277-296, New York: Academic Press.

Tuttle, Russell H. 1981. Evolution of hominid bipedalism and prehensile capabilities. *Philosophical Transactions of the Royal Society, London*, B-292:89-94.

Tuttle, Russell H. 1986. Kinesiological inferences and evolutionary implications from Laetoli bipedal trails G-1, G-2/3, and A, in: *The Pliocene Site of Laetoli, Northern Tanzania* (M.D. Leakey and J.M. Harris, eds.), pp. 503-523, Oxford: Clarendon Press.

Tuttle, Russell H.; Basmajian, John V.; and Ishida, Hidemi 1979. Activities of pongid thigh muscles during bipedal behavior. *American Journal of Physical Anthropology*, 50:123-136.

Tuttle, Russell H.; Cortright, Gerald W.; and Buxhoeveden, Daniel P. 1979. Anthropology on the move: progress in experimental studies of nonhuman primate positional behavior. *Yearbook of Physical Anthropology*, 22:187-214.

Tuttle, Russell H., and Watts, David P. 1985. The positional behavior and adaptive complexes of *Pan gorilla*, in: *Primate Morphophysiology, Locomotor Analyses and Human Bipedalism* (Shiro Kondo, ed.), pp. 261-288, Tokyo: University of Tokyo Press.

Tylor, Edward B. 1871. *Primitive Culture: Researches into the Development of Mythology, Philosophy, Religion, Language, Art and Custom*, London: J. Murray.

Uehara, Shigeo 1981. The social unit of wild chimpanzees: a reconsideration based on the diachronic data accumulated at Kasoje in the Mahale Mountains, Tanzania. *Africa-Kenkyu (Journal of African Studies)*, 20:15-32.

Uehara, Shigeo 1982. Seasonal changes in the techniques employed by wild chimpanzees in the Mahale Mountains, Tanzania, to feed on termites (*Pseudacanthotermes spiniger*). *Folia Primatologica*, 37:44-76.

Uehara, Shigeo 1984. Sex difference in feeding on *Camponotus* ants among wild chimpanzees in the Mahale Mountains, Tanzania. *International Journal of Primatology*, 5:389.

Uehara, Shigeo, and Nyundo, Ramadhani 1983. One observed case of temporary adoption of an infant by unrelated nulliparous females among wild chimpanzees in the Mahale Mountains, Tanzania. *Primates*, 24:456-466.

Umiker-Sebeok, Jean, and Sebeok, Thomas A. 1982. Rejoinder to the Rumbaughs. *Anthropos*, 77:574-578.

Van Horn, Richard N. 1972. Structural adaptations to climbing in the gibbon hand. *American Anthropologist*, 74:326-334.

Vauclair, Jacques 1982. Sensorimotor intelligence in human and non-human primates. *Journal of Human Evolution*, 11:257-264.

Vauclair, Jacques 1984. Phylogenetic approach to object manipulation in human and ape infants. *Human Development*, 27:321-328.

Vauclair, Jacques; Rollins, Howard A., Jr.; and Nadler, Ronald D. 1983. Reproductive memory for diagonal and nondiagonal patterns in chimpanzees. *Behavioural Processes*, 8:289-300.

Vedder, Amy L. 1984. Movement patterns of a group of free-ranging mountain gorillas (*Gorilla gorilla beringei*) and their relation to food availability. *American Journal of Primatology*, 7:73-88.

Verhaart, W.J.C. 1970. The pyramidal tract in the primates, in: *The Primate Brain* (Charles R. Noback and William Montagna, eds.), pp. 83-108, New York: Appleton-Century-Crofts.

Verschuren, Jacques 1978. Les grands mammiferes du Burundi. *Mammalia*, 42:209-224.

Vogel, Christian 1961. Zur systematischen Untergliederungen der Gattung *Gorilla* anhand von Unterschungen der Mandibel. *Zeitschrift für Säugetierkunde*, 26:1-12.

Voronin, L.G. 1962. Some results of comparative physiological investigations of higher nervous activity. *Psychological Bulletin*, 59:161-195.

de Vos, John 1983. The *Pongo* faunas from Java and Sumatra and their significance for biostratigraphical and paleo-ecological interpretations. *Proceedings of the Koninklijke Nederlandse Akademie van Wetenschappen, Series B*, 86:417-425.

de Waal, Frans B.M. 1978. Exploitative and familiarity-dependent support strategies in a colony of semi-free living chimpanzees. *Behaviour*, 66:268-312.

de Waal, Frans 1982. *Chimpanzee Politics: Power and sex among apes*, New York: Harper & Row, Publishers.

de Waal, Frans B.M. 1984. Sex differences in the formation of coalitions among chimpanzees. *Ethology and Sociobiology*, 5:239-255.

de Waal, Frans B.M., and van Hooff, Johan A.R.A.M. 1981. Side-directed communication and agonistic interactions in chimpanzees. *Behaviour*, 77:164-198.

de Waal, Frans B.M., and van Roosmalen, Angeline 1979. Reconciliation and consolation among chimpanzees. *Behavioral Ecology and Sociobiology*, 5:55-66.

Wada, Juhn A.; Clarke, Robert; and Hamm, Anne 1975. Cerebral hemispheric asymmetry in humans. *Archives of Neurology*, 39:239-246.

Wallace, Alfred Russel 1869. *The Malay Archipelago*, London: Macmillan.

Ward, Jeannette P.; Yehle, Arthur L.; and Doeflein, R. Stephen 1970. Cross-modal transfer of a specific discrimination in the bushbaby (*Galago senegalensis*). *Journal of Comparative and Physiological Psychology*, 73:74-77.

Warner, Harold; Bell, Charles L.; Rumbaugh, Duane M.; and Gill, Timothy V. 1976. Computer-controlled teaching instrumentation for linguistic studies with the great apes. *IEEE Transactions on Computers*, C-25:38-43.

Warren James M. 1965. Primate learning in comparative perspective, in: *Behaviour of Nonhuman Primates, Modern Research Trends* (Allan M. Schrier, Harry F. Harlow, and Fred Stollnitz, eds.), Vol. 1, pp. 249-281, New York: Academic Press.

Warren, James M. 1973. Learning in vertebrates, in: *Comparative Psychology: A Modern Survey* (Donald A. Dewsbury and Dorothy A. Rethlingshafer, eds.), pp. 471-509, New York: McGraw-Hill Book Company.

Warren, James M. 1974. Possibly unique characteristics of learning by Primates. *Journal of Human Evolution*, 3:445-454.

Warren, James M. 1976. Tool use in mammals, in: *Evolution of Brain and Behavior in Vertebrates* (R. Masternson, M. Bitterman, C. Campbell, and N. Hotton, eds.), pp. 407-424, Hillsdale, New Jersey: Erlbaum.

Warren, James M. 1980. Handedness and laterality in humans and other animals. *Physiological Psychology*, 8:351-359.

Waterman, Peter G. 1984. Food acquisition and processing as a function of plant chemistry, in: *Food Acquisition and Processing in Primates* (David J. Chivers, Bernard A. Wood, and Alan Bilsborough, eds.), pp. 177-211, New York: Plenum.

Waterman, Peter G.; Choo, Gillian M.; Vedder, Amy L.; and Watts, David 1983. Digestibility, digestion-inhibitors and nutrients of herbaceous foliage and green stems from an African montane flora and comparison with other flora. *Oecologia (Berlin)*, 60:244-249.

Waterman, Peter G.; Mbi, Christiana N.; McKey, Doyle B.; and Gartlan, J. Stephen 1980. African rainforest vegetation and rumen microbes: phenolic compounds and nutrients as correlates of digestibility. *Oecologia (Berlin)*, 47:22-33.

Watts, David P. 1984. Composition and variability of mountain gorilla diets in the central Virungas. *American Journal of Primatology*, 7:323-356.

Watts, David P. 1985a. Observations on the ontogeny of feeding behavior in mountain gorillas (*Gorilla gorilla beringei*). *American Journal of Primatology*, 8:1-10.

Watts, David P. 1985b. Relations between group size and composition and feeding competition in mountain gorilla groups. *Animal Behaviour*, 33:72-85.

Weber, A.W., and Vedder, Amy 1983. Population dynamics of the Virunga gorillas: 1959-1978. *Biological Conservation*, 26:341-366.

Wegener, Jonathan G. 1965. Cross-modal transfer in monkeys. *Journal of Comparative and Physiological Psychology*, 59:450-452.

Weidenreich, Franz 1945. Giant early man from Java and South China. *Anthropological Papers of the American Museum of Natural History*, 40:1-134.

Welker, Wallace I. 1956a. Some determinants of play and exploration in chimpanzees. *Journal of Comparative and Physiological Psychology*, 49:84-89.

Welker, Wallace I. 1956b. Variability of play and exploratory behavior in chimpanzees. *Journal of Comparative and Physiological Psychology*, 49:181-185.

Welker, Wallace I. 1956c. Effects of age and experience on play and exploration of young chimpanzees. *Journal of Comparative and Physiological Psychology*, 49:223-226.

Whitesides, George H. 1985. Nut cracking by wild chimpanzees in Sierra Leone, West Africa. *Primates*, 26:91-94.

Whitmore, T.C. 1975. *Tropical Rain Forests of the Far East*, Oxford: Clarendon.

Whitten, Anthony J. 1980. Arenga fruit as a food for gibbons. *Principes*, 24:143-146.

Whitten, Anthony J. 1982a. Diet and feeding behaviour of Kloss gibbons on Siberut Island, Indonesia. *Folia Primatologica*, 37:177-208.

Whitten, Anthony J. 1982b. Home range use by Kloss gibbons (*Hylobates klossii*) on Siberut Island, Indonesia. *Animal Behaviour*, 30:182-198.

Whitten, Tony 1982c. *The Gibbons of Siberut*, London: J.M. Dent & Sons Ltd.

Whitten, Anthony J. 1984a. The trilling handicap in Kloss gibbons, in: *The Lesser Apes: Evolutionary and Behavioural Biology* (H. Preuschoft, D.J. Chivers, W.Y. Brockelman, and N. Creel, eds.), pp. 416-419, Edinburgh: Edinburgh University Press.

Whitten, Anthony J. 1984b. Ecological comparisons between Kloss gibbons and other small gibbons, in: *The Lesser Apes: Evolutionary and Behavioural Biology* (H. Preuschoft, D.J. Chivers, W.Y. Brockelman, and N. Creel, eds.), pp. 219-227, Edinburgh: Edinburgh University Press.

Wilkerson, Beverly J., and Rumbaugh, Duane M. 1979. Learning and intelligence in prosimians, in: *The Study of Prosimian Behavior* (G. Doyle and R. Martin, eds.), pp. 207-246, New York: Academic Press.

Williams, George C. 1975. *Sex and Evolution*, Princeton, New Jersey: Princeton University Press.

Wilson, Edward O. 1975. *Sociobiology*, Cambridge, Massachusetts: Harvard University Press.

Wilson, Glenn, and Danco, Jeff 1976. Color preference in the gibbon. *Perceptual and Motor Skills*, 43:155-158.

Wilson, Gordon L., and Grunzke, Marvin E. 1969. Apparatus for psychobiological research, in: *Primates in Medicine* (E.I. Goldsmith and J. Moor-Jankowski, eds.), Vol. 4, pp. 94-153, Basel: Karger.

Wilson, Wendell A., and Shaffer, O.C. 1963. Intermodality transfer of specific discrimination in the monkey. *Nature*, 197:107.

Winner, Ellen, and Ettlinger, George 1979. Do chimpanzees recognize photographs as representations of objects? *Neuropsychologia*, 17:413-419.

Wittenberger, James F. 1979. The evolution of mating systems in birds and mammals, in: *Handbook of Behavioral Neurobiology*. Vol. 3, *Social Behavior and Communication* (P. Marler and J.G. Vandenbergh, eds.), pp. 271-349, New York: Plenum Press.

Wittenberger, James, and Tilson, Ronald L. 1980. The evolution of monogamy: hypotheses and evidence. *Annual Review of Ecology and Systematics*, 11:197-232.

Wolfe, John B. 1936. Effectiveness of token-rewards for chimpanzees. *Comparative Psychology Monographs*, 12:1-72.

Wolkin, Joan R., and Myers, Richard H. 1980. Characteristics of a gibbon-siamang hybrid ape. *International Journal of Primatology*, 1:203-221.

Wolpoff, Milford H. 1975. Sexual dimorphism in the australopithecines, in: *Paleoanthropology: Morphology and Paleoecology* (R.H. Tuttle, ed.), pp. 245-284, The Hague: Mouton.

Woo, J.K. 1962. The mandibles and dentition of *Gigantopithecus. Palaeontologia Sinica*, n.s. D, 11:1-94.

Wood, Susan; Moriarity, K.M.; Gardner, Beatrice T.; and Gardner, R. Allen 1980. Object permanence in child and chimpanzee. *Animal Learning and Behavior*, 8:3-9.

Woodruff, Guy, and Premack, David 1981. Primitive mathematical concepts in the chimpanzee: proportionality and numerosity. *Nature*, 293:568-570.

Woodruff, Guy; Premack, David; and Kennel, Keith 1978. Conservation of liquid and solid quantity by the chimpanzee. *Science*, 202:991-994.

Wrangham, Richard W. 1974a. Artificial feeding of chimpanzees and baboons in their natural habitat. *Animal Behaviour*, 22:83-93.

Wrangham, Richard W. 1974b. Predation by chimpanzees in the Gombe National Park, Tanzania. *Primate Eye*, 2:6.

Wrangham, Richard W. 1977. Feeding behaviour of chimpanzees in Gombe National Park, Tanzania, in: *Primate Ecology: Studies of feeding and ranging behaviour in lemurs, monkeys and apes* (T.H Clutton-Brock, ed.), pp. 503-538, London: Academic Press.

Wrangham, Richard W. 1979a. On the evolution of ape social systems. *Social Science Information*, 18:335-368.

Wrangham, Richard W. 1979b. Sex differences in chimpanzee dispersion, in: *The Great Apes* (David A. Hamburg and Elizabeth R. McCown, eds.), pp. 481-489, Menlo Park, California: The Benjamin/Cummings Publishing Company.

Wrangham, Richard W., and Nishida, Toshisada 1983. *Aspilia* spp. leaves: a puzzle in feeding behaviour of wild chimpanzees. *Primates*, 24:276-282.

Wrangham, Richard W., and Smuts, Barbara B. 1980. Sex differences in the behavioural ecology of chimpanzees in the Gombe National Park, Tanzania. *Journal of Reproduction and Fertility*, Supplement, 28:13-31.

von Wright, J.M. 1970. Cross-modal transfer and sensory equivalence—a review. *Scandinavian Journal of Psychology*, 11:21-30.

Wright, R.V.S. 1972. Imitative learning of a flaked tool technology—the case of an orangutan. *Mankind*, 8:296-306.

Wright, R.V.S. 1978. Imitative learning of a flaked stone technology—the case of an orangutan, in: *Human Evolution, Biosocial Perspectives* (S.L. Washburn and E.R. McCown, eds.), pp. 214-236, Menlo Park, California: The Benjamin/Cummings Publishing Co.

Yamagiwa, Juichi 1983. Diachronic changes in two eastern lowland gorilla groups (*Gorilla gorilla graueri*) in the Mt. Kahuzi region, Zaire. *Primates*, 24:174-183.

Yamazaki, Nobutoshi 1985. Primate bipedal walking: computer simulation, in: *Primate Morphophysiology, Locomotor Analyses and Human Bipedalism* (S. Kondo, ed.), pp. 105-130, Tokyo: University of Tokyo Press.

Yamazaki, Nobutoshi, and Ishida, Hidemi 1984. A biomechanical study of vertical climbing and bipedal walking in gibbons. *Journal of Human Evolution*, 13:563-571.

Yehle, Arthur L., Ward, J.P. 1969. Cross-modal transfer of a specific discrimination in the rabbit. *Psychonomic Science*, 16:269-270.

Yeni-Komshian, Grace H., and Benson, Dennis A. 1976. Anatomical study of cerebral asymmetry in the temporal lobe of humans, chimpanzees, and rhesus monkeys. *Science*, 192:387-389.

Yerkes, David N. 1977. Home life with chimpanzees, in: *Progress in Ape Research* (G.H. Bourne, ed.), pp. 5-7, New York: Academic Press.

Yerkes, Robert M. 1916. The mental life of monkeys and apes: a study of ideational behavior. *Behavior Monographs*, 3:1-145.

Yerkes, Robert M. 1925. *Almost Human*, New York: The Century Co.

Yerkes, Robert M. 1927a. The mind of a gorilla. *Genetic Psychology Monographs*, 2:1-193.

Yerkes, Robert M. 1927b. The mind of a gorilla. part II. mental development. *Genetic Psychology Monographs*, 2:375-551.

Yerkes, Robert M. 1928. The mind of a gorilla. part III. memory. *Comparative Psychology Monographs*, 5:1-92.

Yerkes, Robert M. 1934. Modes of behavioral adaptation in chimpanzee to multiple-choice problems. *Comparative Psychology Monographs*, 10:1-108.

Yerkes, Robert M. 1939. The life history and personality of the chimpanzee. *The American Naturalist*, 73:97-112.

Yerkes, Robert M. 1943. *Chimpanzees: A Laboratory Colony*, New Haven: Yale University Press.

Yerkes, Robert M., and Learned, Blanche W. 1925. *Chimpanzee Intelligence and Its Vocal Expressions*, Baltimore: Williams & Wilkins Co.

Yerkes, Robert M., and Nissen, Henry W. 1939. Pre-linguistic sign behavior in chimpanzee. *Science*, 89:585-587.

Yerkes, Robert M., and Tomilin, Michael I. 1935. Mother-infant relations in chimpanzees. *The Journal of Comparative Psychology*, 20:321-359.

Yerkes, Robert M., and Yerkes, David N. 1928. Concerning memory in the chimpanzee. *Journal of Comparative Psychology*, 8:237-271.

Yerkes, Robert M., and Yerkes, Ada W. 1929. *The Great Apes*, New Haven: Yale University Press.

Yoshiba, Kenji 1964. Report of the preliminary survey on the orang-utan in North Borneo. *Primates*, 5:11-26.

Zapfe, Helmut 1960. Die Primatenfunde aus der Miozanen Spaltenfullung von Neudorf an der March (Deninska Nova Ves), Tschechoslowakei. *Schweizerisch Palaeontologische Abhandlungen*, 78:1-269.

Zhang, Yong-Zu; Wang, Sung; and Quan, Guo-Qiang 1981. On the geographical distribution of primates in China. *Journal of Human Evolution*, 10:215-226.

Zhixiang, Li, and Zhengyu, Lin 1983. Classification and distribution of living primates in Yunnan China. *Zoological Research*, 4:111-120.

Zihlman, Adrienne L. 1979. Pygmy chimpanzee morphology and the interpretation of early hominids. *South African Journal of Science*, 75:165-168.

Zihlman, Adrienne L., and Cramer, Douglas L. 1978. Skeletal differences between pygmy (*Pan paniscus*) and common chimpanzees (*Pan troglodytes*). *Folia Primatologica*, 29:86-94.

Zihlman, Adrienne L.; Cronin, John E.; Cramer, Douglas L.; and Sarich, Vincent M. 1978. Pygmy chimpanzee as a possible prototype for the common ancestor of humans, chimpanzees and gorillas. *Nature*, 275:744-746.

Zimmerman, Robert R., and Torrey, Charles C. 1965. Ontogeny of learning, in: *Behavior of Nonhuman Primates: Modern Research Trends* (Allan M. Schrier, Harry F. Harlow, and Fred Stollnitz, eds.), Vol. 2, pp. 405-447, New York: Academic Press.

Index